THE EVOLUTION OF THE VERTEBRAL COLUMN

H. F. GADOW

THE EVOLUTION OF THE VERTEBRAL COLUMN

A contribution to the study of Vertebrate Phylogeny

BY

H. F. GADOW, M.A., PH.D., F.R.S.

Late Strickland Curator and Reader in Vertebrate Morphology
in the University of Cambridge

Edited by

J. F. GASKELL & H. L. H. H. GREEN

CAMBRIDGE

AT THE UNIVERSITY PRESS

1933

CAMBRIDGE
UNIVERSITY PRESS

University Printing House, Cambridge CB2 8BS, United Kingdom

Published in the United States of America by Cambridge University Press, New York

Cambridge University Press is part of the University of Cambridge.

It furthers the University's mission by disseminating knowledge in the pursuit of education, learning and research at the highest international levels of excellence.

www.cambridge.org
Information on this title: www.cambridge.org/9781107633384

First published 1933
First paperback edition 2014

A catalogue record for this publication is available from the British Library

ISBN 978-1-107-63338-4 Paperback

EDITORS' PREFACE

The manuscript materials for this book were left in an unfinished condition on the sudden death of its author. The earlier chapters had been in great part arranged in order, but the later descriptive chapters were separate, though notes were found indicating their probable sequence. The editors are therefore responsible for the final arrangement of the latter and have introduced Chapter xv (A Classification of Tetrapoda) in order to show the scheme under which it has been arranged. We have modified the manuscript by the removal of certain passages which were practically repetitions of passages in other chapters, but we have avoided as far as possible any alteration of either the construction of sentences or the terms used. We are well aware that certain chapters tend to be incomplete and would almost certainly have been altered or expanded by the author when revising the book as a whole, but we have thought it better to leave them incomplete rather than to attempt to expand them. The only corrections we have introduced have been in one or two cases where a statement has been made directly inconsistent with statements elsewhere.

The division into an earlier, general part and a later, systemic, descriptive part was amply indicated by the author, but the earlier part of the manuscript, though in order and to some extent divided up, had not been definitely arranged into chapters. We are therefore responsible for this division and for the chapter headings.

The aim of the book is to put together the evidence of the various lines of development of the vertebrae as a guide to the general morphological scheme of vertebrate evolution: the problem of the evolution of vertebrae had attracted the author more than forty years ago and had remained one of his principal interests throughout. The book is therefore a culmination of investigations and reflections of the greater part of an active life devoted to the morphological problems of the vertebrate phylum.

A number of rough sketches were found with the manuscript illustrating the author's views as set out in the text, but there was no indication of any final selection to be used as illustrations. We have made use of certain of them which had a sufficient description attached to indicate their probable position. Drawings of these have been made by Mrs Gadow, to whom we are also indebted for much help, especially in the composition of the bibliography and index. Fig. 41 has been specially drawn for this book by Dr Bulman. We have also freely made use of illustrations which have already been published in the works of various authors and to which reference is made, more especially in the descriptive chapters. For these our thanks are due to Dr R. Broom, Dr O. M. B. Bulman, Mrs H. S. Hacker, Professor D. M. S. Watson and Dr W. F. Whittard. Acknowledgment is also made to the Akademie der Wissenschaften in Wien (*Der Fauna der Gaskohle und der Kalksteine der Permformation Böhmens*, Fritsch); Messrs Allen & Unwin, Ltd. (*Zoology*, Sedgwick); The Director, American Museum of Natural History, New York (*The Bulletin of the American Museum*); Messrs A. & C. Black, Ltd. (*Vertebrata Craniata*, Goodrich); The Carnegie Institution of Washington (*Revision of the Amphibia and Pisces of the Permian of North America*, E. C. Case); Herrn W. Engelmann, Leipzig (*Festschrift für Carl Gegenbaur*, Goeppert); Herrn Gustav Fischer, Jena (*Handbuch der Vergleichenden und Exper. Entwicklungslehre der Wirbeltiere*, Schauinsland); Herrn Walter de Gruyter & Co., Berlin (*Die Stämme der Wirbeltiere*, Abel); Harvard University Press (*Osteology of the Reptiles*, Williston); Messrs Longmans, Green & Co. Ltd. (*Anatomy*, Gray); Messrs Macmillan & Co. Ltd. (*Cambridge Natural History*, Evans; *Structure and Development of Vertebrates*, Goodrich; *Zoology*, Parker and Haswell); The Council of the Royal Society (*Philosophical Transactions*); Herrn E. Schweizerbart'sche Verlagsbuchhandlung (Erwin Nägele), G.m.b.H., Stuttgart (*Palaeontographica*, von Heune); and to The Council of the Zoological Society of London (*Proceedings*).

J. F. GASKELL
H. L. H. H. GREEN

GREAT SHELFORD
October 1931

CONTENTS

CONTENTS

PART II

SYSTEMIC MORPHOLOGY

CONTENTS

ILLUSTRATIONS

ILLUSTRATIONS

ILLUSTRATIONS

Fig.

ERRATA.

p. 3, fig. 1 B legend, *for* genetal *read* genital.
p. 30, line 12, *for* Abbot *read* Abbott.
p. 53, line 27, *for* Unten *read* Untere.
p. 106, fig. 36 A, The CENTRUM OF PROATLAS should be stippled green.
p. 121, line 12, *for* 1886 *read* 1896.
p. 124, line 9, *for* Leptorhophus *read* Leptorophus.
p. 159, line 10, *for* Mayer *read* Meyer.
p. 166, line 25, *for* Stannis *read* Stannius.
p. 167, line 12, *for* Wedersheim *read* Wiedersheim.
p. 170, line 6, *for* Moody *read* Moodie.
p. 310, last line, *for* transversum *read* transversarium.

PART I

Chapter I

THE AXIAL SKELETON

ITS POSSIBLE EXTERNAL ORIGIN

All vertebrates possess a vertebral column, the evolution of which passes through several successive stages, their phylogenetic causal features being on the whole faithfully repeated by the developing individual; that is to say, the phylogenetic changes produced during evolution by alteration of environment are copied by ontogenetic changes which take place during the development of the individual. Since this ontogeny begins with the earliest germ and lasts to old age, phylogenetic changes must be superposed and the two processes must become blended. The changes which have come to pass by way of accommodation during life mean some alteration in the creature's structure, even though infinitely small and ephemeral. If this repeatedly occurs for generations, the initial cause persisting, the alteration becomes an established feature, gains momentum and henceforth proceeds orthogenetically so that it assumes the dignity of an appreciable visible entity, which the organism may find difficult to undo. If the new invention is advantageous, all the better for the race. Strictly speaking, it was advantageous from the beginning, as it was its mode of reaction to meet a new proposition. In time to come the new feature will in turn be registered as a standing order, and the sooner it is repeated the better for the individual's ontogeny if this is to get through its work of building up within reasonable time. The task will soon be so large and complicated that condensation and even compromises are unavoidable, and many new difficulties will have to be overcome, the results appearing as the so-called cenogenetic features which in turn become old-fashioned or palingenetic. And so *ad infinitum*.

The material for building the whole column comes from the notochord which is of endodermal origin, and from muscular and connective mesodermal tissue which surrounds the chorda, furnishes the walls of the spinal canal and further extends as septa through most parts of the body. As it thus forms the all-pervading framework of connective tissue, this mesoderm received the now somewhat antiquated name of membrana reuniens. Since within this tissue in the vicinity of the chordal sheath appear at an early stage the first clusters of cartilaginous cells as the foundation of the future skeleton, this central and innermost portion of the membrana reuniens is termed the skeletogenous layer. For the embryographer this cartilage arises in situ in order to surround the chorda with a skeletal support from head to tail. Any suggestion of peripheral provenance is mostly resented as devoid of justification.

Kupffer, Klaatsch and Julia Platt have fairly established the fact that cartilage can be of ectodermal origin. Kupffer has made it very probable that the whole branchial skeleton of the Cyclostomes is made of such ectodermal cartilage and it is possible that the so-called outer arches loosely attached to the visceral arches proper of many Elasmobranchs, though generally ignored, are the somewhat modified derivatives of the same outer cartilage. I myself have always cherished the prophetic dictum of Gegenbaur made in 1879: "Aller Knorpel kommt urspruenglich von aussem".

The conviction has gradually forced itself upon me that all the cartilage which ultimately contributes to the formation of the axial skeleton is referable to those cartilaginous elements which arise in the early median fin-folds (Fig. 1). These little rods, originally polymeric, extend from the periphery into the deeper strata, until they come to stop in the neighbourhood of the chordal sheath, where this valuable and superior material leads to the formation of the arcualia. They are still in this condition in the Cyclostomes, entirely cartilaginous, not jointed, but repeatedly dichotomising peripherally, nearly reaching the free edge of the fin. There is no difficulty in following this process throughout the dorsal region and around the tail to the anal region, where the gut naturally puts a stop to this simple arrangement. To account for the ventral arcualia in the trunk we have to choose between several hypotheses:

(1) According to the lateral fin theory the paired fins are the continuation of the median fin. The difficulty here is that most of the little rods in these lateral fins are pledged to the formation of the free fins and their girdles. But we can imagine that similar rods, by working their way along the intermuscular septa, thereby lay the foundation of the ribs. Beyond doubt ribs are the distal portions of the ventral arches and therefore may be founders of the ventral arcualia instead of themselves being the elongated derivatives of the latter.

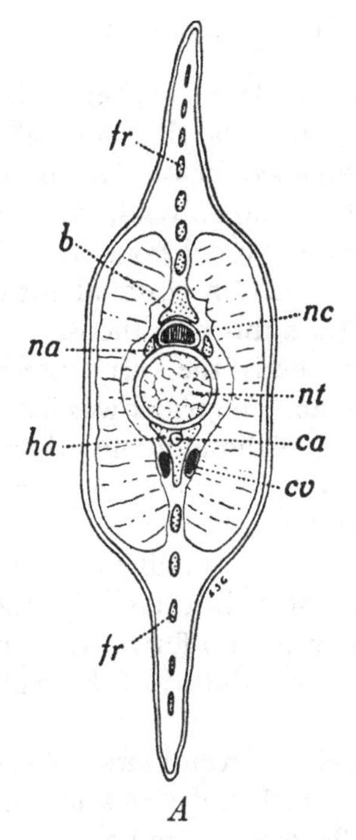

A

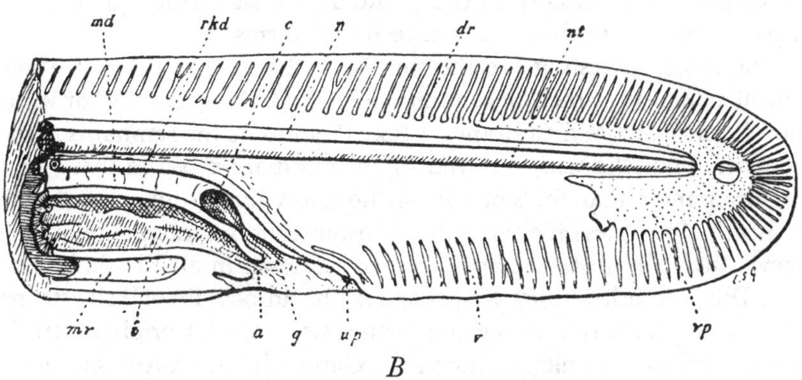

B

Fig. 1.

A. *Petromyzon marinus*. Transverse section of tail. *b*, base of dorsal fin-radial; *ca*, caudal artery; *cv*, caudal vein; *fr*, cartilaginous fin-ray or radial; *ha*, base of ventral fin-radial; *na*, neural arch; *nc*, nerve cord; *nt*, notochord with sheaths.

B. Tail of *Myxine glutinosa* L., cut so as to show the skeleton and the opening of the intestine, etc., left-side view. *a*, anus; *c*, gap behind mesentery leading from right to left coelomic cavities; *dr*, cartilage radials of dorsal median fin; *g*, median opening through which the genetal cells escape; *i*, intestine; *md*, dorsal mesentery; *mv*, ventral mesentery; *n*, nerve cord; *nt*, notochord; *rkd*, left kidney duct; *up*, urinary papilla; *v*, cartilage radials of ventral median fin; *vp*, cartilaginous plate. (Goodrich, *Vert. Craniata*, 1909.)

(2) Alternatively, the caudal ventral arcualia, having arrived at the anal region, have diverged to the right and left and by a later process of proliferation have extended forwards. This is supported by the fact that the paired ventral haemapophyses, the counterparts of the dorsal neurapophyses, are serially homologous with the caudal ribs of some fishes. The former existence of a postanal gut extending into the tail may support or complicate this hypothesis.

(3) As the third hypothesis, the vertebrate consisted at an archaic stage only of head and tail. The trunk was formed later by interstitial growth for the reception of those organs which were crowded out of the head, as this itself was being built up by concentration and fusion of the already more or less metameric skleral units. The tail is obviously older than the trunk. It still retains various ancient normal characters; whilst, being so old, it has had time during its reduction, which is progressing from its end forwards, to produce many pseudo-primitive features later reproduced in the trunk, such as the paired ventral arcualia.

The *fin-folds* have received several kinds of supports or stays, for the general description of which the terms fin-fibres, dermal rays and cartilaginous, eventually osseous, radials are sufficient, but there is a plethora of more than a dozen technical terms.

The most interesting are the fin-fibres, the so-called horny fibres, Hornfaeden, etc., which Gegenbaur took for an 'Abscheidung' or secretion from the basal membrane. I take them to be the remnants of degenerated muscles which moved the primordial fin before these were anything more than folds of skin, without any inner support, but with a sheath of muscles on either side. Various considerations support this view: these fin-fibres are ancient, being known in all Elasmobranch and Dipnoan fishes: they are present in the adipose fin of Teleosteans, and were probably the sole supporting stays of that original tail fin which vanishes completely during the Gano-Dipnoan-Crossopterygian and Teleostean developments. The contractile substance, the muscle proper, has become lost; the perimysium, the so-called non-cellular connective tissue, is preserved. These selfsame horny or "elastic" fibres are not elastic tissue, not being composed of elastine, but of some other related matter, which has been called elastoidin. Krukenberg, who worked out this question many years ago, concluded that both elastine and elastoidin were referable to the same kind of collagen mother substance. According to Goodrich these fibres or ceratotrichia seem to originate at the growing distal edge of the fin, immediately below the basal membrane; but later on they sink into the connective tissue, and have attached to their proximal ends the radial muscles of the fin. In fact these fibres seem to be continued into the very muscle fibres of which they are the degenerated remnants. If the degeneration began at the distal end, which must have been the oldest part of the raised fin-fold, then it is to be expected that ontogeny would repeat this process. On the outside they are naturally overlaid by the denticles, the originators of the dermal rays or dermotrichia, which most likely covered the skin before the fold was raised and muscularised.

If my interpretation of the significance of these ontogenetic processes is correct, the previous phylogenetic existence of a muscularised fin-fold without an inner skeletal support must be assumed, although such a structure, according to Gegenbaur, is unknown in the vertebrates and was therefore used by him as the strongest objection to the lateral fin theory.[1]

1 The structure of the fin-rays is well shown in a specimen in the Cambridge Museum of the dorsal fin of a Dogfish, picked up on the shore where it must have been rolling in the surf, countless times dried and baked in the sun. It consists of a thick mass like a brush or tassel composed of the innumerable split and resplit "horny fibres", giving a demonstration of their structure and arrangement not easy to reproduce.

Chapter II

SEGMENTATION AND RESEGMENTATION

THE FORMATION OF METAMERIC VERTEBRAE

At a very early stage of development there appear to the right and left of the chorda the Urwirbel, that is, Ursegmente or Protovertebrae; cubic segments arranged in the long axis of the body. They continue to appear from near the presumed end of the head backwards; their number becomes considerable, and may reach two or more score according to the length of the body to be. Each cube contains a cavity, and its wall is the material which will give rise to a segment of the body proper, minus the ectoderm and the organs of the vascular and alimentary systems. These cubes are usually called somites and are distinguished according to later differentiation as myo-, sklero-, nephro- and gonotomes, all segmentally arranged. Even this segmentation is certainly not the primordial condition but an acquisition, since at the very earliest stage the sum total of the somites appears as one continuous right and left band, a condition which must apply also to their myotomic portion. It may be concluded that this, the only contractile string, must have contracted peristaltically and like a vibrating cord with nodes and internodes which, when fixed by repeated action, divided the pair of right and left muscular strings into contractile and non-contractile (let us call it mesenchymatous) matter, the latter forming the transparent interprotovertebral septa, the first indication of segmentation, to which necessarily the other "tomes" must be adjusted; in other words, all the tomes must become segmented, even the neurotomes, as being the peripheral portion or complement of the myotomes. It is a crude notion to start with the passive sklerotomes for whose benefit the muscles were ordered to make the future column movable. Such a notion appeals to a *causa finalis* instead of to a *causa efficiens* which the very nature of a muscle demands, for function must marshal and organise the available material and thereby build organs.

It is obvious that if each myotome corresponded with, and was restricted to, its sklerotome, there would be no progress of motion. Two neighbouring sklerotomes must be made to act as the *points d'appui* for one myotome. This is effected by a simple adjustment; each sklerotome becomes split at the level of the node, and the anterior and posterior halves of neighbouring sklerotomes are fused together, the bridging myotome assisting this to occur in the region which is not under strain. Thus we have now a rearrangement, a resegmentation of the column, the new segments or units being henceforth metameric skleromeres or vertebrae (Fig. 2). Each skleromere is the result of a fusion of the cranial half of the split tome with the caudal half of the split tome in

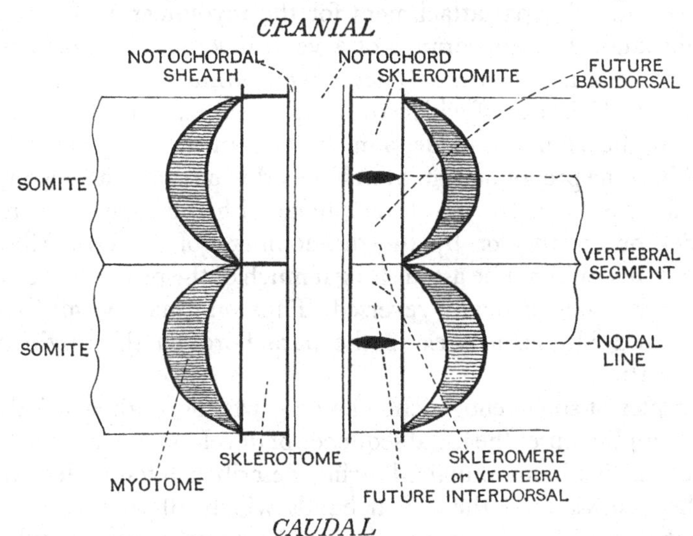

Fig. 2. Diagrammatic horizontal section of two primitive body segments. On the left the primitive somites are shown with their corresponding sklerotomes and myotomes; on the right resegmentation with the formation of skleromeres or permanent vertebral segments. The skleromere is formed from the caudal half of the anterior sklerotome, the future basidorsal, and from the cranial half of the posterior sklerotome which forms the future interdorsal.

front. Some authors, e.g. Piiper, called these sklerotome halves sklerotomita. Schauinsland seems to be the first to have recognised that the dorsal arcualia arise in such a way that the basidorsal cartilaginous blocks belong to the caudal, and the interdorsals to the cranial sklerotomita. And since it has unavoidably come to pass that in this diction the caudal sklerotomite makes the anterior half, and the cranial sklerotomite makes the posterior half of the vertebra, confusion is liable to occur. It seems therefore advisable to avoid cranial and caudal and to speak only of the anterior and posterior half or end of a vertebra, when it is once understood how the vertebra arose.

There is still some difficulty in tracing the origin of the ventral arcualia. Schauinsland and Howes however have independently been able to show that a portion of the basidorsal mass undoubtedly grows downwards around the chorda, and on the latero-ventral side forms the basiventral. The behaviour of the interdorsals with reference to the interventrals seems to be analogous.

It stands to reason that the first function of the skleromeres is protection of the spinal cord; by uniting as neural arches they afford a suitable and principal attachment for the myotomes as observed by Schauinsland. The appearance of a ventral series of arcualia would therefore be a later feature. They have nothing to protect, but are simply the blocks out of which the whole or greater part of the axial portion of the column is made, namely the centra. If this ventral mass should become preponderant, even if only during some passing phyletic stage, for instance, by the development either of a great centrum (pseudo- or gastro-) or by the development of powerful ribs and chevrons, it would not be astonishing if much of the process of vertebral building is cenogenetically reversed. Thus one may be misled, like Marcus and Blume, to seek in the parachordalia the origin of all skleral matter.

Examples of such cenogenetic changes, which are then strictly adhered to and so upset the usual sequence of development, are the hypochordal cartilaginous unpaired string described later under Anura, and the behaviour of the dorsal bands which afford indications of segmented neural arches before they turn into continuous bands.

Chapter III

THE COMPOSITION OF THE STANDARD VERTEBRA

ITS BILATERAL ORIGIN FROM FOUR PRIMITIVE CARTILAGES ON EACH SIDE

The fundamental scheme of the composition of the ideally complete vertebra, which actually exists in the Sturgeon (Fig. 3*B*), and from which any known vertebral modification can be derived, is so simple, and the nomenclature that I proposed in 1895 is so logical and easily mastered, that it was only a question of time for it to be generally accepted. It suffices to use nine or, including ribs, ten terms, or perhaps, if we include the arcualia and interarcualia, twelve terms instead of at least forty or more, which have been used at various times, and not a few of which are undefinable.[1]

Many morphologists have since given valuable help by suggestive criticism of the whole scheme, notably Branson, Schauinsland, Schwarz, Abel and Piiper; many others have assisted, perhaps unintentionally, by descriptions of hitherto unknown combinations of the units, mostly fossil.

The following are the units involved in this scheme (Fig. 3*A*):

(1) Basidorsals with (2) supra-basidorsals and (3) dorsospinalia.

(4) Interdorsals with (5) supra-interdorsals.

1 An example is afforded by the synonyms for the interarcualia:
Intercalaria corporum or Intercalaria superiora.
Schaltstuecke, dorsal intercalaries.
Neurapophysen.
Pleurapophysis superiora or Pleurapophysis dorsalia.
Pleurapophyses.
Pleurocentra.
True centrum.
Hypocentrum pleurali; to distinguish it from the Hypocentra of Gaudry (basiventrals).

Hypocentra and Pleurocentra are terms which have caused confusion, centrum being used in a new sense, as the centre of ossification, instead of its true meaning of body or Koerper. This loose nomenclature has culminated in the invention of neurocentra (Williston) for the neural arches. The term pleurocentrum itself is due to the crude notion that it was the original bearer of the ribs (pleura), on to which distal portions of the basiventrals have crept; the ribs are thus often borne by the parapophyses of the amniotic centrum. For other lists of the appalling number of synonyms of the arcualia see *Phil. Trans.* 1885, Fishes, pp. 170–7; Amniota, pp. 23–5.

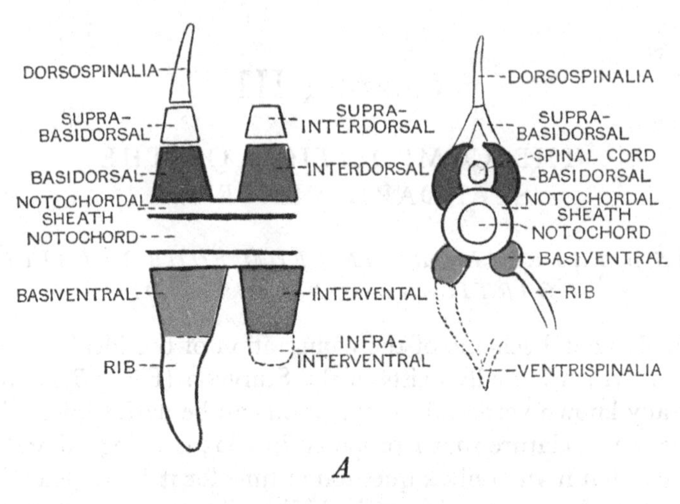

A

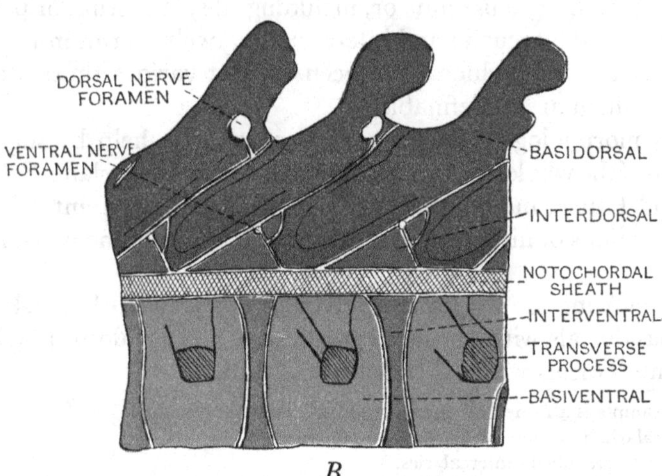

B

Fig. 3.

A. Diagram of the units forming the complete vertebra of the general scheme. On the left a side view of the elements is shown, the cranial end being to the left; on the right a transverse view of the basal elements with their relationships to the central nervous system and the notochord.

The colour scheme of this diagram will be adhered to throughout.

Blue = Basidorsal, BD.	Red = Interdorsal, ID.
Orange = Basiventral, BV.	Green = Interventral, IV.

B. Drawing from a plaster model of three vertebrae of the Sturgeon, showing the four primitive elements which compose them. The relationship of the foramina for the exit of dorsal and ventral spinal nerves to the basidorsal block is shown. The vertebrae are seen from the left side.

(6) Basiventrals with (7) ribs and (8) ventrispinalia.

(9) Interventrals with (10) infra-interventrals.

These ten cartilaginous, or later ossified, units are the components of the arcocentrous vertebra. The basidorsals with their epimeres (2) and (3), and the basiventrals with their epimeres (7) and (8), make up the system of the principal arcualia and form the anterior half of the vertebra. Numbers (4), (5), (9) and (10) collectively make up the system of the interarcualia and form the posterior half of the vertebra.

The basiventral (BV) forms the Pseudocentrum of Amphibia.

The interdorsal (ID) forms the Notocentrum of Amphibia.

The interventral (IV) forms the Gastrocentrum of Amniota.

Every one of the possible combinations and permutations of these fundamental blocks has materialised in some vertebrae of some individual, species, genus, or larger group; and it is the morphologist's business to trace and try to arrange these variations in lines, and so to find out whether these lines are true indications of affinity and descent, or whether they are instances of convergent or parallel evolution. One of the main tasks of the present work is the attempt to bring order into chaos, with a view to arrive at, or to help towards, a natural classification of the many thousands of living and extinct vertebrates. The amplitude of these variations ranges from the possession of all four arcualia, each of which may be differentiated distally into epimeres such as the supradorsals and infraventrals, to the reduction of the whole vertebra to a single pair of basidorsals. In many cases it is difficult to ascertain what has happened. Interpretation of a particular bony unit may be helped by ontogenetic observation and, if backed by fossils, may be of supreme importance. There are, for instance, vertebrae composed of only two large pieces, BV and BD, as in the Labyrinthodont *Mastodonsaurus*, but the same bipartite effect is produced in many Amniota by the BD and IV. The clue to the fact that these strikingly similar vertebrae are fundamentally different we owe to some early fossil specimens.

It is absolutely wrong to base a classification solely upon one organic structure, in the present case the vertebrae; as wrong as to base it upon any other organ, which seems promising, because it happens to possess differential characters easily expressed, whether it be the temporal arches, now in vogue, the convolutions of the gut, the characters of the carpus and tarsus, the nervous system or the sacrum. Every one of these attempts is bound to break down, the respective organs ceasing to be trustworthy, because, owing to evolution, any two or more clearly diverging lines must start from some indifferent condition, from which intermediate links radiate. We consider ourselves lucky when we can discern the trend of a new character, and can find justification to follow up this trend without meeting with any insuperable obstacle. Some we can follow at length until we reach some still existing terminus, but there are many more which after a fairly long run come to a dead stop. The latter cases are really the more valuable, because by a process of elimination they reduce the number of candidates for the ancestry of any particular group of creatures.

One absolute NO is sufficient, no matter which organic system provides it. However, our search should not be hopeless, since there *is* a road to follow.

To account for the four pairs of arcualia I originally assumed that the basiventral grew upwards and, beyond the level of the chorda, separated off its apex as the interdorsal; the same notion was applied to the down-growing basidorsal, which separated off its apex as the interventral. In order to account for the necessary movability of these blocks and vertebrae I was forced logically to reconstruct the whole vertebra (Fig. 4*A*), putting the interdorsal 39, for example, in front of the basidorsal of vertebra 39. This plausible arrangement became widely accepted and in many instances accounts for the erroneous statement that the ID lies in front of its BD. Since the interdorsal is the same as the pleurocentrum of some authors, and the same as the notocentrum of others, the mischief is complete. The corollary follows that instead of issuing behind the basidorsal to which it belongs, the spinal nerve issues in front of its vertebra, which cannot be the case.

Perhaps partly confused by Cope's speculations, I mistook in *Eryops* for the interventrals what American palaeontologists subsequently proved to be the interdorsals. Consequently the inclusion of *Eryops* among the Reptiles was wrong. In my *Classification of Vertebrata*, 1898, I proposed the subclass Proreptilia for *Eryops* and *Cricotus*, as the lowest of Reptiles. Curiously enough, Cope himself called *Cricotus* a Reptile, not without reason, because its large posterior axial block certainly contains the Amniotic centrum. For obvious reasons this erroneous inclusion of the Embolomeri with the Reptiles has been allowed to lapse, but nevertheless my error with regard to *Eryops* has been retained as a test case to prove that my fundamental system of vertebral composition was wrong. Since the American Rhachitomi have nearly made a centrum out of their interdorsals (therefore a notocentrum), the same would appear to apply also to the Amniota. This, however, is wrong,

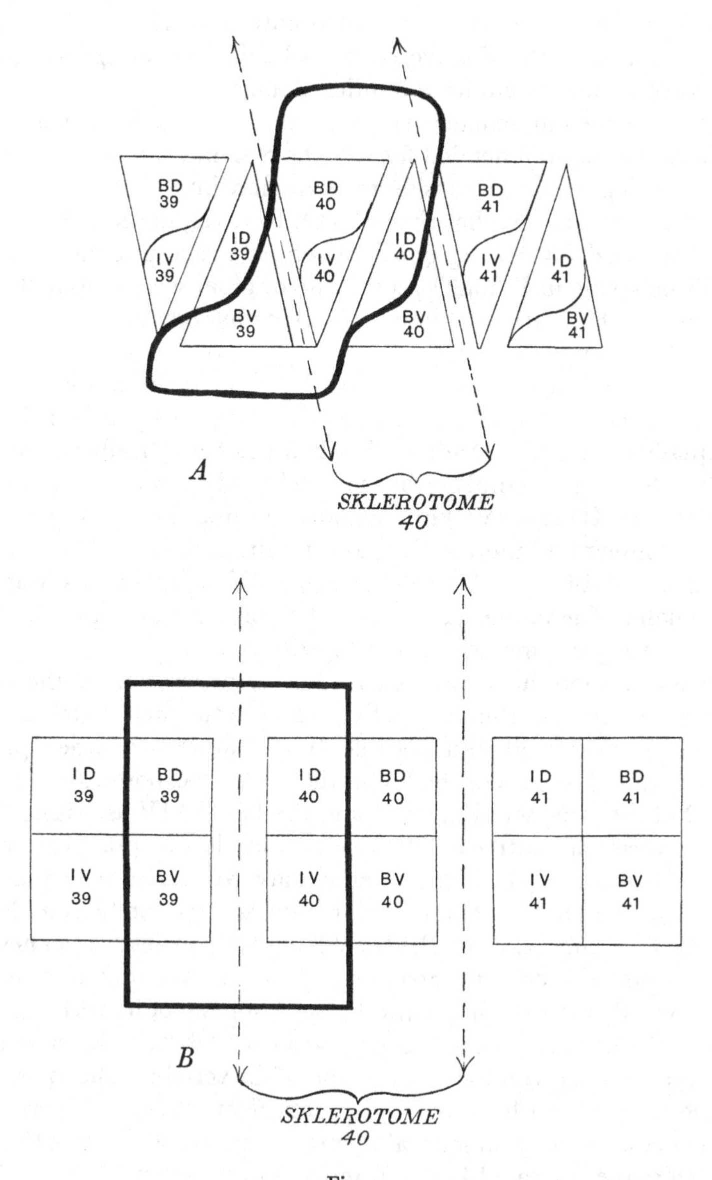

Fig. 4.

A. Old scheme of the composition of vertebra 39. The various blocks are labelled, the sklerotome is indicated between the dotted lines, the vertebra by the thick outline.

B. The same according to the amended scheme.

as the Amniota have a very large gastrocentrum besides small vestigial interdorsals, the latter discovered by Schauinsland in *Sphenodon* and later corroborated in embryos of other Amniota.

The fundamental scheme which is about to be described is the rehabilitation in an amended form of the resegmentation, or Neugliederung, which in 1895 I considered extremely improbable. It is now generally accepted. At that time also O. Hertwig, in the 1888 edition of his *Lehrbuch d. Entwicklungsgeschichte*, refused to accept Umgliederung der Wirbelsaeule in Remak's sense. The original scheme, that I then put forward (Fig. 4*A*), was simply a working hypothesis to explain the action of the myotomes upon the skleromeres, the actively contractile myotomes being held to cause an overlapping of the passive sklerotomes. The correction which must be made in this hypothesis is that the equally numbered BD and IV blocks are not genetically connected, but that the intervertebral cartilage is solely and entirely formed by the interarcualia, ID and IV. This amended scheme (Fig. 4*B*) gives now a much simpler and more satisfactory result, as follows: Vertebra 39 is composed of BD 39 + ID 40 + BV 39 + IV 40, and arises from the caudal half of sklerotome 39 + cranial half of sklerotome 40. The final and the wrong scheme are figured together.

Various authors have suggested other arrangements of the components so as to alter their respective shares in the final vertebra. The new scheme is the simplest possible and is satisfactory when put to practical test, if we accept the BV as the ventral portion which carries the BD of the same serial number and the ID and IV as arising from the intervertebral cartilage. The question has lost its practical value, and has become of academic interest only. We know now that the numerical combinations of the fundamental scheme can be completely upset, as for example, when the basal half of BV 40 fuses as a chevron on to the posterior end of centrum 39; or when it is optional, as in the Anura, whether the ID fuses with the posterior end of its vertebra as a knob, and therefore produces a procoelous vertebra, or when it does the reverse, fusing with the anterior end of the vertebra following, thus making it opisthocoelous. For all practical purposes such vertebrae are of course homodynamous, all being composed of BD and ID, but carrying in the one case BD 39 + ID 39 and in the other ID 39 + BD 40. These vertebrae are physiological, not solely morphological conceptions.

There is another somewhat puzzling consideration, viz. the numbering of the sklerotomes and skleromeres. If skleromere or vertebra I consists of the caudal half of sklerotome I and the cranial half of sklerotome II, there must be the cranial half of sklerotome I to spare, probably forming the proatlas. Another illustration can be taken from the Amniota: according to the scheme, vertebra 4 consists of BD 4 and IV 5, the ID 5 having dropped out; if its BV 4 fuses on to the preceding vertebra as a chevron, the complete vertebral mass 4 then consists of BD 4 and centrum 5 + BV 5. If the whole BV vanishes, this vertebra consists only of BD 4 and centrum 5. The vertebra is here again only a physiological entity. Birds will also fall into line, for in their case Piiper has made it possible to bring order into chaos.

An account of the embryology of the vertebral column, in the careful style of Howes or Piiper, is needed for the Gecko, Lizard and Crocodile, above all for the first. All the rest will then fall into line, Man and Tortoises included.

Chapter IV

THE FIVE ONTOGENETIC AND PHYLOGENETIC STAGES OF THE VERTEBRAL COLUMN

A broad view of the principal tendencies in the evolution of a physiologically adequate axial skeleton reveals the successive, repeated supersession or substitution of the original central axial material by sheaths or mantles which are placed more peripherally.

First the chorda, next the chorda with a thick sheath made by it, then a truly perichordal mantle beyond this sheath containing cartilaginous units. This third stage develops into stage 4, in which the cartilaginous elements invade the chordal sheath. This invasion may become the chief development and so give rise to the side group 4*a*, in which it culminates in the formation of chordacentrous vertebrae. But in the main line of descent, stage 5, this invasion becomes retrogressive and is overshadowed by the development of the cartilaginous elements of the skleromere external to the chorda and its sheaths (Fig. 5).

FIRST STAGE

This consists of a simple notochord surrounded by a single layer of cells. It still survives in the fully grown animal in the case of *Amphioxus*.

DESCRIPTION OF FIGURE 5

Stage 1. Simple notochord with peripheral layer of cells.

Stage 2. Notochord with fibrous sheath and elastica externa.

Stage 3. Four longitudinal cartilaginous blocks become differentiated in the membrana reuniens.

Stage 4. The blocks have formed the dorsal and ventral arcualia, and cartilage cells have ruptured the elastica externa and are invading the chordal sheath. They are also growing round the outside of the external elastic membrane. The arrows indicate the directions of migration.

Stage 4*a*. Chordacentrous vertebra. The arcualia have now fused dorsally and ventrally to form the neural and haemal arches. The chordal sheath is extensively invaded by cartilage cells. The complete development of this stage forms thė chordacentrous vertebra. Growth of cartilage cells from the arcualia round the outside of the elastica externa has not materialised.

Stage 5. Arcocentrous vertebra. Growth of cartilage cells from the arcualia forms a perichordal mantle, the invasion of the chordal sheath is negligible. The chordal sheath is on the wane. Neural and haemal arches are fully formed. Full development of this stage gives rise to the arcocentrous vertebra.

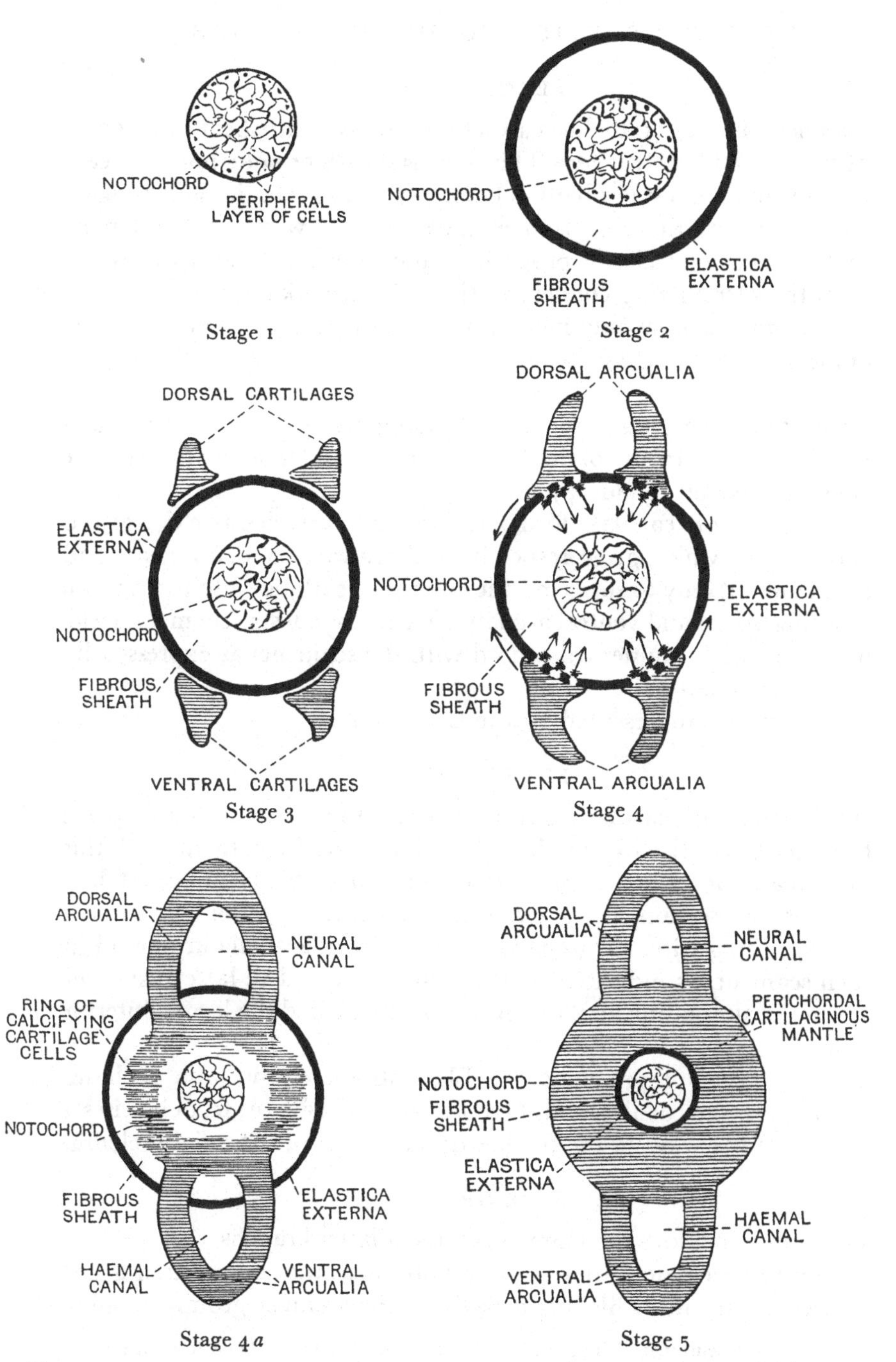

Fig. 5. Diagram of the five ontogenetic and phylogenetic stages of the vertebral column.

SECOND STAGE

The notochordal epithelium secretes a non-cellular sheath composed of successive fibrillar layers. This fibrous sheath or notochordal sheath can assume a considerable thickness. It is itself elastic, and is again surrounded by the very thin elastica externa, which has still more highly developed elastic properties, and which separates the whole from the surrounding connective tissue framework of the body.

This stage is now only met with as a transitory phase of vertebrate ontogeny.

THIRD STAGE

Nests of future cartilage appear in the supporting tissue or membrana reuniens in the vicinity of the elastica externa, and thus form a structure called the skeletogenous layer.

The clusters of cartilage form four longitudinal series, two dorsal and two ventral, which grow respectively downwards and upwards, and thus tend not only to surround the chordal sheath but also to form the dawning dorsal and ventral arcualia. These are not strictly metameric, not yet being in numerical accord with the segments as expressed by the neuromeres.[1]

This stage is represented by the Cyclostomes.

FOURTH STAGE

Immigration of cartilaginous cells from the arcualia takes place through the elastica externa into the thick chordal sheath, and fills this to a greater or less extent with cartilage. This ingress can be watched in transverse sections of the embryo proceeding from the four corners corresponding with the arcualia. Four pairs of arcualia are found in each segment, as indicated by the spinal nerves. The latter issue between the alternating arcualia, immediately behind the larger anterior pair.

This stage is represented by Elasmobranchs and also includes Chimaeras, the cartilaginous Ganoids and the Dipnoi; from it arises a parting of the ways in the further development of the axial skeleton.

SIDE GROUP 4*a*

This group includes the Chimaeras and Elasmobranchs.

Whilst in the older groups of Elasmobranchs the chorda itself is not interfered with in its lifelong growth, in the younger groups its thick

1 The term "neuromere" is used both here and throughout this book to indicate that area which lies between two successive issuing nerve roots.

sheath, strengthened already by the immigrating cartilage, becomes infiltrated with calcareous deposits which are then arranged in concentric layers, asterisks or crosses, or combinations of these (Fig. 6). These calcifications restrict the further growth of the chorda itself: a constriction begins in the middle and, proceeding fore and aft, produces the well-known hourglass shaped blocks into which the chordal sheath is broken up. Where the successive hourglasses meet the chorda is not interfered with, and together with the connective tissue of the skeletogenous layer acts as an intervertebral rather primitive joint.

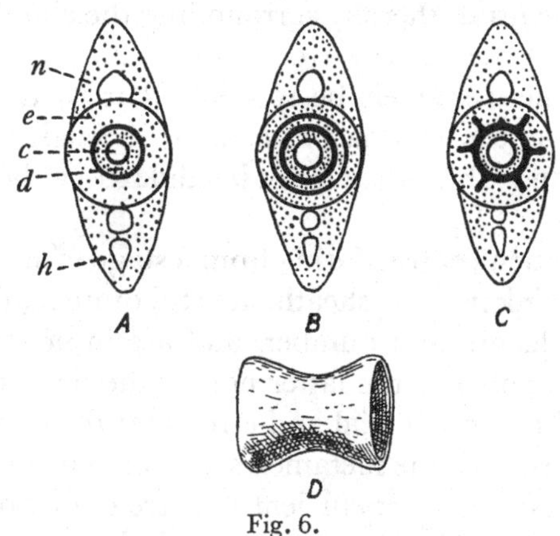

Fig. 6.

A—C. Diagrammatic transverse sections of vertebrae of Selachians. *A*, cyclospondylous; *B*, tectospondylous; *C*, asterospondylous condition (after Hasse, from A. Sedgwick, *Zoology*, 1905). *c*, notochord; *d*, central calcareous ring; *e*, elastica externa; *h*, haemal arch; *n*, neural arch.

D. Calcified cylinder of a centrum of *Squalus vulgaris* (Goodrich).

Thus are formed and introduced vertebral chordagenous chordacentra (Fig. 5, stage 4*a*).

The arcualia, which are at first but indirectly concerned in this process by having contributed cartilage cells for the chordal sheath, become in the less archaic Elasmobranchs more and more implicated. Whilst in the oldest genera the main mass of the cartilaginous arcualia lies outside the general level of the elastica externa, later ingrowth has, so to speak, gained momentum, so that most of the elastica is destroyed, at first perhaps having been driven in, and squeezed; so that eventually the chief arcualia, the basidorsals and basiventrals, become deeply embedded in the compound chordal sheath. Since this sheath naturally also grows in size and thickness, it thereby engulfs the basal parts of the arches and these sink in as growth proceeds. Thus it comes to pass that a well-macerated vertebra of a recent Elasmobranch shows in transverse section its solid calcified material in the shape of an upright cross, whereas in a Teleostean vertebra the solid material appears in the shape of a slanting cross or saltire, since it is formed in this instance by ossification of the arches which form the bulk of the vertebra.

The originally strictly chordagenous centra have indeed become a compound of chorda sheath and purely cartilaginous arches. This process goes still further in various Sharks and especially in the Skates. The dorsal arcualia tend to meet and to fuse with each other and also to fuse with the ventral, thereby surrounding the chordal sheath with a cartilaginous mantle. It is a trend of evolution to which we shall return later. It need scarcely be repeated that by the above process the importance of the chordacentra as a distinctive feature of the Elasmobranchs and Chimaeras is not invalidated, although externally obscured.

The Elasmobranch centra, dating from a stage when the arches had not yet begun to sink into the sheath, are still quite independent of the arcualia in both length and number, and are in no way determined by the size and number of the latter nor by the neuromeres. It is by a long-continued process of trial and error that these centra or bodies have come to agree with the metameres proper. The problem requires the establishment of the most efficient "centre of motion", in accordance with the way in which the particular kind of fish swims, bends or wriggles. In *Chimaera* the whole axial portion of the chorda is composed of a continuous series of narrow disks of which from three to five correspond with a neuromere.

The ultimate adjustment of the vertebral column, which becomes the normal and usually final one, is such that each vertebra corresponds to one neuromere and one skleromere. Since the latter is composed of a pair of basidorsals and a pair of basiventrals, and paired interdorsals and interventrals, and since the basiventrals often form pseudocentra, and the interarcualia produce autocentra, it follows that two kinds of centra can be present to each "mere".

A number of terms to describe the various conditions found have been invented: for instance, Polyspondylous, Diplospondylous and Monospondylous, which unfortunately convey the notion that each disk has the dignity of one vertebra or Wirbel. The confusion is shown by the fact that some authors call the diplospondyle a Doppelwirbel (meaning two vertebrae), while others speak of two Halbwirbel, which is more reasonable.

It would be difficult to reconcile these changeable centra with the splitting and recombining of the sklerotomes, since these chordacentra are independent formations, although naturally they must have got their material from the sklerotomes.[1]

When correlation has at last been fixed, diplospondylism may rightly be explained by referring the posterior half to the cranial part of the split sklerotome and the anterior half to the caudal. Schauinsland and others have made it highly probable that originally all the arcualia were alike in size, and that later the offspring of the cranial sklerotome half has undergone reduction in size, while the caudal half has become the important basidorsal pair forming the neural arch. Abundant traces of these changes exist in Urodeles and even in Amniota. Perhaps the fact of greatest importance is the discovery of independently ossifying interdorsalia in the Amniota, the very units which we supposed to have been lost in the gastrocentrous vertebrae. That they must have existed is a postulate of reason. That they are repeated more or less in the embryos of Reptiles and Birds, somewhat vaguely in Mammals, and therefore have enabled us to recognise them in fossil reptiles, is in agreement with onward evolution and a welcome proof of the soundness of our fundamental scheme of the composition of the vertebra.

FIFTH STAGE

The chordal sheath, once thick and powerful, is now on the wane. It may be reduced to a very narrow tube so that it contains few traces, if any, of immigrated cartilage, and is not strengthened by calcareous deposits; if the anyhow very thin elastica externa has been destroyed, it passes imperceptibly into the skeletogenous layer. Some authors discern and describe a transitory zone, the thickness of which varies according to whether they attribute more material to the former chordacentrous or to the coming arcocentrous formation.

This vestigial representative of the chordal sheath shows an unmistakable tendency to hurried and condensed ontogenetic development. Cells naturally "perichordal" surround the chorda with a continuous

[1] The following shows the danger of arguing from preserved specimens. The vertebral column of an adult *Alopecias* (Cambridge Museum) was prepared and carefully sketched. After the specimen had been kept in alcohol for some weeks, the sketch seemed so full of faults as to be almost unrecognisable. The well-calcified central columns and the more numerous arches had shrunk at different rates. The arches had actually shifted away from those centra upon which they had been lodged in the fresh state, and the numbering of corresponding parts was upset. A diagram of the present state would give an erroneous impression.

mantle, in which appear soon, to the right and left, dorsal and ventral clusters of the forming cartilaginous arcualia. These are metamerically repeated. Instead of a stiffening by means of calcareous matter which infiltrates or plates the cartilage in the style of the Elasmobranch's cyclospondylous type, henceforth both endochondral and ectochondral ossifications assert themselves. The preponderance of the newly invented bone of the arcualia seems to be responsible for the gradual suppression of the chordal sheath, and soon for the various constrictions of the chorda itself, which is now doomed to disappear.

This type of vertebra, the axial portion of which is predominantly formed by the arcualia, is called Arcocentrous and is that of bony Ganoids, Teleosts and all Tetrapoda (Fig. 5, stage 5).

OTHER VIEWS AND MODIFICATIONS

Goette invented the term "primaerer Wirbelkoerper" for the tube-like, apparently not yet metamerised, cartilaginous column, which, becoming surrounded by the bases of the arches, is converted into the "secundaere Wirbelkoerper". Most embryologists have accepted these terms, occasionally with some modifications. Miss Abbott and myself had especially stated that it would be wrong to assume that the so-called membranous stage is followed by a non-segmented continuous cartilaginous stage. Schauinsland agrees with this warning. He states, however, that the chordal sheath of the Elasmobranchs and Dipnoi "entspricht dem primaeren Wirbel" of the Amniota, to which Goette's terms refer. "Entsprechen" is a vague word; it may mean "answer up to", homologous or analogous representation, continuity or substitution.

The young cartilaginous column exhibits many cenogenetic features. For instance, the originally right and left bands of the basiventralia appear in *Xenopus* (see Anura, Chap. XVIII) as one median band below the chorda and remain so as long as they last, without metamerism, a simple case of summary condensation. But at the same time the equally continuous but paired dorsal bands show from the beginning prominent indications of the basidorsal units.

Although the chordal sheath of the Elasmobranchs is not always the same entity as the arcualia-containing mantle of the Amniota, the error of confusing them has crept in. Further, since the sheath produces chordacentra, whilst the amniotic centra owe their origin to the interarcualia, the mistake has arisen that the bodies, centra, or Koerper have come to be considered as quite different from the "arches".

Henceforth the vertebrae consist of "Koerper und Bogen". As the psychology of this argument involves also the all-important question of Notocentra *versus* Gastrocentra, we may sum up this idea in another way: body or centrum is that portion of a complete vertebra which is surrounded by or eventually carries its arches. Therefore the mammalian centrum, i.e. the centrum or body by priority of name, cannot be composed of arches. It must therefore have arisen from something predestined, preformed between the chorda and the bases of the arcualia. This confusion is due to using the term "centrum" and the vernacular "body", or main mass, promiscuously and in two different senses. Many Sharks have developed the unique feature of chordacentra. The main mass or body of the pseudocentrous vertebra (e.g. Urodeles) is made of the neural and haemal arches (basidorsals + basiventrals). The main mass of the Amniotic gastrocentrous vertebra is its "centrum or body", made of the ventral interarcualia or interventrals as carriers of the neural arch (Fig. 8).

Just as the chorda may be almost completely suppressed in arcocentrous vertebrae, so the formation of a bony carapace may lead to partial reduction and suppression of the vertebrae themselves. In some of the Plectognathi (Coffer and Trunk fishes) this causes the reduction of a portion of the vertebral column of the trunk to an almost ineffective condition, the spinal cord being sufficiently guarded by a no longer ossified tube of connective tissue, whilst the muscles have gained stronger and broader attachments to the inside of the shell. The same applies to Tortoises, where their vertebral column is fused with, and partly replaced by, the carapace (see Chap. XXII).

It is worth remembering that the notochord throughout its existence retains the archaic character of being devoid of blood vessels, so that we have here an "organ" nourished entirely by osmosis and yet capable of attaining a length of many feet and a thickness of more than one inch. It comes well under the definition of an organ, a structurally self-contained entity, which in this case does not merely passive work by support, but also active work by returning to the straight, thanks to its elasticity.

Chapter V

PIIPER'S THEORY

THE EVOLUTION OF THE VERTEBRAL COLUMN IN BIRDS

Very recently Piiper has proposed the following ontogenetic and phylogenetic stages of the vertebral column.

For comparison, and as an indication of the wide scope of Piiper's paper, a condensed account is here given of his chapter headed "Ontogenetical Stages and Phylogeny".

He distinguishes four main stages in the membranous and cartilaginous condition of the vertebral column. His principles of classification are (1) the form of the notochord, (2) absence or presence of the perichordal and vertebral rings, (3) the intrasklerotomic fissure, (4) the intercentra.

Stages I and II are in the mesenchymatous and prechondral stages, III and IV the phases of the column in the later prechondral and in the cartilaginous condition.

I. Prespondylous. Notochord in the original cylindrical condition. Perichordal rings not yet formed, but single mesenchyme cells are detached from the sklerotomes and begin to surround the notochord: this is our dawning immigration into the chordal sheath, the ontogenetic hazy formation of the sheath as shown in Elasmobranchs and Gano-Dipnoi.

Sklerotomes divided by the intrasklerotomic fissure into cranial and caudal "sklerotomites".

II. Protospondylous. Notochord with intrasklerotomic dilatations and intersklerotomic constrictions. The dilatations are enclosed by mesenchymatous perichordal rings which later fuse into the perichordal tube, and the ring is flanked fore and aft by a dorsal and ventral pair of sklerotomites: on the ventral side these flanking masses fuse longitudinally.

III. Mesospondylous. The notochord now displays intersklerotomic (i.e. intravertebral) dilatations and intrasklerotomic (i.e. intervertebral) constrictions.

The column consists of vertebral bodies (centra + arches), intercentra and intervertebral bodies. The outer portion of the centrum comes from the dorsal and ventral arcualia. The intercentra (Birds) are present throughout the column, and consist of the fused ventralmost portion of the IV and BV. The intervertebral bodies consist of the primary intervertebral bodies flanked by the caudal portion of the IV: there is here apparent confusion.

IV. Metaspondylous. Secondary dilatations appear at both ends of the centra, the main dilatation in the middle of the centra becoming reduced.

These dilatations are of no value. Dilatation is the expression of the fact that the chorda grows when it is not prevented by other growths, or what is the same thing, when it is not protected by a ring; the presence of a ring, flanked by the sklerotomites, for a while restricts the active increase of the chorda. The fact is undoubted that the chorda of Birds becomes intervertebrally dilated and intravertebrally constricted, that is, amphicoelous. This happens because the chorda is pressed everywhere else and can grow unimpeded only when the basiventral mass is being reduced. When the chorda is eventually reduced to a meniscal pad, the joint becomes plane.

Piiper continues: The centra and neural arches become elongated and undergo final chondrification. The intercentra, except those of the atlas, axis and tail, are completely reduced.

The author's intervertebral apparatus begins to qualify for supremacy and intervertebral ligaments are formed. Interdorsals and interventrals seem to be neglected, and the avian trunk vertebra consists apparently only of a neural arch and a centrum *sui generis*.

The author then gives and discusses examples of his stages. The prespondylous stage is paralleled by the embolomerous type of fossil and recent *Amias* and some carboniferous Stegocephali. He practically means my acentrous stage, somewhat mixed up with the embolomerous. The vertebral centra are decidedly arcogenous, with a cranial and a caudal half. The anterior arch-bearing half of the embolomerous vertebra is comparable with the lower or chordal portion of the caudal sklerotomite; the caudal half of the centrum is similarly comparable with the cranial sklerotomitic portion below the spinal ganglion. He thus includes the interdorsal with the interventral arcualia. *Cricotus*, for example, according to Piiper's view, has not yet quadripartite vertebrae, the ID and IV of the trunk being not yet present as separate bony units.

The author combines here the prespondylous with the early stages of his protospondylous stage. The real protospondylous stage produces four prechondral arcualia, as for example in the quadripartite Archegosauri, which are the prerhachitomous vertebrates (Pseudocentrous Type, Chap. XVI). He assumes that (as indicated partly by Birds) the supra-interdorsals join the basidorsals. He makes the acute remark that probably a layer of pliable tissue (the perichordal rings, later the perichordal tube) was interposed between the osseous arcualia and the notochord. In fact, these creatures retained a perichordal sheath still sufficiently thick to be turned into chordacentra if strengthened by calcification and immigrating cartilage; this, however, has not taken place.

In the mesospondylous stage intercentra are developed from the subchordal portions of the sklerotomes, as in Rhynchocephalia and some Cotylosaurs; they are very likely homologous with the intercentra and chevrons of Birds. Further, the centra of *Sphenodon* and Birds are homologous.

Finally, the metaspondylous stage is characterised by the absence of trunk intercentra, and by the amphicoelous centra, which recall the Mesozoic and Tertiary toothed Birds, which are amphicoelous and lack the intercentra. We ask how can the metaspondylous stage, which presumably is that of the modern Birds too, be characterised by amphicoelous vertebrae, considering that fossil vertebrae with opisthocoelous, procoelous and plane joints are known; again, the usual joint is now saddle-shaped, a modification which combines procoelous and opisthocoelous characters. Our author seems to have neglected the salient feature that in the Amniota the bulk of the axial portion is tending (1) to absolute preponderance, (2) to become the sole carrier of the neural arches, (3) to become eminently subchordal.

Piiper's conception of the composition of Amniotic vertebrae is extremely complicated, and it seems to contain an error which is difficult to trace and to explain. It concerns chiefly his intervertebral ring or primary intervertebral body. This arises according to him in the "intervertebral spalte" or in the vertebral fissure, and then the chorda becomes constricted by it. The ring is said to differentiate into three zones, which we will mark as α, β and γ (Fig. 7). The anterior zone α is called the opisthospondylous zone, because it coalesces with the posterior portion of the adjacent "primary vertebral ring or body". The posterior or prospondylous zone coalesces with the anterior portion of the other adjacent primary vertebral ring, which by this fore

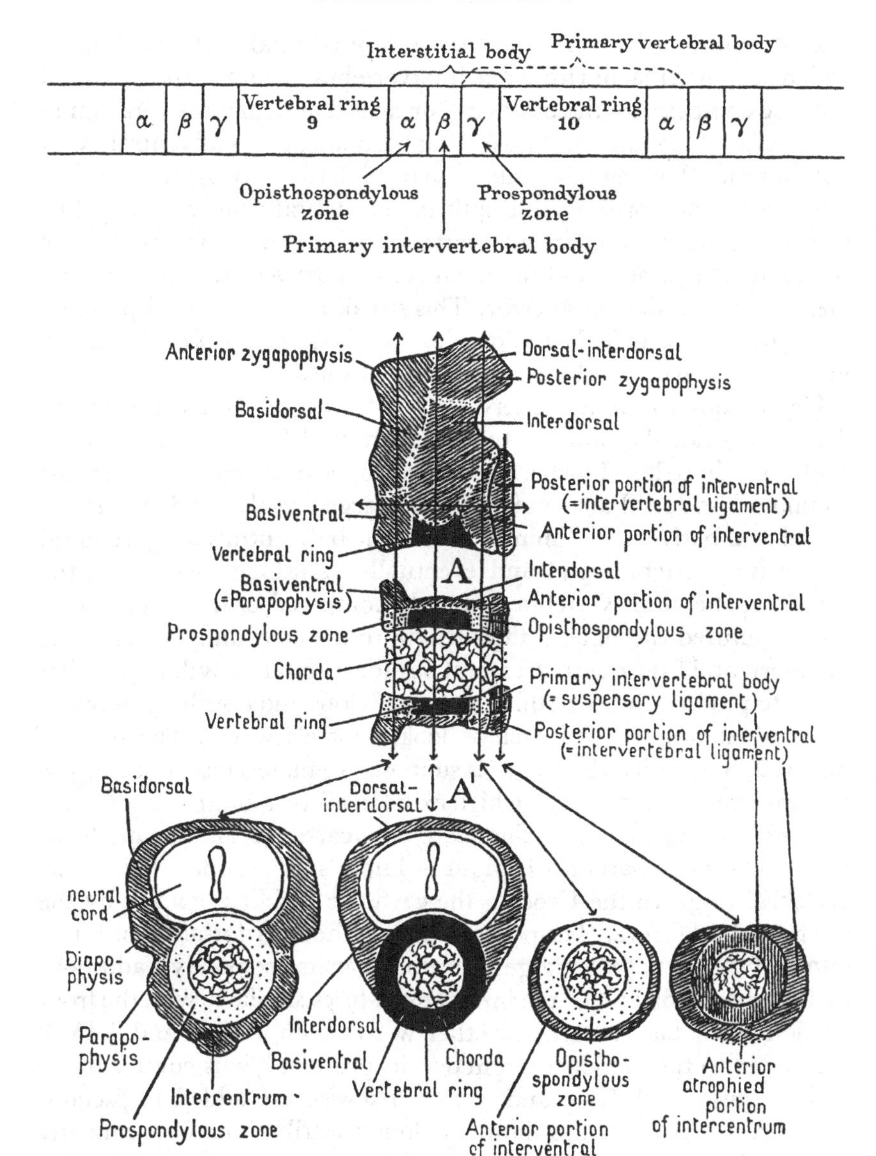

Fig. 7. Diagram (from Piiper) illustrating the composition of a cartilaginous bird vertebra. *A*, side view of vertebra and intervertebral body; *A*ꞌ, frontal section of vertebra *A* in the direction indicated by horizontal double arrow; *B–E* represent transverse sections of vertebra *A* through regions indicated by vertical double arrows. (*Proc. Roy. Soc.* vol. 216B, p. 349.) Piiper's theoretical scheme is shown above.

and aft addition becomes the "secondary or final vertebral body", which of course reaches from one intervertebral fissure to the next, and must be composed of the basidorsal or neural arch and the centrum in the final adult condition. We ask what becomes of the basiventral. Our author somewhat vaguely assumes that this basiventral is formed out of the middle zone of the triple split intervertebral ring, which middle streak is spoken of as an intervertebral ligament, giving rise to the suspensorial ligament and the meniscal pad between the adult centra. Here lurks a fundamental error. This streak has not escaped previous embryologists, who looked upon it as a formation *sui generis*, instead of identifying it with the last axial vestiges of the dwindling basiventrals.

Fig. 8 shows that the intervertebral fissure of the gastrocentrous Amniota is not the same as the intervertebral fissure of the pseudocentrous Urodela. Further, we have to bear in mind that in the Amniota basidorsal and interventral enlarge greatly, while the interdorsal and basiventral diminish in size, the basiventrals being reduced to the intervertebral pad and eventually vanishing. Again, in the Amniota, the adult vertebra is mainly composed of the large neural arches, sutured throughout their length to the centrum. Further, the intervertebral fissure as such cannot possibly produce anything, and if it seems to produce out of nothing prospondylous and opisthospondylous zones, their origin had better be looked for elsewhere. The mass of material, into which the fissure is supposed to materialise, must be the true intervertebral cartilage which exists in all vertebrates between the successive basal arcualia. Therefore this cartilage alone can throw light upon the apparent behaviour of Piiper's three zones of his intervertebral ring. In the Urodela the cartilage is split vertically by the intervertebral fissure. The posterior half of the interdorsal is continued into, or fuses with, the enlarged basidorsal as its prespondyle addition, and the posterior half of the interventral fuses similarly with the front portion of the basiventral. In other words, both $\frac{1}{2}$ ID 8 and $\frac{1}{2}$ IV 8 are added to the pseudocentrum 9 as its prospondylous contribution; and the anterior $\frac{1}{2}$ ID 9 and $\frac{1}{2}$ IV 9 likewise are added to pseudocentrum 9, but as its opisthospondylous contribution. Consequently the pseudocentrum is at either end flanked and augmented by half of the intervertebral cartilage.

The third or middle portion, β, does not exist either in Newts or Amniota. In the latter what looks like a middle zone is the attentuated BV 9, which originally lying behind, caudally to the intervertebral fissure, comes to lie as a pad in the precise space of the fissure. It, the meniscal pad, is in reality zone γ and is thus confused with a fictitious β, which owes its existence to a confusion of the original Urodele vertebra, which is holospondylous, strictly speaking sexpartite and has the formula $\frac{1}{2}$ ID 8 + BD 9 + $\frac{1}{2}$ ID 9 dorsally and $\frac{1}{2}$ IV 8 + BV 9 + $\frac{1}{2}$ ID 9 ventrally, with the central axial portion of the Amniotic vertebra, which has the formula BV 9 + IV 9, the latter breaking up, as Piiper states, into three nuclei IV_{I}, IV_{II} and IV_{III}.

It follows that, as Abel first pointed out, the axial bulk or centrum in both Urodela and Amniota is a physiological conception only. The pseudocentrum is augmented fore and aft by intervertebral cartilage. The Amniotic autocentrum, on the other hand, is made entirely from the intervertebral cartilage. The intervertebral fissure is likewise a physiological conception and is not morphologically homologous. This is as it should be, for function builds up the organ out of available material, the organ is not ready made, having then to find out how it can best be turned to use. There is no mystery in this, since the material consists of living cells and life itself is a function. When protoplasm first originated owing to some unknown physico-chemical combination, life originated too.

Chapter VI

CHORDACENTRA AND ARCOCENTRA

The term "chordale Wirbelkoerper" is Koelliker's, 1860. The Koerper or centra are formed by the chordal sheath, to the inside of the elastica externa. It was accepted by Gegenbaur and Klaatsch, and now is agreed to by the majority of morphologists. Klaatsch called the other Wirbelkoerper *perichordal*. For reasons given in 1895, I adopted this distinction, suggesting the terms Chordacentra and Arcocentra, which are now generally used (Fig. 5).

Hasse, 1879, was the first to show that the chordacentra are formed by cartilaginous cells which immigrate into the chordal sheath from the arches. Miss Abbot and myself followed this up, and our paper in 1895 gives a long discussion of their formation and subsequent changes. We then stated that chordacentra are possessed by all Elasmobranchs, which was a great exaggeration and bad mistake, contradicting the previous text. The mistake was the greater seeing that Gegenbaur had given a well-nigh complete list of Elasmobranchs arranged in five categories or stages ranging from complete centra to none, and even to a stage when they are replaced by the arches. Moreover, we pointed out that the abolishing of the chordacentra could be, and has been, brought about either by the gradual suppression or by the conversion of the sheath, the suppression of the primitive centre being the more complete victory, and that this struggle for supremacy showed itself first by the conversion not of the chorda itself but of its product, the sheath. The struggle ended with the supremacy of the arcualia, which after all had introduced conversion by sending cartilaginous cells through the elastica; we need not therefore be surprised when in a round-about way the chordacentrous type may sometimes come partly to resemble the second or arcocentrous type. Klaatsch had lucidly shown that a tendency exists (illustrated by the *Raiidae* in comparison with *Squalodon*) to transfer the immigration of cartilaginous cells into the sheath to such early ontogenetic stages, that the recapitulation of a cell-less sheath becomes more and more obscured, implying at the same time the bursting of the elastica. A grand conception of fundamental importance, rarely grasped by the embryographer.

Further, we stated that in numerous Selachians the dorsal and ventral arcualia grow towards each other, meet and even fuse, ultimately surrounding the true chordacentrum by a ring of cartilage. If this process is transferred back to an earlier ontogenetic stage, it might lead to suppression of the chordal sheath and result in the direct formation of arcocentra. We added that Amphibia and Amniota might possess such arcocentra. Since 1895, when the mixed formation of the centra was still unknown, many traces of transitional remnants have been discovered: such as Schauinsland's interdorsals and even supradorsals in Amniota, remnants of the chordal sheath in the Tetrapoda, and even amalgamation of this sheath with the destroying cartilage.

In the case of the Elasmobranchs we clearly see the following stages. The invention of the chordacentra has led the many genera possessing them to form still larger and larger calcified blocks, which finally form a unique feature; in other genera these apparently most successful structures declined, giving way to the superior arcocentra. We therefore find that the column is acentrous in the lowest Elasmobranchs, chordacentrous in the lower Sharks, arcocentrous in the newer members, which are in the majority; that is to say, invention, success, decline and substitution of something still better. It is one of many instances that an invention (something "come-upon" perhaps accidentally by emergency), if useful, gains momentum and rapidly develops to the utmost limit. Still "das Bessere ist des Guten Feind", "Le mieux est l'ennemi du bon."

Ridewood failed to appreciate this chapter of evolution. Instead of realising the fact that the fully developed chordacentrous vertebra is a unique, perfect structure, and found only in Elasmobranchs, he relegated it to a side issue. He was attracted, like Hasse, by the almost endless modification of the rings and stars of the cyclospondylous and asterospondylous types, single, multiple and in combination. Being calcified so as to form such exquisite objects for fossilisation, and being often the only remnants of their bygone owners, they are a boon to the systematist, while to the morphologist these pictures, beautiful in transverse section, are of less value. Ridewood became a champion of the arcocentrous vertebra.

Marcus and Blume find fault with the division of the Wirbelkoerper into Bogencentra and Chordacentra, which they say has done much mischief, and they blame the many adherents of this view, for instance Abel, who does not accept the homology of the amphicoelous vertebrae of Sharks, Teleosts and Ichthyosaurs! On p. 30 of their paper they call their parachordalia (the half-moon cartilages on the side of the chorda) "Chordascheidenverstaerker", and therefore hold that they are to be contrasted with the arcualia. But these very arcualia "verstaerken" the chorda by sending cartilage cells into it and thereby show the chordacentra to be a side departure, a product of the previously existing arcualia. To this they have several objections:

(1) In *Hypogeophis* they have found cells entering the very interior of the chorda. Therefore their vertebrae are Chordawirbel. Perhaps we are expected to conclude that the cartilaginous septa of the Geckos warrant the same conclusion.

(2) Their parachordals are the first stage towards vertebral formation, whilst the Bogenelemente appear later. And now comes the surprising statement "Wahrscheinlich giebt es gar keine Bogenwirbel", which is reason enough for the collapse of Gadow's scheme.

(3) The vertebrae of cartilaginous fishes are also partly to be derived from Bogenelemente.

(4) *Torpedo* is an interesting intermediate form between those with vertebrae made by the sheath with immigrated cells, and those whose chorda contributes nothing and leaves the development entirely to the perichordal cells which Marcus and Blume derived from the sklerotomes. Now if such an intermediate form exists, the sharp division into Chordawirbel and Bogenwirbel (whatever these terms mean) is not permissible. Unless our authors mean centra, we are inclined to ask "Wahrscheinlich giebt es ueberhaupt gar keine Wirbel?" Schauinsland and other morphologists make much of the fact that such transitional conditions do exist between the two types of centra.[1]

Marcus and Blume wind up as follows: We condole with the systematist who, because of defective and misunderstood ontogenetic observation, tries upon such a differential character to bring order into a group of animals. May we, on the contrary, sympathise with the

1 Schauinsland states that I compare the "chorda-cartilages" with the cartilaginised chordal sheath of Selachians, and that the Urodeles possess a cartilaginous chorda-centrum like the Selachians! This is rather reading too much between the lines. I said, *Phil. Trans.* p. 11, "if we assumed for argument's sake that the Urodeles possessed a thick chordal sheath (which I stated immediately before to be practically absent), the intravertebral portion of their vertebrae would undoubtedly be made up of chorda-centra...when an intra-vertebral septum is completed this portion of the vertebral body is practically a chorda-centrum (and all the more if Howe's views of this chorda-cartilage should be correct) but if, as he admits for lizards, this cartilage is formed from BD and BV, then it would be a pseudo-centrum; and this would be a most significant feature connecting *Sphenodon* and lizards with the Urodele condition".

embryographers who, with deficient knowledge of their stock-in-trade, misinterpret what they find because they cannot rise high enough to see the wood of evolution for its trees.

The view well-nigh generally accepted that the axial bodies, the Wirbelkoerper, arise through fusion of the arcualia or Bogenkoerper is safe enough, but is to be modified for the Amniota by the fact that their interarcualia lose their independent status of cartilaginous blocks, which eventually ossify, because their material, which forms the centra, amalgamates with that of the prechondral cartilage and with that of the hitherto not pledged mesenchyme. The process of their formation is therefore an instance of shortened development, a summarising by direct action of the various stages or arrangements by which the vertebra has been laboriously built up, by physiological means. It is a condensed recapitulation by ontogeny, together with many cenogenetic features, some of which are now so ancient as to have themselves become phyletic.

Proofs of this are ample. The fossil vertebrae can be arranged in series connecting the lowest with the highest modern forms, while also showing a satisfactory radiation of Phyla. A fossil find is absolute proof that its structure represents an actually working stage. The embryo is full of promising notes, the value of which depends upon our acumen, or credulity. Comparative anatomy alone is able to perceive what these tokens are meant for, and that they have been issued to tide over some temporary emergency. Whatever is found in the embryo is of no phyletic morphological value, unless it can be shown to have materialised into an organic part in some adult creature of the group we are studying. For instance, in *Cricotus* (see Chap. xvi) the basiventrals form a large stout ring which acts as a pseudocentrum, articulating intravertebrally and intersklerotomically with the true centrum, which is the equally stout posterior ring. The whole vertebra shows that it must have arisen from two different sklerotome halves, and further research promises to show that this embolomerous condition has left its traces in the embryonic vertebrae of most Amniota.

Schauinsland remarks that in embryos of Geckos a peculiarity appears which recalls the vertebral formation of Amphibia and diverges from that of other Reptiles. In the latter a considerable development of perichordal cartilage takes place, which calcifies and thereby forms the first appearance of the centrum; on the other hand, in Urodeles and Geckos fibro-cartilaginous lamellae appear in the immediate vicinity of the midvertebral portion of the chorda.

Chapter VII

FURTHER DEVELOPMENTS OF THE FUNDAMENTAL SCHEME

The ideally complete vertebra being limited to four right and left pairs of skeletal blocks or units, any further modification by variation in shape and size of any given pair can only be effected in correlation with the others and eventually at their expense. Relative increase in size of the basidorsals and basiventrals implies a corresponding reduction of the intervertebral mass, and vice versa. Increased size of the basiventrals affects that of the interventrals, eventually causing their suppression; and vice versa, preponderance of the interventral mass leads to the diminution of the basiventrals. Upward or dorsal extension of the ventral arcualia affects the dorsal units, and vice versa.

As pointed out elsewhere in detail, any combination and permutation in reason has materialised somewhere (Fig. 8), and not a few cases of convergent or parallel development are discussed.

The causes initiating, and the impetus given to, these changes have to be looked for in the environment. Restricting ourselves here to the Eotetrapoda, the process dates back into the Lower Carboniferous times, when these creatures were sorting themselves out according to the preponderance of aquatic or terrestrial life. This sorting out, implying trial and error, had its ups and down owing to countless vicissitudes. Some went but a short way along a certain line, their plastic frame yielding to the prevailing environment until a new proposition asserted itself to be overcome by a new yielding or accommodation. This could be done, provided established rights allowed of a compromise; in other words, if the material still contained the wherewithal, and plasticity was not yet impaired. Otherwise they came to grief.

Others, more lucky, gained momentum, thanks to the undisturbed inheritance of the newly acquired feature, and established longer, perhaps very long orthogenetic lines. The successful flourished, with future possibilities if they happened to avoid the lure of hypertely, that is to say overspecialisation with exaggeration of a good thing, a formation running wild and getting beyond control.

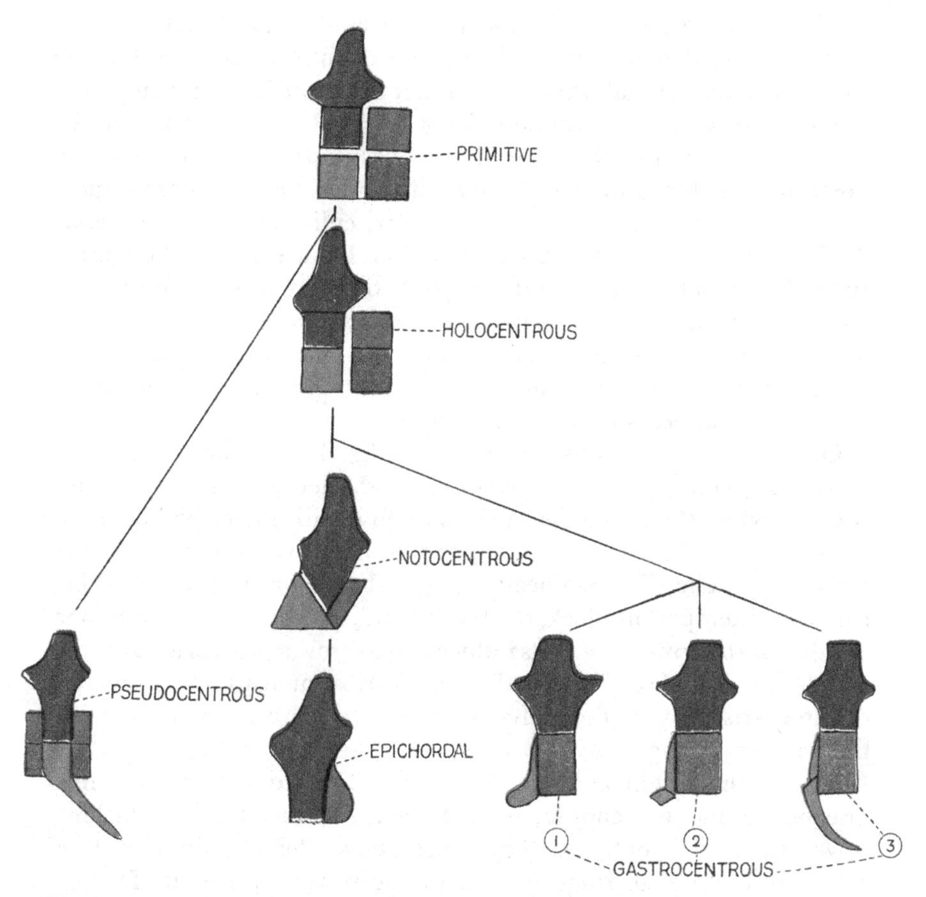

Fig. 8. Diagrammatic tree showing the evolution of the various types of vertebrae from the four primitive blocks. Colours as in Fig. 3.

In the primitive type the four blocks are separate as in the Sturgeon.

In the pseudocentrous type of the Urodela the interbasalia split and fuse to the basalia in front and behind.

In the holocentrous type the respective dorsal and ventral blocks fuse, but the vertebra remains in two separate pieces. This is found in the Embolomeri.

In the notocentrous type the interventrals disappear and the interdorsals grow ventrally around the chorda. This is found in the Rhachitomi.

The epichordal type is an exaggeration of the notocentrous type, in which the basiventral also disappears and the vertebra is formed entirely of dorsal elements. This is found in the Anura.

In the gastrocentrous type the interventral preponderates and the interdorsal disappears. Three forms in which the basiventral may occur are shown: (1) as intervertebral fibro-cartilage only; (2) with the addition of a small intercentrum; (3) with the addition of a larger chevron.

The game was, and is still, played incessantly and at many places; and if played long enough it will produce countless cases of isotely, no matter whether we call them convergent or parallel, heterogeneous or homogeneous, analogous or homologous. Haeckel rightly said that the whole art of tracing the natural true classification of any group of creatures is reduced to being able to distinguish between homological and analogical features. It is a big order, enlivened by cenogenesis. And it must be remembered that if we had knowledge of all the creatures of a given time, we might group together such as seem to run into each other; but if a few scattered imperfect samples are all we can follow up into successive horizons and these in widely separated regions, one in South Africa, another in Texas and a third in Bohemia, the problem appears hopeless of solution.

Of the four pairs of constituent blocks (Fig. 8), any one, two or even three, except the basidorsal which produces the neural arch, may be suppressed, while the remaining units may fuse with their neighbours in various ways. The neural arch may shift from one ventral on to another for support. The ribs, two-headed or one-headed as the case may be, can leave their parent block, the basiventral, and by shifting on to one or other of the remaining units induce this to grow a process to carry the wandering homeless rib. Further, these ribs, at least their chevron counterparts, may fix themselves on to the vertebra in front, to which they do not belong.

The original right and left basiventrals tend to form a semi-ring, grasping round the chorda, with the horns of this meniscus looking upwards, until eventually they meet above the chorda and thus establish a complete ring, or disk, the dorso-ventral height of which equals the thickness of the axial column, if we exclude the neural arch.

Whilst the interventralia are in the cartilaginous Fishes and in the Urodela easily recognised as still separate cartilaginous blocks, the reverse is the case in the majority of the Tetrapoda. Here in the adult stage they are known as separate entities only in a few Stegocephalia, and even then mostly in an unmistakably reduced condition, for instance in the trunk and tail of *Archegosaurus*, in the trunk only of *Chelydosaurus* and *Sphenosaurus*, in the cervical or rather anterior trunk region only in *Eryops*.

Ontogenetically, however, both interdorsals and interventrals can easily be ascertained in Urodela as paired pieces which, combining into dorsal and ventral semi-rings and thus grasping the chorda, fuse into complete cartilaginous rings and thus form the so-called intervertebral cartilage. This fusion is the first hint that in later higher groups the condensing or shortening processes of cenogenesis may show very early stages of the intervertebral cartilage in a state of amalgamation. Thus in the Amniota the actual composition of the centra and the respective shares of the interdorsal and interventral units is obscured. The realisation of this is of the greatest phyletic importance, as it proves the difference between the tripartite notocentrous vertebrae of all those Amphibia (most Stegocephali and the Anura) which are not pseudocentrous like the Urodela, and the also tripartite but gastrocentrous vertebrae of the Amniota.

There are three possibilities for the formation of true centra in contradistinction to pseudocentra.

I. Both interdorsal and interventral blocks join in the making of one centrum. Examples are the posterior axial disks of the Permian *Cricoti* (Fig. 8).[1] For convenience sake I had called this disk the postcentrum, and most palaeontologists and others are inclined to take this view of its composition. It would be the morphologically complete equivalent of the whole intervertebral cartilage of the Urodeles, only with the difference that it becomes more like an independent disk, which seems to tend towards joining the anterior half of the vertebra. This latter cannot be called a precentrum, since it is composed of the neural arch and the rib-bearing or chevron-bearing basiventral.

II. Only the interventral units form the true centrum, which is therefore called the Gastrocentrum (Fig. 8). In its ontogeny it seems to have started as a ventro-lateral mass grasping the chorda from below. Later this mass, in an analogous manner to the basiventrals, is developed into a thick tube which, by surrounding and squeezing, reduces and destroys the centrally placed chorda. This happens partly at the cost of the interdorsals, which we had until recently to assume to have vanished without a trace in the typical Amniotic vertebra. Fortunately they have since been discovered by Schauinsland in *Sphenodon* and quite recently even supra-interdorsals have been found in Coeciliae, and recognised as such in Birds. Even Mammals are said to have

1 *Cricotus* (from the Greek κρικωτός = made of rings) is the type of Cope's embolomerous vertebra, a correctly built term but scarcely appropriate, since "embolus" is the recognised pathological term for something thrown-in, which causes a blockage in a structure to which it does not belong. In the formation of the vertebra, nothing is thrown-in; whether it be a dorsal, ventral or compound element, it is an essential part of the vertebra. Another creature with a ventral block, obviously basiventral, has been described by Cope as *Embolophorus*, but this name and the creature with it seem for some occult reason to have fallen into abeyance. The vertebra has been figured and described in *Brit. Mus. Cat.*

traces. The obvious suppression of these shortlived entities has been dated too far back.

III. Only the interdorsal units form the centrum, which is therefore called Notocentrum (Fig. 8). It is of the greatest significance that this trend to make a tripartite vertebra out of two dorsal and only one ventral pair of the original units has never produced a proper axial block to act as the carrier of the neural arch. As chiefly American palaeontologists have shown, these centra, which are being evolved from the interdorsals, still show in the various Stegocephali every stage from a dorsal pair of osseous blocks, the halves still separate and resting upon the chorda or just uniting above it, through a stage in which they are fused into a semi-ring grasping the chorda with its downward extending horns, until ultimately the mass reaches so far down as actually to wedge itself in between the successive basiventrals. On the strength of this intrusion, these interdorsals may in a way be likened to emboli, but the embolus of *Embolophorus* is the united pair of basiventrals; hence the confusion. How far down these interdorsal wedges reach depends, however, also upon the size of the basiventrals, which may be so much enlarged that they form a continuous series, thereby precluding any further progress of the intruders. Their attempt to make a complete dorsoventral disk is foiled.

In an apparently very well-preserved trunk vertebra of *Eryops*, figured by Matthews, also described in other specimens, there is a gap left between the downward extending horns of the interdorsal. This gap was not filled by chorda as has been suggested, but more probably by the still rather large interventral pair, which if still cartilaginous would not be preserved. This interpretation tallies with the existence of undoubted vestiges of bony interventrals in those still earlier creatures, *Archegosaurus*, *Chelydosaurus* and *Sphenosaurus*, and with the presence of osseous interventrals in the cervical region of *Eryops*. Fusion of these with the downward extending interdorsals would bring the really quadripartite vertebrae of the Rhachitomi,[1] which according to definition should be tripartite, very near the embolomerous type. It was a mistake to imagine that the Rhachitomi were the forerunners of the Embolomeri. Naturally the quadripartite condition is the oldest, and the temnospondylous stage precedes that in which fusion of the units has taken place. But it is easy to show that the Eryopes missed this

1 The type specimen of the rhachitomous vertebra was actually called *Rhachitomus*, but this genus was found to be identical with *Eryops*, both having been described by Cope in the same year.

chance, and instead allowed the interventrals to decrease steadily, with the inevitable result of qualifying for the notocentrous type and thus ending in a cul-de-sac, when these large Amphibians died out as stereospondylous Mastodonsaurs. However, before this happened they founded as a still flourishing side-branch the Anura, which represent the most exaggerated terminus of the notocentrous development by the epichordal type of the Aglossa and other low Anura (Fig. 8). The Aglossa, *Pipa*, *Xenopus* and the somewhat intermediate tiny *Hymenochirus*, may be nicknamed the Ratitae of the Anura. They are the most archaic of all, and yet in some respects are highly specialised and have had time to lose characters, which their immediate ancestors had already acquired and have transferred to the rest of the order. All this may have happened because they have returned to exclusively aquatic life, there being no doubt that the general structure of the whole Amphibian class is due to that most important of all steps ever taken in the evolution of vertebrates—the step from aquatic to terrestrial life, which made the Tetrapoda and culminated in Amphibia, Reptiles, Birds and Mammals.

We have to return to the first possibility materialised by the Embolomeri. The question is whether the same fusion of ID and IV applies to a limited extent to other extinct and even recent creatures. There is indeed one vertebral metamere which in all Tetrapoda still seems to represent this truly archaic stage, namely the Atlas with its odontoid centrum.

It is a bit of special pleading to assume that, because the odontoid as the first centrum of the Amphibia is formed by the ID, and because the Amphibian first centrum must be identical with that of the Amniota, therefore the Amniotic odontoid is formed from the ID. It is a nice syllogism, logical but nevertheless wrong, because one premise is wrong, at least not proven. The assumption that the ID alone forms the odontoid or first centrum is not necessarily correct. We can at least try to show that it is the amalgamated compound which will form in other groups both the ID and IV, this condition dating from a stage before the intervertebral cartilaginous mass was cut up by centres of ossification into separate ID and IV units. The recent Elasmobranchs with all component units in the cartilaginous stage allowed us by analogy to reconstruct the whole fundamental scheme of the complete ancient vertebra. Howes and Swinnerton describe the metameric segmentation of the vertebra of *Sphenodon*, "whereby a series of segments becomes recognisable, each comprising a centrum, an intercentrum and a pair of neural arches; and the facts justify our regarding the centrum as of paired origin,[1] a view which lends support to Gadow's conclusion that it represents the fused 'interventralia' of the lower vertebrate forms". The same authors describe certain ossifying tracts as extending into the lateral cartilaginous expansions, which are then rapidly absorbed, this absorption proceeding from the ventral side upwards; "exactly the same process takes place in the ossification of the centrum of the atlas as of an ordinary vertebra". The above points by themselves prove nothing, but they indicate the completeness of the composition of the first centrum. On the other hand, both authors have shown that the atlas-centre fuses with BV_2, and by peripheral expansion also with centre II, which is an indication that BV_2 is not yet alienated from the centre I mass, and that on the contrary both proclaim their close affinity as ventral elements. Further, there is the preponderant position of centre I and centre II on the ventral side of the chorda, whilst farther back in the trunk the chorda is transferred more and more into the middle, not because the chorda sinks in, but because the ventral mass extends upwards and surrounds it. Lastly, the ventral preponderance in the bulk of the first and second vertebrae of the Urodeles is enormous, and their odontoid seems to be formed almost exclusively by ventral elements, the detail of which is discussed in the chapter dealing with the atlanto-occipital joint (Chap. x).

To sum up. In the embolomerous and pseudocentrous types both interdorsals and interventrals form the intervertebral cartilage. Other Amphibia have become notocentrous, because the interventrals have been suppressed, whilst in the Amniota the interdorsals have become vestigial or absent. The first centrum is homologous in all Tetrapoda. The conclusion is that this first centrum must contain both dorsal and ventral elements. How far this double composition can be traced in other vertebrae, with reciprocal preponderance of dorsal elements in one group, of ventral elements in another, must be left for a special paragraph. The atlas need not be the oldest and first formed of all vertebrae; on the contrary, it is more likely that it is the latest of all, not to be compared in age with the tail. It proclaims its relative youth as the vertebra which still most clearly retains the old temnospondylous condition. The presence of vestigial interdorsals within it is now known in some Amniota.

The ontogenetic changes in the vertebral column of the Fishes can

1 This important point has escaped Schauinsland in his contemporary work on *Sphenodon*, in which the morphology of the centrum is treated in a more easy way.

all be checked by the direct method of comparison with living adult forms, since, notably in the cartilaginous fishes, all the constituent parts are available, and all the phyletic stages are still represented in some living group or other. In the Tetrapoda this is no longer the case, because most of the important phyletic stages are represented by fossils only, in which naturally the cartilage is not preserved and calcareous incrustation is rare. With few exceptions the extent and shape of the cartilage, sometimes its mere existence, can only be surmised by circumstantial evidence. Only such parts as are more or less ossified, or calcified, afford positive evidence. It is not unlikely that the process of ossification was initiated by deposition of calcareous matter in the perichondrium, where it was dealt with by such potentially cartilaginous cells as had learned how to deal with the mineral infiltration, how to organise it, and thereby how to convert themselves into histioblasts of a higher order, that is to say, from chondroblasts into osteoblasts. The whole process may be called severe from a histological point of view, since it implies the existence of histioclasts, cells which both destroy and live upon others and also ingest the indigestible waste products; they excrete the digested mixture peripherally and this forms the wherewithal to make the superior tissue, bone.

There is a deep significance in the fact that a considerable amount of dead mineral substance is relegated into the skin, where it has proved extremely useful as enamel, dentine, vaso-dentine, which is incipient bone, and osteoderm. All this commotion is taking place in the skin, whilst the inner skeleton has remained during long phyletic stages in the purely cartilaginous condition. Milky infiltration of calcareous matter into cartilage is common enough; the advantage to the cartilage as such is not obvious, in ourselves it spells gout. In the lowest known vertebrates, like the Ostracoderms, the preponderant mass of their skeleton lies on the outside, as in invertebrates; the higher the status the more the cartilaginous, or bony, inner skeleton preponderates. This process can be traced through the weird Silurian Ostracodes to the armoured fossil Amphibia, whose recent members are nearly all naked. On the other hand, the bony Teleosts have mostly cutaneous, non-bony scales. Minor examples among recent genera are the yellow patches of *Salamandra maculosa*, which are full of mineral matter, the whitish parts of the skin of many Anura, which are likewise thus densely infiltrated; this is carried to an extreme in various species of *Hyla*, in which individual patches of such hard concretions appear, being sometimes five or more millimetres broad, the patch remaining pure white, any pigment being absolutely excluded. There is an analogy to the above principle of utilisation of waste products which is presented by the pigments; these, with rare exceptions, like the black peritoneal pigmentation of some Lizards, are also relegated to the skin.

Recapitulation becomes more difficult to trace as we ascend from the lower to the higher Tetrapoda, because of the unmistakable fact that one or more vertebral units are liable to drop out, or to fuse with their neighbours; amalgamation may even take place in the so-called prochondral stage. The whole process can, however, be reconstructed by circumstantial evidence. For instance, newts start and proceed ideally, like diagrams. In the Anura, however, all the ventralia are laid down either as a paired, or as an unpaired, string of cartilage. In the Coeciliae this has led Marcus and Blume to the discovery of their "parachordalia" with a hopeless misinterpretation of the facts. They have also discovered in *Hypogeophis* cartilaginous elements which they had to call supradorsalia, the term provided in my general scheme. But our authors take this opportunity only to jeer at the fact, that thirty years ago I mentioned the possession of supradorsal elements as the only differential character of the vertebral column between Fishes and Amphibia. I rejoice that the Coeciliae take their share in bridging that gulf.

What is more important is that interdorsals had already been discovered by Schauinsland in *Sphenodon* and quite recently similar units have also been discovered by Piiper in Birds, which he has more cor- correctly recognised as supra-interdorsals. That such elements must have existed, connecting the various classes of vertebrates with each other, was a postulate of reason. That they are now known in embryos of Amphibia, Reptiles and Birds speaks well for the soundness of the fundamental scheme. Above all, it shows how faithfully phylogeny can be repeated by ontogeny, and last but not least that, what in cartilaginous fishes are solid and separate building blocks, are but temporary units in some members of the higher classes; while in other members they are difficult to recognise and ultimately no longer distinguishable from the rest of the building material.

Our serious attention has to be drawn to Piiper's recent beautiful embryological study of the whole vertebral column of Birds, mostly dealing with *Struthio* and *Larus*, illustrated by nearly fifty clear figures of sections, showing even the histological detail from the early prochondral to the ripe stage of his objects, so that we can easily reconstruct the whole. He traces the changes of our BV, BD, ID and IV in the various vertebrae, from atlas to tail, and pays special attention to the difficult interpretation of the intervertebral cartilage with its derivatives. Almost needless to say he assigns its proper vestigial value to the short-lived traces of the chordacentrous tissue, and considers the centra of the Amniota as arcocentrous. The work is sure to occupy other morphologists, being full of new observations and reflections, not a few unexpected discoveries having been made. One of the most important conclusions is that he dislikes the existence of the gastrocentrum of Amniota, favouring apparently the notocentrum.

Fortunately his descriptions and especially his Fig. 29 (see p. 323) are so clear, that we cannot wish for a better proof of our own contention. According to the author the whole interventral mass is the subchordal half of the cranial sklerotomite, the dorsal half containing the interdorsals, which in passing he states to be larger than the subchordal half; perhaps a slightly exuberant statement, which he does not press further. But the important point is that he divides the interventral mass into three portions: a cranial division in front, which eventually fuses with the basiventral; a caudal division above, and in the middle; and a ventral unpaired or subchordal portion of the cranial sklerotomite, which forms the posterior projection of the whole vertebra. His Fig. 29 of a ten days' Gull embryo makes it quite clear that:—

(1) The interventral mass represents a very large part of the centrum, certainly its whole ventral half, the dorsal half being formed by the fused ID and BD.

(2) That if ID and BD fuse, BD thereby now comes to carry the centrum (we may ignore the author's suggestion that the "dorsal interdorsal", read our supra-interdorsal, is reduced to the zygapophysial pad). At the same time we know the BV can ultimately vanish in the trunk, and in this case the centrum would consist of the ID and IV and therefore be a holocentrum like that of the quadripartite Cricoti. Indeed the unexpected and unavoidable conclusion, amounting almost to a proof that *Cricotus*, which Cope had declared to possess the true Amniotic centrum, represents the ancestral condition of the Amniota, is that vestiges of the holocentrum still exist in Birds, more than probably in Reptiles and perhaps even in Mammals.

(3) Whatever has happened, the intercentrum of Birds, part of our BV, according to Piiper fuses with the anterior portion of its interventral mass. The latter therefore must form the caudal or posterior portion of the intercentrum.

We have learned another lesson from Piiper's observations, namely that in the prochondral stage the posterior half of the vertebra may be likened coarsely to a mass of porridge with three pats of melting butter, the pats being the vestiges of a once united interventral cartilage. It is already confluent even with the dorsal half, and thus by a cenogenetic short-cut produces a centrum, which is, as it were, a single cast of the cartilaginous and other mesenchymatous material. There is nothing against the possibility that this process of condensed and shortened development may in time to come produce the whole vertebra in once cast. Such a vertebra would be truly stereospondyle, composed only of a large centrum which carries the large neural arches.

There remains the justification of the term Gastrocentrum, the distinctive character of the Amniota. We know now that the dorsal and the ventral halves of the intervertebral mass have ceased to be pure, and that where both coexist the vertebra might be held to be classifiable as holocentrous; this would, however, be pedantic and just as useless as declaring that there is no hard and fast line to be drawn between chordacentrous and arcocentrous vertebrae. We might as well give up speaking of Amphibia, Reptiles, Mammals, etc., since these run into each other. The test of a term is the practical and commonsense one. Our criterion is the trend which leads to the final stage of an organ reached by the majority of a group of creatures. The centra of the Amniota cannot be called Notocentra, which are, in their full development, characteristically Amphibian. The centra of the Amniota are as characteristically developed in the opposite direction, therefore we need the new term Gastrocentrum.

The following is the proof of this. In the trunk the chorda passes through the middle of the centra, but further forward it comes to lie nearer the spinal canal, which it almost reaches in the pointed odontoid process, that is to say, the centrum of the atlas, not because the chorda has shifted, but because the ventral portion of the cartilage is still preponderant. Another significant feature is that in large Lizards and various fossils the neurocentral suture lies so far dorsal that the centrum forms the bottom portion, or even more, of the spinal canal, the rest of which is made of the neural arches and of what vestiges there remain of the interdorsals, which, as Marcus and Blume and Piiper have found, fuse with and therefore contribute to the bulk of the neural arches. This means reduction of the interdorsal and reciprocal growth of the ventral elements. Howes and Swinnerton accepted without hesitation my interpretation that the centrum is formed from the interventrals, when they discovered in *Sphenodon* that it arises from a right and left unit on the sides of the chorda, a point omitted by Schauinsland. On the contrary, analogy shows the trend of dorsal preponderance of the ID in the Stegocephalia, culminating in the hypertely of the neural elements in the epichordal type of the Anura. Much is to be gathered from the reciprocal development of the arcualia. When, as in the Eryopes, the ID becomes larger, the IV tends to be reduced to vanishing point, and the BV occupies the vacated space. *Mastodonsaurus*, a terminus of the Labyrinthodonts, is a clear case. The vertebrae have become stereospondylous, being made out of a ventral pair of arcualia which have greatly increased, the IV being in process of being suppressed, while the large triangular wedge-shaped ID points downwards and almost completely separates the successive basiventrals, which carry the BD. The vertebrae have thus become entirely pseudocentrous and are a beautiful instance of parallelism with the Amniota with their IV centrum carrying the BD. The resemblance of the two bipartite vertebrae becomes still more striking, when the ID is lost, as usually happens in the Amniota, and the BV or intercentrum is reduced to a meniscal cartilaginous pad. An indirect proof by circumstantial evidence and analogy that the Amniotic centrum is the IV.

Cope came very near to an understanding of the value of the gastrocentrum, when he made the rather involved statement that the pleuracentra of the Stegocephali remain to be the true centra of *Cricotus*, and as such homologues of the reptilian and therefore true (Amniotic) centra. This is partially true because he had in mind Fritsch's hypocentra pleuralia (our interventrals) and forgot, or mixed them up with the notocentra of the Amphibia, hitherto simply called pleuralia.

The results of this long discussion show that three types of autocentra, or true centra, exist (Fig. 8):

(1) *Holocentrum.* This is composed of ID + IV. It is typically developed by the Cricoti or so-called Embolomeri. Further differentiation of this holocentrum leads to types (2) and (3).

(2) *Notocentrum.* The dorsal half, ID, is preponderant. This development is reserved for Amphibia.

(3) *Gastrocentrum.* The ventral half, IV, is preponderant. This is characteristic of the Amniota.

This shows the biradial evolution of the autocentrous division of the arcocentrous vertebrae from a common stem.

Chapter VIII

THE VERTEBRAL PROCESSES IN THE TETRAPODA

THEIR DEFINITIONS, SYNONYMS AND DISTRIBUTION

The nomenclature in the main centres round the term apophysis, which means an outgrowth from some basal unit, therefore a secondary additional feature or process. Correct examples would be Owen's apophyses combined with prepositions like dia-, para-, hypo-, outgrowths used for muscular attachments or for skeletal articulation. But he also compounded apophyses with nouns, like neuron, haema, pleuron, for arch-like growths which tend to surround, enclose or protect. Such noun-combined apophyses are the neurapophyses and their ventral counterparts, the haemapophyses; both of these are, however, not outgrowths from the centrum or body of the vertebra, as was supposed erroneously, but are the whole growths of the entire original units themselves. Moreover, there was the difficulty presented by his pleurapophyses, which he intended for the ribs. Treating these as belonging to the same category as the neurapophyses, he applied the term haemapophyses to the sternal portion of the rib, thereby causing endless confusion.

It is scarcely worth mentioning that epiphyses and apophyses belong to totally different categories, although it may not always be easy to distinguish separately ossifying epiphyses from hypapophyses or from supradorsals.

Not a few of these numerous terms, although well intended and now carelessly accepted, are based upon faulty idealistic conceptions.

The processes fall into two main groups, arising respectively from the neural arches, that is to say, the basidorsals, and from the centra, that is to say, either the basiventrals or the interventrals (Fig. 9).

A. Processes arising from the neural arches or basidorsal units:

(1) Processus spinosus.
(2) Zygosphenes and Zygantra.
(3) Prezygapophyses.
(4) Postzygapophyses.
(5) Metapophyses.
(6) Anapophyses.
(7) Metasphene and Metantrum.
(8) Centrosphenes.
(9) Diapophyses.

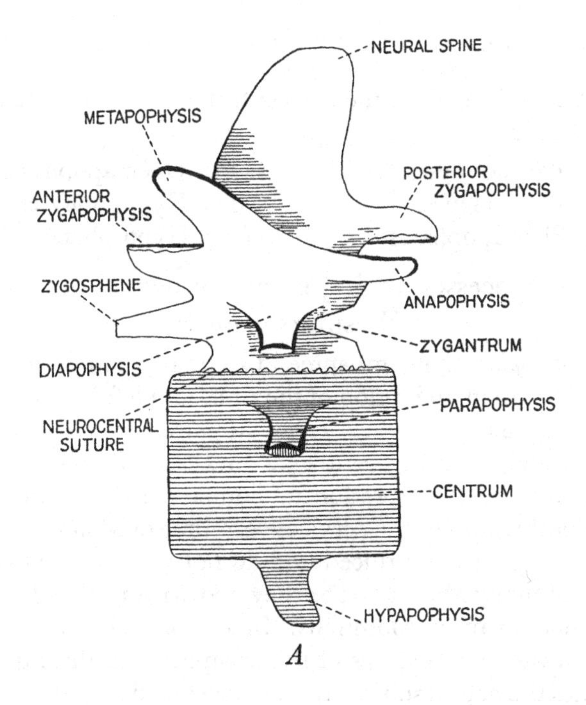

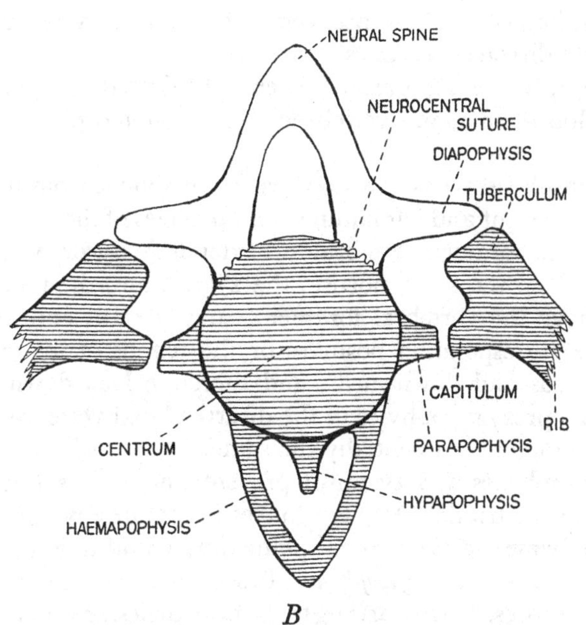

Fig. 9. Diagrams of vertebral processes. *A*, side view; *B*, end view. The dorsal elements and their derivatives are left plain, the ventral elements and their derivatives are hatched.

B. Processes arising from the basiventral or from the centre-forming interventral units:

(10) Parapophyses.	(13) Haemapophyses.
(11) Parasphenes.	(14) Hypapophyses.
(12) Pleurapophyses.	(15) Hyposphenes.

A. Processes arising from the neural arches or basidorsal units.

(1) *Processus spinosus*, neural spine. Originally paired, produced by the dorsal elongation and fusion of the right and left basidorsals. Probably the occasional occurrence of an unpaired subterminal element with a separately ossifying centre, as in *Dimetrodon* and Man, is the vestigial homologue of a supradorsal. Sometimes the neural spines are reduced to nothing, as in the cervical vertebrae of Monotremes; or the right and left halves are reduced and do not meet as, for example, in the caudals of many Mammals; or they end with terminal bifurcation as in the trunk of various Mammals. In Tortoises their distal ends fuse with some of the neural scutes of the carapace. In the vast majority of vertebrates the neural spines are well developed; excessively in some Theromorpha, where they are very long and carry several tiers of transversely directed processes.

(2) *Zygosphenes and Zygantra* (Owen). Indicated in Aistopoda, and well developed on many vertebrae of Pythonomorpha, Snakes and Iguanidae.

The zygosphene arises as a stout wedge or sphene from the anterior surface of the right and left laminae and pedicles of the neural arch, at the level of the foramen vertebrale; it extends forwards, and in Snakes fits into a deep recess or mortise, the zygantrum, the inner walls of which, above the vertebral foramen, carry a right and left pair of facets; the corresponding positives or the tenons of the zygosphene look outwards and downwards, and therefore look down upon the facets of the prezygapophyses in the disarticulated vertebra. The conditions in Iguana are essentially the same.

The zygosphenes and zygantra prevent undue dorsal and ventral excursion of the trunk. Only the snakes in pictures by popular artists proceed by waves in the vertical plane, and climb trees in spirals!

(3) *Pre- or Anterior Zygapophyses.* Processus articulares or Processus obliqui superiores. Paired, arising from the pedicles, or, when higher up, from the laminae, and extending horizontally forwards; with articular facets which are dorsally overlapped by the postzygapophyses. As a

rule the prezygapophysial facets are horizontal, looking upwards; sometimes slanting obliquely, looking inwards.

(4) *Post- or Posterior Zygapophyses.* Processus articulares or Processus obliqui inferiores. Paired, arising from the pedicles or laminae; since their downward facets fit those of the anterior zygapophyses, their bulk lies in a more dorsal horizontal plane.

In the parlance of human anatomy (Fig. 10) the prezygapophyses and postzygapophyses are called respectively superiores and inferiores, regardless of the fact that the prezygapophyses are the more ventrally placed elements. Both kinds of zygapophyses are often rather more in

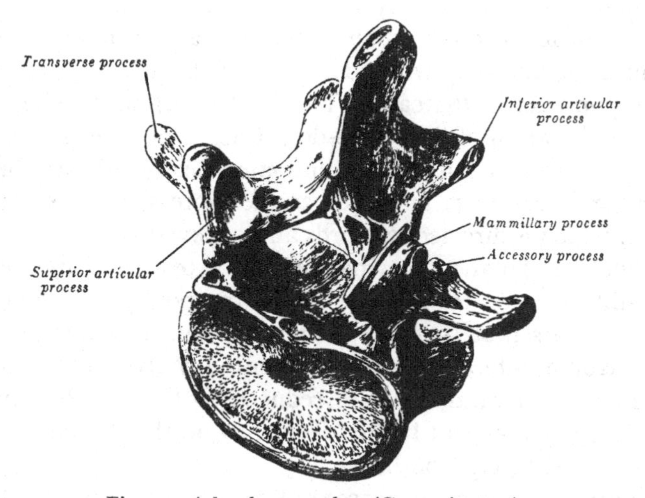

Fig. 10. A lumbar vertebra (Gray, *Anatomy*).

the nature of facets than prominent processes, but they are the oldest paired interlocking articulations and almost universally present.

(5) *Metapophyses* (Owen). Paired swellings or outgrowths from the latero-anterior aspect of the neural arch, arising with broad bases at a transverse level between the spinous process and the prezygapophyses. As processes they are directed forwards and upwards. They owe their origin entirely to muscular attachments. They are restricted to the Mammalia and show great variations. By extending their bases they may be merged partly with the prezygapophyses; or they may occur in the shape of thick, rounded knobs; or the process may be bulky, directed forwards and important enough to form "additional facets" for the reception of the posterior extensions of the preceding vertebra, such as the postzygapophyses. Lastly, in some cases they

may touch the anapophyses (which are shortly to be described), although not articulating with them.

The metapophyses reach their greatest size in the Armadillos as obliquely extending upward struts which act as supporters of the carapace. In Man they are reduced to the processus mammillares, recognisably present only on the posterior thoracic and the lumbar vertebrae, arising between the superior processus obliquus and the transverse process, and touching the processus accessorius, which is the vestige of the anapophysis of the same vertebra (Fig. 10).

(6) *Anapophyses* (Owen). These paired processes are rather elusive, since when present at all they vary much in position, shape and size. They are restricted to the posterior thoracic and lumbar vertebrae of the Mammals, and may with advantage be described as representing the metapophyses of the caudal half of the vertebrae. They are the caudal continuation of the latero-dorsal horizontal swelling, which is sometimes traceable; it then culminates headwards as the metapophysis and reappears tailwards as the more slender, shorter anapophysis, which is directed upwards and backwards. Frequently they lie in a horizontal plane somewhat below that of the metapophyses and extend still more ventro-laterally below the postzygapophyses. Thus it comes to pass that they may approach and may merge with the other horizontal lateral ridge which eventually gives rise to the transverse process. In Man they are reduced to the so-called processus accessorii. They are most highly developed on the thoracic vertebrae of various rodents, e.g. the Beaver.

Originating as muscular processes like the metapophyses, they are intimately correlated with the muscles, and further profoundly influenced by the shape, size and position of the diapophyses, which themselves are outgrowths of the neural arches and owe their origin to the articulation with the ribs. An example of this is shown in *Manis* and *Orycteropus*: on the posterior thoracic vertebrae a small tubercle appears upon the diapophysis which farther back, on the lumbars, divides into the beginning of the metapophysis and anapophysis, the former increasing progressively in length until the last lumbar is reached; on the thoracic vertebrae the metapophysis has a small facet on the under and forepart of its base, articulating with the anapophysis of the preceding vertebra. It is the beginning of the posterior middle accessory joint, described in the American Edentates in the next chapter.

When, as in the Anteaters, the anapophysis, thus restricted to the posterior half of the vertebra, is merged into the mass of the postzygapophyses, which moreover carry the diapophyses, called here "transverse processes" by Flower, the whole compound bulky mass is grasped between the metapophyses and prezygapophyses of the succeeding vertebra and carries three articular facets: the postzygapophyses, the posterior middle accessory, and a third, which overlaps the anterior edge of the next following "transverse process". They remind us of the analogous lateral, intervertebral, additional joints of the lumbars of Perissodactyles.

(7) *Metasphene and Metantrum.* On the lumbar vertebrae of Dinosaurs Broili has described a vertical wedge or process extending backwards from between the postzygapophyses and fitting into a notch between and below the prezygapophyses of the next following vertebra, just above the spinal canal. This sphene or wedge had been named Hyposphene, a term anticipated by Wiedersheim for an entirely different structure in the Gymnophiona, and the notch or mortise was called Hypantrum. To correct this confusion the new terms Metasphene and Metantrum are herewith established.

The metasphene, arising from the posterior half of the vertebra and fitting into the front end of the next following, presents conditions exactly the reverse of those of the zygosphene, nor can it be homologous with either the hyposphene or with the parasphenes.

"Hyposphene" and "Hypantrum" articulations have also been described on the lumbar vertebrae of the theromorph *Diadectes.*

(8) *Centrosphenes.* Case's term for a pair of outgrowths below the anterior zygapophyses on a vertebra, apparently the epistropheus, of the theromorph *Naosaurus.* They would, by position, agree with zygosphenes, but are supposed by Case to have fitted on to a facet of the first neural arch.

(9) *Diapophyses* (Owen). Apophyses transverses of Cuvier, the Radix posterior processus transversi of Soemmering, or the Obere Querfortsaetze of Joh. Mueller. These paired right and left processes of the neural arch, below the level of the zygapophyses, owe their origin entirely to the tubercular attachment of the ribs. If the tubercular portion is lost, the remainder of the rib may nevertheless be carried by the diapophysis; if the whole rib is lost, the process carries nothing, ending free. Frequently the diapophysis is merely a facet; on the other hand, it may help dorsally to enclose the foramen transversarium; moreover, it may serve for muscular attachments. Often it is of considerable length, as in the lumbar region.

On the last thoracic vertebra of Man apparently the whole diapophysis is called processus transversus; on the lumbars the processus transversus ends in three processes: (1) Processus lateralis, the most ventral and main continuation of the whole processus transversus. This processus lateralis, in other mammals, may be absent and a vestigial rib takes its place, as in the first lumbar vertebra of *Equus Przewalski*, Cambridge Museum; (2) Processus accessorius, the remnant in Man of the anapophysis; (3) Processus mammillaris, the remnant of the metapophysis, which on the lumbar vertebrae has shifted from the base of the "transversus" towards the processus articularis superior; on the second lumbar it comes to form a tuberosity upon the articular process itself.

Since the ribs of the Amniota tend to shift their attachments further tailwards and dorsally with a rotation of their long axes, it often comes to pass that both their capitulum and tuberculum articulate with or fuse with the diapophysis, so that this carries the whole rib although reduced. In this case we may speak of a transverse process in a special sense.

B. Processes arising from the basiventral, or from the centre-forming interventral units.

(10) *Parapophyses* (Owen), the Untere Querfortsaetze of Joh. Mueller. These processes or facets are also paired, and owe their origin to the necessity for a capitular attachment for the ribs. They are, however, not always serially homologous with each other, if the vertebral unit which produces them is taken as the criterion. In most cases they arise from the latero-ventral side of the centrum, well below the neurocentral suture. If, however, the capitulum crosses the suture, the resulting facet is formed by both the centrum (interventral) and the neural arch (basidorsal); it may develop into a very short process and thus become a true, compound transverse process: this condition has not materialised, however, in any surviving forms, but may have existed in certain fossil groups. The parapophyses are in fact the direct response of that part of the vertebra which the capitulum of the rib needs for its support.

The lumbar and caudal vertebrae of various groups of vertebrates appear at first sight to carry long blade-shaped transverse processes, which however reveal themselves in immature individuals as composed entirely of the ribs, with an oval facet directly upon the centrum.

The nature of the transverse processes, whether diapophysial or parapophysial, will be discussed in detail under Crocodiles, Tortoises, Lizards, Snakes, Birds and Mammals.

(11) *Parasphenes*. This name merely represents the extreme development of the parapophyses which is found in the posterior region of the trunk of the Gymnophiona.

(12) *Pleurapophyses* (Owen). This is the antiquated term for the ribs, in conformity with the notion that the ideally complete vertebra consisted of (1) neural arches for the enclosure of the spinal cord, (2) pleural arches enclosing or protecting the body cavity with the viscera, and (3) haemal arches, enclosing the blood vessels in the tail and partially protecting the dorsal aorta in the trunk of Elasmobranchs.

The next two terms, Haemapophyses and Hypapophyses, for want of proper definition have been so much confused with each other and used so promiscuously, that many authors consider them as homologues and hold that one of these redundant terms ought to be dropped. Yet they are not the same elements and they can both occur in the same individual, for example, in the cervico-thoracic region of Birds.

(13) *Haemapophyses* (Owen), the Untere Bogen, Haemal arches, or Blutbogen. These are the ventral counterparts of the neurapophyses, being outgrowths of the basiventrals, lying ventro-laterally and originally paired; they therefore belong to the anterior end of their vertebrae. Their reduction to unpaired nodules will be described in the cervical region of Lizards. Their relation to the ribs in the Elasmobranchs will be detailed in Chap. XIV. They reach their fullest development as the chevrons of the Amniota. Their vestigial remains form Cope's intercentra.

(14) *Hypapophyses*, the Processus spinosi inferiorés, or Unten Dornfortsaetze of Rathke. He calls the ventral half of the atlas ring, BV, the first hypapophysis. Leydig concluded that in Snakes the haemapophyses of the trunk pass imperceptibly into the hypapophyses of the tail.

The hypapophyses arise as median, unpaired longitudinal ridges from the ventral side of the vertebral bodies or centra, that is to say, they are in the Amniota clearly of interventral origin. They tend to arise rather from the caudal half or even the end of the vertebra as a single, unpaired pedicle which tends to bifurcation: they show these characters most prominently in the tail of Snakes, and in the thoracic vertebrae of various Birds, for example, in Penguins and Colymbidae or Divers. Other examples of long hypapophyses are found in the neck or anterior trunk vertebrae of Snakes, notably Cobras and the egg-swallowing kinds, and in the cervical vertebrae of the pythonomorph *Clidastes* and Crocodiles. With one exception these processes always serve for muscular attachments, never for articulation.

(15) *Hyposphenes* (Wiedersheim). The only case where the hypapophyses are used for the purpose of articulation occurs in the lumbar region of Gymnophiona: they here reach their greatest development and are grasped by and articulate with the parasphenes. They are then distinguished by the name of Hyposphenes.

It is often difficult to trace the exact homologies of these numerous processes, but they deserve our interest as illustrations of the fact that similar requirements have evolved them independently in totally different groups of creatures, for example, Snakes and Iguanas, Gymnophiona and Aistopoda have followed the same principles of interlocking of their vertebrae, though they use different material to bring this about. Dinosaurs and Iguanas interlock with material from the zygapophyses but in reverse position. The problem of interlocking was always present. How to satisfy this "besoin" (Lamarck) was a question of detail; it would not be satisfied until in some one of the many possible ways a development had taken place, perhaps accidentally, which answered so well, that the employed material could be improved into a highly specialised piece of machinery. Lamarck's "besoin" is not a yearning or desire: it is the test of living. So long as there is possibility of improvement the organism will not be stationary, "He won't be quiet till he gets it." Whether he is conscious of it is a transcendental question.

Chapter IX

ARTICULATION OF THE VERTEBRAE

THE FORMATION OF ACCESSORY ARTICULATING SURFACES

All the various paired and unpaired processes growing from the vertebrae have been stimulated into existence by irritation, due either to contact with neighbouring skeletal parts or to attachment of muscles. Hence they are ultimately all referable to motion, and it is reasonable to suppose that the first segmentation arose from stabilised, rhythmical, peristaltic contraction of a primordial continuous muscular mantle, budding in the direction of the longitudinal axis. First myotomes and nerves, then sklerotomes, of which the four pairs of basalia by variable arrangement laid the foundation of the central axial jointing, culminating in the formation of the intervertebral joints. These later primary articulations need no further discussion, since they are not formed by "processes".

The formation of accessory articulations between the vertebrae can be traced through the following phyletic stages:

(1) The earliest condition still survives in cartilaginous fishes. In these there are no processes, with the occasional exception of dorsal spinous elongations united into an unpaired carrier of a movable more or less jointed spine, for example, *Chimaera*.

(2) Attempts to form paired zygapophyses are frequent in bony fishes. Owing to the backward slant of the elongated spinous processes these come into contact with one another. This leads to a swelling of the anterior aspect of the spinous process, at first unpaired, and lying somewhat distally from the base. The thickening further develops into a right and left forward growth, which eventually grasps the spinous process in front, giving the first indication of a pair of prezygapophyses. The corresponding postzygapophyses formed on the spinous process grasped by the prezygapophyses at first remain rudimentary, but later, in more advanced fishes, become improved so as to form fully developed postzygapophyses, which are overlapped from behind by the prezygapophyses. Both are eventually placed considerably nearer the centres of the vertebrae. In some Teleostei both pairs of

zygapophyses are drawn out into long spines directed respectively forwards and backwards, which no longer touch each other; in other cases, for example, in Xiphias, the Swordfish, the prezygapophyses may grasp the large upright blade-like dorsal spinous process and limit its excursion, although without skeletal contact. An interesting analogy is presented by various Whales, in which the prezygapophyses of the anterior trunk region are typically overlapped dorsally by the post-zygapophyses; in mid-trunk and tail the prezygapophyses are turned into vertical blades, grasping the spinous process of the vertebra in front (Fig. 11), whilst the paired postzygapophyses become confluent and vestigial in the posterior lumbars and still more in the tail.

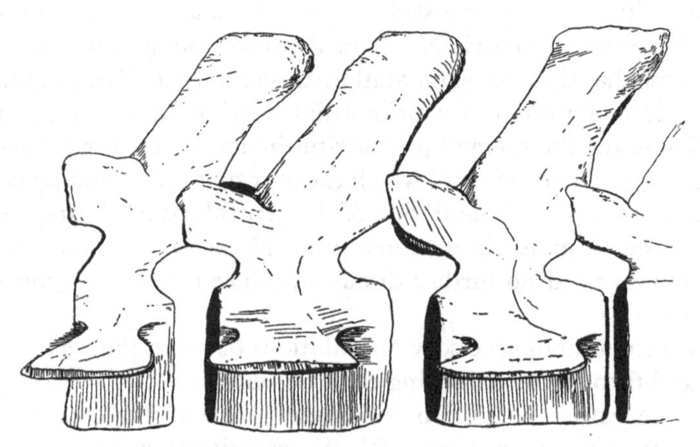

Fig. 11. Left side view of three mid-trunk vertebrae of a Whale showing the prezygapophyses grasping the spinous process.

(3) Prezygapophyses and postzygapophyses have become the normal feature in the Tetrapoda (Fig. 12). The facets of the prezygapophyses look upwards and inwards, being overlapped from in front by the post-zygapophyses, the facets of which are turned downwards and outwards. This rule should be remembered; it enables one to determine whether a given vertebra is procoelous or opisthocoelous.

These three stages are clearly repeated in the ontogeny of the Newts, though in the adult stage of some the prezygapophyses are much the better developed, the postzygapophyses remaining rudimentary.

Considerable modifications are introduced by the ribs with their development of bifurcated upper ends, which gain in variable ways attachment to either the centra, thus giving rise to the parapophyses, or to the neural arches, causing the formation of diapophyses.

(4*a*) The limbless snake-shaped Amphibia represent a side departure. Both Aistopoda and Gymnophiona possess the peculiar parasphenes, which are best developed in the latter group, grasping the unpaired hyposphenes of the preceding vertebra. Both groups have the anterior and posterior zygapophyses, which in the Aistopoda are connected by a lateral ridge which carries the diapophysis for the rib's sole attachment. Short, delicate zygosphenes, fitting into zygantra, exist in the Gymnophiona, but are merely indicated in the Aistopoda.

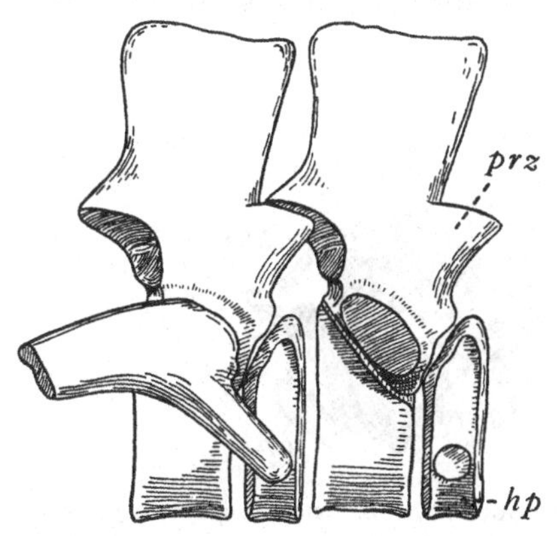

Fig. 12. Trunk vertebrae of *Eogyrinus*, right-side view, drawn by D. M. S. Watson. *hp*, hypocentrum; *prz*, prezygapophysis.

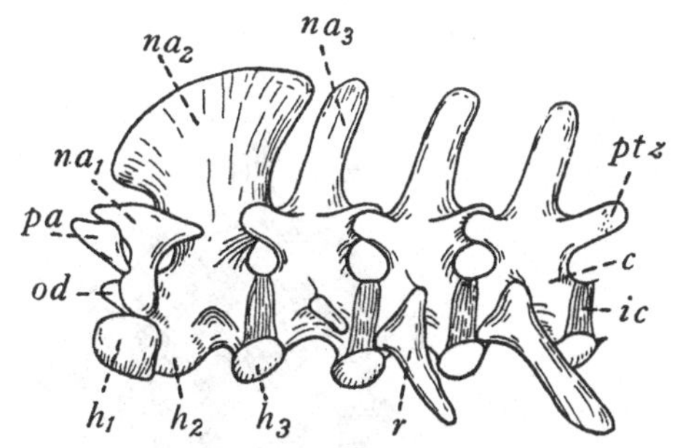

Fig. 13. *Sphenodon punctatum*; left-side view of anterior vertebrae, diagrammatic. *c*, body of vertebra (pleurocentrum); *h*, hypocentrum; *ic*, intervertebral disk; *na*, arch; *od*, odontoid process or first pleurocentrum; *pa*, proatlas; *ptz*, postzygapophysis; *r*, rib. (Goodrich).

This combination of paired and unpaired vertebral articulations by processes from the neural arches and from the vertebral centres is restricted to these apodous Amphibians.

(4*b*) The prezygapophyses and postzygapophyses alone form the articulations, and reach a high stage of development. The condition is present in the recent Reptiles (Fig. 13), excepting Snakes and Iguanidae, and in the Birds. The Chelonia are peculiar; in the Turtles all cervical joints are much reduced in conformity with the short scarcely retractile neck. In those with a fully retractile neck the eighth or last cervical vertebra forms elaborate joints: its centrum fits with a knob into the centrum of the ninth vertebra and its postzygapophyses form broad, curved, concave facets for the reception of the prezygapophyses of the latter vertebra, which is immovably fixed. This arrangement has led to a unique extreme in the Trionychidae: the centres of the eighth and ninth do not touch each other, becoming separated by partial resorption, so that the vertebral articulation is effected exclusively by the correspondingly elaborate zygapophyses. The Snakes and Iguanidae have developed an additional pair of joints: zygosphenes and zygantra (Fig. 14).

(5) The height of perfection of the articulation of the vertebral column is reached by the Mammalia. Several features deserve special notice. As has already occurred in the Anura and in most Sauropsida, the trend of evolution is to elaborate the intervertebral articulations, and restrict them to the neural arches, to the exclusion of any possible use of hypapophyses for this purpose. With regard to other possible

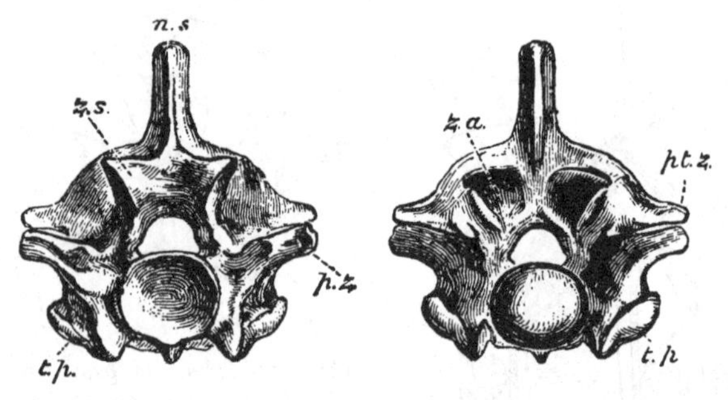

Fig. 14. Vertebra of Python; anterior and posterior views. *n.s.*, neural spine; *p.z.*, prezygapophysis; *pt.z.*, postzygapophysis; *t.p.*, transverse processes; *z.a.*, zygantrum; *z.s.*, zygosphene. (Huxley.)

processes, only the lumbar "transverse processes" of the Perissodactyles and certain Edentates form serviceable contact facets, but even these tend to co-ossify with age; this is an accident, which, not satisfying a Lamarckian "besoin", but being impossible to undo, has to be patched up.

Many long-necked Mammals have a well-developed ligamentum nuchale, which extends from the occiput to the elongated thoracic spinous processes. Being very elastic and strong, it saves labour as carrier of the heavy head, and elevates head and neck into their normal position when the cervical depressor muscles relax: an example is a horse when grazing.

The spinous processes of the thorax are usually directed backwards, and those of the lumbar region forwards, and there is generally one with a straight spine, the anticlinal vertebra, corresponding with the so-called centre of motion (Fig. 15 *A*).

A short bifurcation at the free end of the spinous process, for muscular insertion, is of frequent occurrence. This is a reminiscence of the paired nature of the process, and is carried to the extreme in the caudal vertebrae of many Mammals, the right and left halves not meeting each other. Thanks to the cauda equina there is no spinal cord to protect.

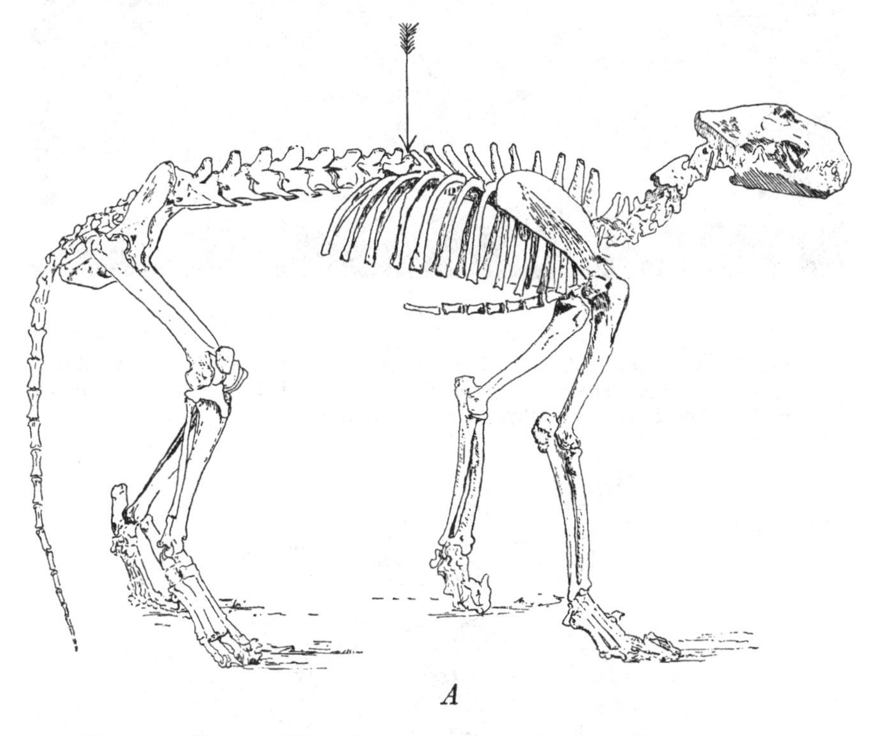

Fig. 15 *A*. Skeleton of Lion showing the centre of motion. The arrow points to the anticlinal vertebra.

A most important and characteristic feature is the development in the lumbar region of an anterior and posterior pair of often large processes, the metapophyses and anapophyses, neomorphs caused by muscular attachment (Fig. 15 *B*, *C*). In the American Edentates they are further developed into pronounced basal processes which carry additional articular facets. They are worth special attention, as a little side chapter of evolution which can be traced from insignificant muscular ridges, or tuberosities, to interlocking joints, "tenon and mortise" fashion, perplexing by their numbers and elaboration. They are secondary facets or processes of the metapophyses and anapophyses, therefore of tertiary

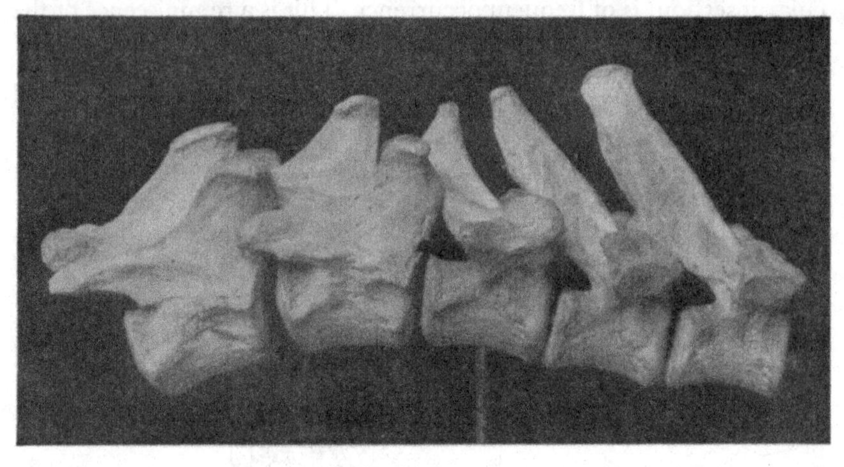

B

Fig. 15 *B*. Enlarged view of the last five thoracic vertebrae of a Lion from the right side, from a photograph. The anticlinal vertebra is in the middle, and the two posterior vertebrae show large metapophyses and anapophyses.

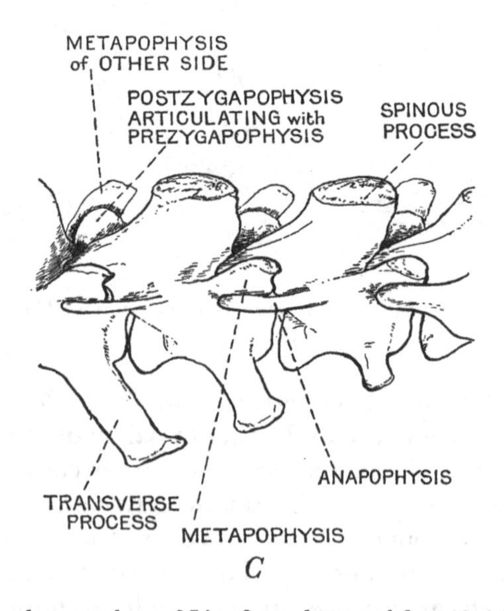

C

Fig. 15 *C*. Two lumbar vertebrae of Lion from above and from the right side, showing metapophyses and anapophyses and the articulation formed by prezygapophyses and postzygapophyses. The lateral articulation on the near side is hidden by the metapophysis.

standing. Flower describes them well, selecting the greater Anteater, *Myrmecophaga jubata*, in which they are perhaps best developed.

The first ten thoracic vertebrae are quite normal, the only paired articulations being those between the prezygapophyses and the postzygapophyses: total facets four, all quite horizontal (Fig. 16 *A*). On the eleventh thoracic vertebra appears a new facet, situated upon the dorsal aspect of the amalgamated mass of the postzygapophysis and anapophysis. This posterior facet, pz^1, in Flower's terminology, which is really the end of the anapophysis, looks upwards and plays against a down-looking process az^1, which arises from the front portion of the metapophysis of the next vertebra (Fig. 16 *B*). This same anterior process, with the protruding prezygapophysis, forms a deep recess for the reception of the double-faceted postzygapophyses. A reversed posterior recess is enclosed between the postzygapophyses and the centrum. There are therefore two anterior and two posterior pairs of facets, giving a total of eight on the eleventh thoracic vertebra. On the twelfth and thirteenth thoracic vertebrae (on the thirteenth and fourteenth in a specimen in the Cambridge Museum) appears a third pair of facets on the hinder margin of the uniting laminae just above the spinal canal. These facets, pz^2, look outwards, and fit against corresponding facets, az^2, on the inner side of the metapophyses of the next vertebra (Fig. 16 *C*). The metapophysis now has two facets, and projects forwards into the space or notch between pz^1 and pz^2 of the antecedent vertebra. There are now three anterior and three posterior pairs: total twelve facets. On the penultimate and last presacral vertebrae, the first sacral being always the twenty-sixth of the whole vertebral series whether there are three or only two lumbars, there appears a fourth pair of facets on the under surface of the hind edge of the transverse processes, which overlap the upper surface of the next following transverse processes. The same happens between the last lumbar and the first sacral vertebra. The penultimate lumbar vertebra, having such a fourth facet only on the posterior end, has therefore three anterior and four posterior pairs of facets: total fourteen. But the last presacral or last lumbar has both pz^3 and az^3, making a total of sixteen articular facets (Fig. 16 *D*): these articulations permit dorso-ventral bending, but disallow rotation. Fig. 17 shows the lumbar vertebrae articulated together.

Essentially the same arrangement exists in Megatherium and in the Armadillos, of which, however, the large *Glyptodon* with its thick carapace is distinguished by the complete ankylosis of most of the trunk vertebrae. These three groups have been classified as Xenarthra,

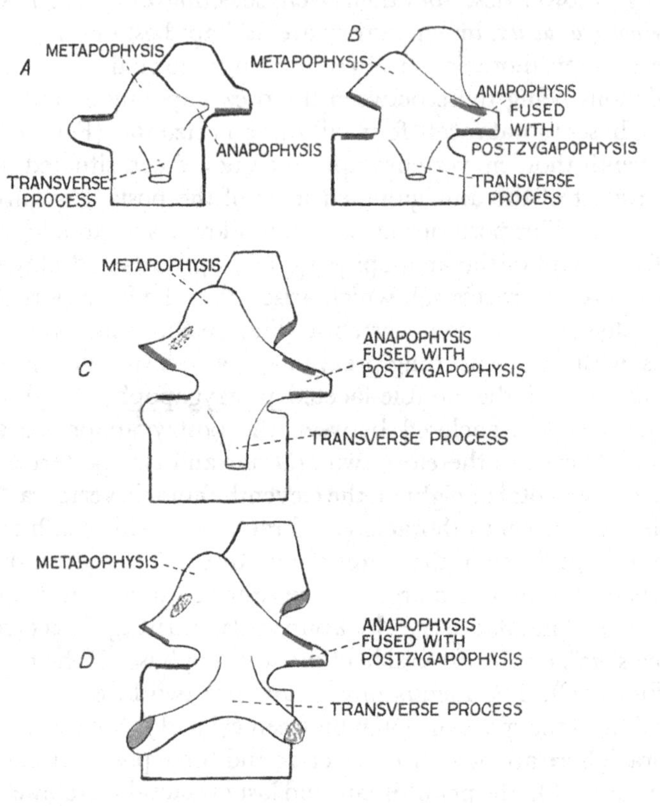

Fig. 16. Diagram of the articular surfaces of the vertebrae of *Myrmecophaga jubata*.

A. Tenth thoracic vertebra. The only articulations are between prezygapophyses and postzygapophyses, the facets being coloured blue.

B. Eleventh thoracic vertebra. Additional facets az^1 and pz^1 coloured in red, lying respectively on the under surface of the metapophysis and the upper surface of the anapophysis which is fused with the postzygapophysis.

C. Thirteenth thoracic vertebra. Showing the third pair of facets az^2 and pz^2 in green, lying respectively on the inner side of the metapophysis and the hinder margin of the united laminae.

D. Last lumbar vertebra. A fourth pair of facets az^3 and pz^3 shown in orange are lying respectively on the antero-superior and the postero-inferior aspects of the greatly widened transverse process.

"with strange joints", in contradistinction to the Old-world *Manis* and *Orycteropus* which have been called Nomarthra, "with customary joints".

The Sloth's lumbar vertebrae are quite aberrant from those of any other Mammals. According to Flower, the Sloths slightly indicate that disposition which is carried out to a greater extent in the Xenarthra. This view has been accepted without question, but it can be shown that the Sloths are neither incipient nor directly reduced Xenarthra.

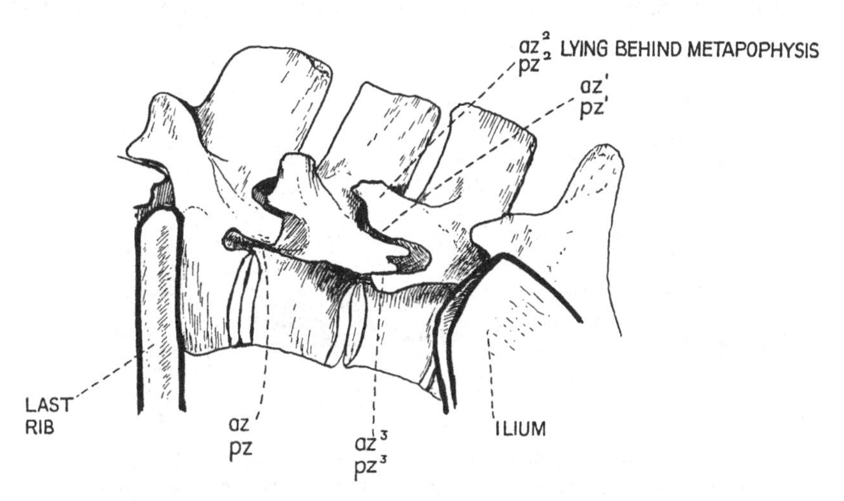

Fig. 17. The lumbar vertebrae of *Myrmecophaga jubata* articulated together, showing the positions of the articulations. Seen from the left side.

Fig. 18 shows my interpretation of the condition in the Sloth. The roof of the neural arch of the last lumbar vertebra forms a broad horizontal plane, the laminae having been depressed into the horizontal. It has a wide anterior and posterior recess. The pedicles extend sidewards into broad horizontal wings, the transverse processes of Flower, which include the anapophyses. They carry on their posterior inner side large facets, almost vertical, but looking slightly downwards. These facets are the pz^1, borne by the anapophyses, which are amalgamated with the transverse processes instead of with the postzygapophyses as in Armadillos and Anteaters. The deep fissure of the Sloth's lumbar vertebra is therefore the remnant of the wide space between the anapophyses and the postzygapophyses of normal Mammals with well-developed and free ending anapophyses. The postzygapophysial facets lie on the under side of the backwardly extending flanges.

The anterior half of the vertebra has undergone changes less difficult to understand. The metapophyses are short and thick low knobs, lying in the transverse plane of the posterior rim of the very wide and deep anterior recess. The combined metapophysial and prezygapophysial mass protrudes but little beyond the centrum. The usually bulky mass between the prezygapophyses and the az^1 facets is in the Sloth reduced to a small sharp ridge, so low that the prezygapophysial and az^1 facets meet. Consequently the articulations of the Sloth are reduced to a total of four pairs: namely, prezygapophyses and postzygapophyses;

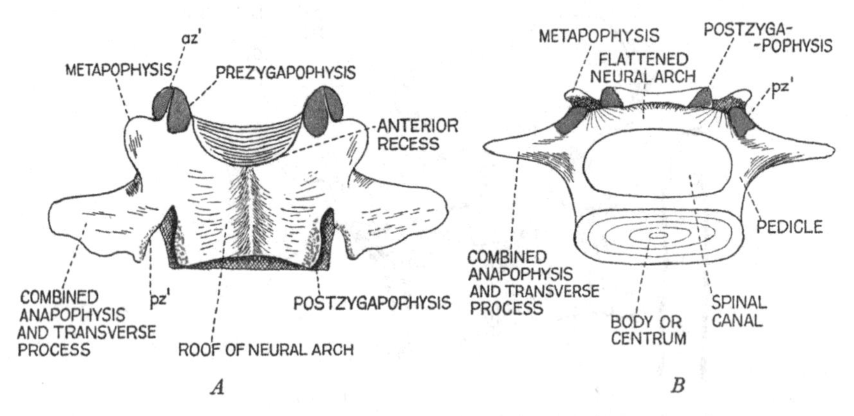

Fig. 18. Lumbar vertebra of a Sloth; *A*, from above, *B*, from behind. They show the great flattening of the neural arch and the anterior and posterior recesses around which lie the articular facets, the prezygapophyses, the postzygapophyses and the accessory facets az^1 and pz^1. In *A* the postzygapophysis and pz^1, being really on the under aspect of the vertebra, are dotted; they are shown in *B* lying on each side of the posterior fissure.

anterior and posterior additional facets az^1 and pz^1. The additional pz^2 and az^2, and the pz^3 and az^3 on the transverse processes, are not present. Everything on this vertebra has been placed high above the very large vertebral centrum, which forms an unusually flat horizontal bottom to the spinal canal. Further, the whole superstructure has been flattened downwards and outwards; thus it has come to pass that the originally wide recess at the hinder aspect of the pedicles has been reduced to a narrow fissure. Lastly, the anapophyses have remained with and amalgamated with the transverse processes instead of with the postzygapophyses, and they lie in the same horizontal plane as the latter instead of below them. These aberrant features suggest correlation with the upside-down attitude which the Sloths assume when moving about and feeding.

Chapter X

EVOLUTION OF AN OCCIPITAL JOINT

The formation of a stabilised cranio-cervical joint is the result of many trials. Every imaginable combination has materialised and is still present in some fish. Every stage or attempt has been ontogenetically repeated. The occipital end of the cranium was in a state of flux just as much as the vertebrae behind. The additions of dorsal and ventral portions to the cranium do not keep step. The formation of a basis cranii by amalgamation of ventral cartilaginous arcualia is already well advanced, while the dorsal portions still retain their separate condition, so that there is the possibility of not merely one, but of several joints in this region, which will give a limited multiple flexibility. Phyletically it has taken a long time to get these discrepant features into accord, like the analogous settlement of the older spinous processes and the younger interspinalia of the Teleostean dorsal fins.

The chordacentra of Elasmobranchs, with their deeply amphicoelous shape, are not compatible with the formation of occipital condyles, which are arcocentrous formations. For instance, in *Hexanchus*, the calcified centra are pierced by the chorda and are gradually continued into the cranium; there is no indication of an occipito-cervical joint, its place being taken by several consecutive joints. The chordacentra are free on the ventral side, but to right and left the fused ventral arcualia form a low ledge, while the dorsal arcualia are large and still recognisable as separate units. In the Tetrapoda the remnants of the chordacentra are reduced to the chordal sheath and as such are continued far into the cranium. The latter and the cervical vertebrae are made out of the dorsal and the considerably enlarged ventral arcualia, both furnishing the material for the postoccipital joint. To understand the difference we have to look for a chordacentrous centrum combined with a postoccipital joint. Examples of this are rare, but are represented by *Chimaera* and the Skates. In the Silurian *Macropetalichthys*, the end of the cranium proper is easily recognisable, but ventrally and behind the vagus it is continued into a long bar, without any nerve holes, which ends dorsally in a high and strong spinal crest, ventrally in a strong condyle, which must be taken to be the anterior half of a distinct cranio-cervical joint. The whole unique structure must have been formed in direct correlation with the equally unique division of the large dermal plates which form the "Arthrodire" joint. Little can be deduced from this, but with the help of *Chimaera* and analogies we may perhaps come nearer an understanding of the principles which shaped the cranio-cervical joint of the archaic bony fishes and thence the Tetrapoda.[1]

In *Chimaera* the basal median portion of the occiput is saddle-shaped, concave from right to left, convex dorso-ventrally. It articulates only with the first chordacentrum and borders the base of the medullary canal. A pair of large flanges stretching out behind the saddle to right and left are undoubtedly neural elements, each with a large facet for the spinal mass behind, which is a complex of several fused neuralia as shown by the three pairs of dorsal and ventral nerve holes. Farther back this complex is continued into the normal trunk region. The stout spinal crest recalls that of *Macropetalichthys*. There is thus formed a saddle-shaped intercentral joint and a right and left interneural joint, the lateral occipital facets fitting upon what may be called the prezygapophyses of an atlas complex.

In the Skates the cranio-cervical articulation is also threefold. (1) A spout-shaped nozzle arises from the spinal column and plays upon the basioccipital mass, which forms the base of the medullary canal; this nozzle is itself perforated by the chorda. (2) and (3) A pair of large flanges arise from the base of the spout and extending upwards articulate extensively with the lateral occipital region.

In the Tetrapod, *Cryptobranchus*, the general appearance of the cranio-cervical joint is strikingly like that of the Skate, the articulations being likewise threefold. (1) A knob-shaped nozzle with a slight cavity on its dorsal surface arises from the anterior end of the spinal column: its apex fits in between the two lateral occipitals, touching both by small facets, or rather resting slightly upon the median backward projection which looks like a shortened basioccipital but is in reality the product of the joined bases of the lateral occipitals. (2) and (3) The lateral occipitals articulate each by a broad facet with the processus anteriores of the

1 For more than forty years I have referred to "Proto-Gano-Dipnoi" as the ancestral group of what at that time was spoken of as Pentadactyloides; this does not prevent Ganoids proper from being ancestral to the Teleostei. The Elasmobranchs are not of this ilk and have acquired their typical characters comparatively late. But ever since Gegenbaur's epoch-making work on the cranium and cranial nerves of Elasmobranchs as the foundation of the origin of the vertebrate head, which was followed by Balfour's embryological researches and discoveries based upon Elasmobranchs, it was taken for granted that these, to the exclusion of all other fishes, represented the ancestral stage of all Gnathostomata.

first vertebra, the facets or processus anteriores being formed by a shoulder behind the base of the knob; the shoulders themselves being the equivalents of ventral arcualia, basiventrals with or without their interventrals, and as such homologous with the centrum of the atlas, or more strictly speaking its pseudocentrum.

The recent Amphibia, Coeciliae, Urodela and Anura are characterised by the absence of basioccipital bones. The cranial extension of the chorda vanishes during embryonic life, and in the Urodela and Coeciliae this chorda lies above the so-called occipital or basal plate, which is supposed to be made by the fusion of the right and left parachordals, while behind the cranium these elements are serially represented by the hypochordal plate, which forms the ventral half of the atlas ring. These parachordals also vanish without producing any permanent cartilage; their room either remains empty or it is taken up by the lateral occipital elements, which, as Peter has shown in Coeciliae, fuse with their basal counterparts below the brain, reminding us of the same procedure as that which Watson has found to take place in the Labyrinthodonts. It is worth noting that this fusion produces, at least in *Cryptobranchus*, the shelf or backward process upon which partly rests the odontoid-like process of the atlas; it is a substitute for the missing basioccipital, a significant analogy to the large paired flanges of the Skates, which are outgrowths of ventral elements extending upwards to meet the lateral occipital cartilage. It is not difficult to recognise the absence of a basioccipital mass as the result of the more or less complete reduction or loss of the hypoglossal region so characteristic of the Amphibia: *Laccocephalus*, one of the oldest Labyrinthodonts, had still a distinct hypoglossus and therefore still possessed a bony basioccipital.

The homologies of the atlanto-occipital region of *Cryptobranchus*, a good type of Urodele, stand as follows: atlas composed of BD_I and BV_I, pseudocentrous, probably containing the IV_I in a fused state. The odontoid-like nozzle is the ventral half of the proatlas[1]; it represents the ventral half of the first sklerotomite, which literally remains to spare. Its dorsal portion, ID_0, would be pleurocentrum O of some authors. The odontoid-like process, fusing with centrum I, stands therefore in exactly the same relation to the atlas as does the true odontoid process, IV_I, to the epistropheus, and its last lingering vestige

1 I myself rejected the existence of such a proatlas, for the reasons given in *Phil. Trans.* 1895, p. 13; and Gegenbaur dismissed it scornfully and tersely in his *Lehrbuch*. We were under the ban of the several segments being added to the cranium by the Amniota. Since then the researches, chiefly embryonic, of Howes, Schauinsland, and others have firmly established the proatlas.

is the ossiculum terminale, a dens or rather denticle of the dens. It follows that, if the first sklerotomite has given rise to the proatlas, the lateral occipital region abutting against the atlas BV_I must be the serial homologue of BD_I. Gaudry's equation that the lateral occipital equals the pleurocentrum has therefore to be given up. Theoretically the neocranium of Gegenbaur may be composed of arcualia like the vertebral column, practically they can no longer be recognised; it is now a purely academic question.

A proatlas is necessary, and various ventral and dorsal remnants of it have been discovered. What is equally valuable is that we have evidence that what befell the proatlas or vertebra O applies to the atlas. The latter has given its centrum to the epistropheus and the atlas ring is in some Marsupials reduced to the neural arch. We can say that, reduced as it is to a ring-pad, it is in contact with occiput and epistropheus; on the other hand, in the Hornbills the whole atlas is fused with the second vertebra into one mass. The atlas is thus liable to lose, so to speak, its personal identity or independence.

The paired condyles of the Urodela and Anura can be satisfactorily explained, although the Anura show one remarkable difference. The vestigial chorda ends in the cranium as in the Urodeles, but runs above the occipital plate and becomes surrounded by the meeting bases of the lateral occipitals, thereby separating the chorda from the medulla, as, for example, in the Labyrinthodonts. The Anuran occipital plate is therefore not homologous with the hypochordal cranial plate, the basal arcualia, of the Urodela, but is in strict conformity with the notocentrous type of the Anura with their tendency to epichordal development with suppression of the ventral arcualia. This is another demonstration that an organ, if wanted, is made from the material which happens to be available; the ventralia in the Coeciliae and Urodela, the dorsalia in the Anura produce the occipital basal plate. The Anura are in this respect in agreement with Watson's admirable analysis of the conditions prevailing in the Labyrinthodonts; he proved that in them the lateral occipitals gradually increase in size and produce the entire right and left condyles, while during this phyletic evolution the basi-occipital gradually decreases. The Anuran condylar flanges articulating with the atlas are consequently not strictly homologous with those of the Urodela. The Frog's atlas is, however, not an entirely dorsal structure, since, as was shown by Ridewood, the hypochordal mass is still preserved, uniting with the BD_I. The ventral arcualia being fused into longitudinal bands, it is impossible to ascertain the share of the interventrals in the atlas-centre, nor how much, if anything, remains of their proatlas.

The existence of a basioccipital in the older Labyrinthodonts is of the greatest importance, since it forms a link with the triple condyle of the Sauropsida (Fig. 19). Watson shows that the large occipital condyle is formed by the basioccipital. The condyle is nearly circular; the lateral occipitals may contribute to it, making it triple with three facets. The lateral occipitals are small, articulate with the basioccipitals, and their right and left bases almost meet, excluding thereby

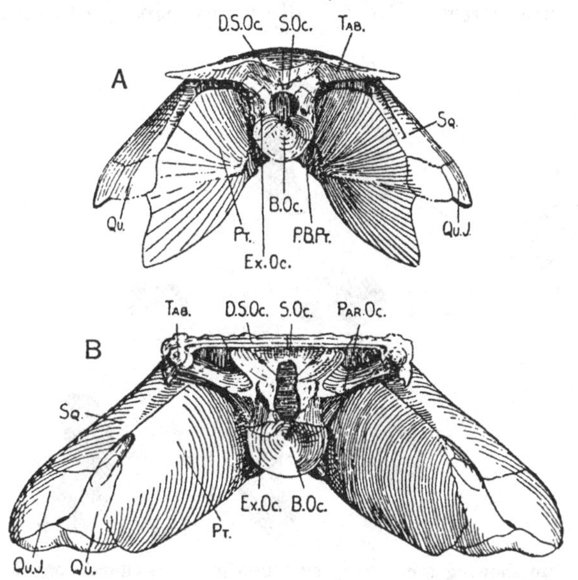

Fig. 19. *A*, *Palaeogyrinus decorus*. Reconstructed from type. *B*, *Orthosaurus pachycephalus*. Reconstruction of the occipital aspect of the skull. (Watson.) *D.S.Oc.*, dermo-supra-occipital; *S.Oc.*, supra-occipital; *Tab.*, tabular; *Sq.*, squamosal; *Qu.J.*, quadrato-jugal; *P.B.P.*, basipterygoid process; *B.Oc.*, basioccipital; *Ex.Oc.*, lateral occipital; *Pt.*, pterygoid; *Qu.*, quadrate; *Par.Oc.*, paroccipital.

the spinal cord from the basioccipital. The condyle is concave, according to Watson, for the reception of the atlas. In *Palaeogyrinus* exoccipitals, that is to say the lateral occipitals, remain separate; in *Orthosaurus* they almost meet, and the author figures well above the basioccipitals a right and left articular facet which undoubtedly must fit upon the neural arches of the atlas. The basioccipital articulates largely with the basiventral; the small exoccipitals with the basidorsal. For the very variable composition of the condylar joint the reader is referred to the numerous figures (Fig. 20), which save tedious description. It is sufficient to state here that any intermediate stage occurs in some creature or other; for instance, first there is the primitive triple condyle with all three units of equal size, followed by either preponderance of the lateral units, or the gradual enlargement of the median unit, combined with recession of the lateral ones, until the single condyle of Birds is reached, which is imitated by the sham monocondyle of the Snakes, effected by one epiphysis covering the three units. The two termini of the reptilian stem, Birds and Snakes, have returned to the primordial monocondylar condition. This is a most remarkable instance of evolution in a circle, perhaps better called spirally progressing evolution.

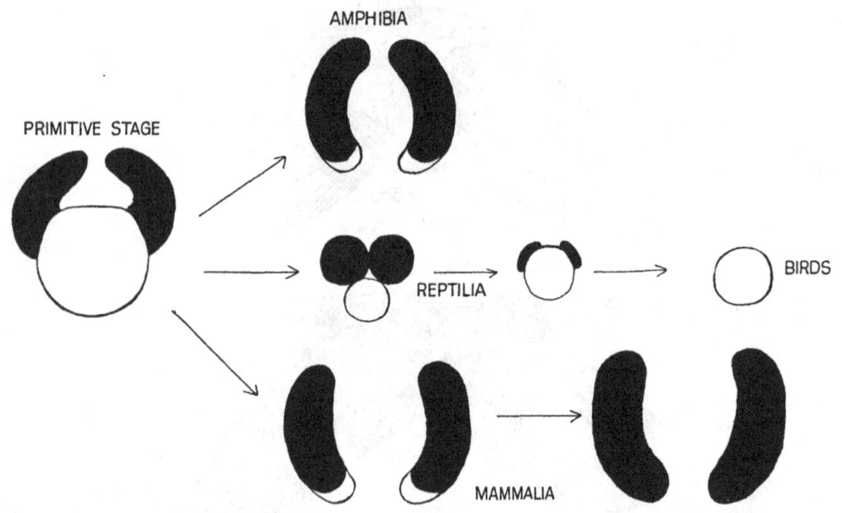

Fig. 20. Diagram showing the lines of evolution of the occipital condyles. The lateral occipital elements are black and the basioccipital white. The three divergent lines are shown leading respectively to the Amphibia, to the Reptilia and through them to the Birds, and thirdly to the Mammalia.

Evolution of lower to higher organised creatures means progress, and the occipital joint has returned to a condition whence it started; a final condition that has been attained by suppression of the two lateral elements, which it has taken untold ages to turn into a double paired joint. Was this then after all a mistake by evolution? Further, the triple joint of Snakes is hidden, turned into a single joint, a make-belief. Does evolution after all flounder about without a purpose? Why not stick to the original single middle joint and improve it? Even biconcave joints can be turned into plane joints, etc. Why then flounder about?

The case seems hopeless, but it is not. The floundering does it and this is induced by the hundred different ways in which creatures use their head, and each habitual motion reacts upon the articulating units. Radiation, and orthogenetic cumulative inheritance continued to the utmost, come into the question. Hundreds of variations have been tried; some of them have been preserved because they answered well some special case. There is further the old "law", perhaps involving parsimony, that it is easier to improve one instead of several units, and the avian condyle is after all one of the most perfectly constructed cup-and-ball joints. Evolution has done well after all.

The Mammalia are gastrocentrous Amniota, and are derived from triple-condyled theriodont Reptiles. Yet they have large paired slightly convex condyles, while the basioccipital has mostly withdrawn from the odontoid. We can safely dismiss the bicondylar Anura as not being in the direct line of mammalian descent; on the contrary, they are an interesting case of convergent analogy. The triple-condyled early Labyrinthodonts might be thought of as ancestors, but they are likewise excluded by their eminently notocentrous nature; the holocentrous *Cricoti* would refer the common meeting ground unnecessarily far back. There remain the three condyled Reptiles of the Cotylosaurian-Theriomorph stage. Withdrawal, shortening of the median occipital, due to increased growth of the lateral occipitals, would require only a slight alteration, no greater in degree than the lengthening of the basioccipital and withdrawal of the lateral elements which produces the monocondylar type. Both possibilities have been tried and brought to perfection. The bicondylar arrangement suits the mammalian constitution. Two pivots, right and left, naturally restrict movement, and when, as in many Mammals, the condyles fit into deep cups, movement of the head is limited to bending up and down, while turning of the head is made possible by the seven-jointed neck. But the practically monocondylic Tortoises and Crocodiles turn their heads as little as the Frogs. The mechanical exigences at play are beyond our ken.

Seeley was the first who derived the typical dicondylic from the monocondylic type, monocondylic being here taken in the physiological sense, by the reduction of the basioccipital and enlargement of the lateral occipitals. Osborn, after a survey of the range of variation of the condyles of the various orders of Mammalia, came to the same conclusion, particularly mentioning the resemblances existing between the theriodont *Cynognathus* and Mammals. He distinguishes four stages which, he is careful to remark, are morphological and not genetic. He starts (1) from a median solely basioccipital condyle and postulates two intermediate stages, namely (2) monocondylic tripartite, most Reptiles, and (3) dicondylic transitional, *Cynognathus*, until he arrives at (4) the typical dicondylic mammalian formation. *Echidna* is supposed to conform with the third stage. Clearly he fell into the same error as myself, when in this third stage he introduces a split basioccipital attached to the lateral elements. And we ask which are the Reptiles with the solely basioccipital condyle of stage (1), since he himself makes the unexpected statement that "no Amphibian, so far as known to the writer, exhibits any participation of the basioccipital in the formation of the condyle". A study of the condyles of the various groups of Reptiles led me to the conclusion that the single basioccipital condyle of Crocodiles, Birds and Snakes from a physiological standpoint is a terminal, not starting stage. I therefore looked for help from the Urodeles as representatives of a distinctly lower level, and fell into the error mentioned above, instead of starting the required series with the tripartite Reptiles themselves which so obviously split into those which retain, and eventually enlarge the basioccipital, and those whose basioccipital is overpowered, partly suppressed by the lateral occipitals.

It was Haeckel who first opposed Birds and Reptiles as monocondylia to the amphicondylous Amphibia and Mammalia. Huxley later also stressed the fact that "the Amphibia are the only air-breathing vertebrata which, like the Mammalia, have a dicondylic skull". That was a cautious remark, showing that Huxley was aware that sweeping categorical morphological terms cannot be trusted as taxonomic witnesses. As mentioned elsewhere in the text they usually break down.

To sum up. The triple condyle represents a very important stage full of further possibilities. It consists of three pieces, an unpaired ventral and a pair of dorsal pieces, all fit for condylar function. Each separately can be enlarged or reduced, and all three together can form one larger unit, which then acts as a so-called single condyle, placed ventrally with practically one single facet for articulation with the atlas. Any reasonable combination has been tried and has materialised in some Tetrapod or other. These various morphological trends or tendencies are not predestined, but are the result of trial and error as determined by the chief function of the joint required by the mode of life of the creatures. However, there is one purely morphological tendency amounting to a natural law that organs which start by being multiple become reduced to fewer, though more highly developed, units. One well-made "universal joint" is better than three bundled together, and if the joint must be multiple, a well-made right and left facet is better than three separate facets in the same transverse plane. It was an unfortunate error to emphasise the difference between monocondylism and dicondylism, represented respectively by Amphibia and Mammals versus the Sauropsida, and to take this as a valuable feature of taxonomic value, for both are referable to the same triple condyle. But it took long to appreciate this fact by more careful study of the recent forms, and fossil forerunners, the Stegocephali, brought the proof. The so-called single occipital condyle of the Permian Eryopes, with basidorsal paired and basiventral unpaired, has changed into the double condyle of the Triassic descendants. The fact that such terrestrial Amphibia had acquired such a paired condyle, made out of the lateral occipitals, was cherished by those who advocated the Amphibian ancestry of Mammals, their earliest fragments likewise dating from the Trias.

Chapter XI

THE HOMOLOGIES OF THE ATLAS AND SECOND VERTEBRA OF THE AMPHIBIA

The last nerve which issues from the cranium of the Fishes is the vago-accessory complex, and as the Fishes have no tongue in the proper sense, it had long been held that they are devoid of a nervus hypoglossus, differing thereby from all Amniota, which have an intracranial postvagal XIIth or hypoglossal nerve. Since the Amphibia have a tongue supplied by nerves which are postvagal and at the same time postcranial, their agreement with Fishes with regard to cranial nerves was hailed as strengthening the value of classification of vertebrates into Ichthyopsida or Anamnia, and Amniota.

We owe to Credner the more satisfactory division into Fishes and Tetrapoda, the necessary link between them having to be looked for in his Eotetrapoda. Gegenbaur, by his fundamental work on the "Kopfskelett" (1872), revealed to us that the whole postvagal portion of the Amniote skull is formed by the addition of several originally spinal or cervical vertebrae and nerves. It is therefore called the neocranium and gives rise to the occipital region of the Amniota. The Amphibia were in this respect still at the same stage as the Fishes in which such an addition had not taken place. If this be true, the conclusion is unavoidable that if the nervus hypoglossus of the Amniota is composed of three metameres, which in the Anamnia are still cervical, the atlas of the Amphibia must be the homologue of the foremost of the three vertebrae, which in the Amniota have fused on to the skull, and that the strictly serial homologue of the Amniotic atlas must be the fourth vertebra of the Amphibia. To mitigate this uncanny conclusion various attempts have been made to prove that the Frog's atlas has resulted from the fusion of two vertebrae, or that part of the first vertebra has joined the cranium.

Van Wijhe broke new ground by his study of the mesoderm segments and the development of the cephalic nerves of the Selachians. Rosenberg discovered that in some Selachians, notably *Carcharias* and *Mustelus*, at least one complete vertebra with all its components and nerves has been added to and partly fused with the postvagal cranium.

Gegenbaur drew attention to similar but much further advanced conditions in various Ganoids, and in the same year published a most welcome critical summary of the metamerism of the head and the vertebral theory of the cephalic skeleton.

Meanwhile Fuerbringer contributed a series of papers on the muscles and nerves of the shoulder girdle of Amphibians, Reptiles and Birds, a study which naturally led him to the hypoglossus. He crowned his researches by his contribution to Gegenbaur's *Festschrift*, under the title "Ueber die spino-occipitalen Nerven der Selachier und Holocephalen und ihre vergleichende Anatomie". This monumental work of 440 quarto pages and eight double plates contains a well-nigh exhaustive morphology of the whole postvagal and cervical regions from Lampreys to the Mammalia. An able summary by Gaupp is to be found in Hertwig's *Compendium*.

One of Fuerbringer's conclusions is that the Amphibia, because of their postcranial hypoglossus, must have already branched off the vertebrate stock at the stage of the early Selachians, and that the true homologues of the Frog's atlas and next cervical vertebrae are contained in the occiput of the Amniota.[1]

It is difficult to follow him. Even after Waldschmidt had discovered, in *Ichthyophis*, a separate nerve leaving the cranium between the vagus and the occipital joint, and after it had been corroborated by Peter, I personally felt that the usually accepted account of the occipito-cervical region left an unsatisfactory feeling. Fuerbringer brushed the significance of this nerve aside. But since von Huene described and figured a hole in the critical region of a Stegocephalian skull, the nerve

1 Fuerbringer (*Festschrift*, III, p. 485) condenses his arguments as follows: "The so-called occipital arch of the Amphibia, presupposing its phylogenetic reality, seems to represent rather the vestige of a multiple of primary occipital vertebrae, and thus to be comparable with the sum of the occipital vertebrae of the Selachians, although in much further advanced reduction. According to Gegenbaur, emphasised by Gaupp, many agreements exist between the cranio-vertebral articulation of the Selachians and Amphibians. I therefore homologise the posterior limit of the Amphibian cranium with that of the Selachians. Accordingly, the primary addition of occipital vertebrae to the Amphibian cranium appears to me decidedly possible, even probable, although none of the ontogenetic investigations amount to proofs, nor do they sufficiently elucidate the phylogenetic process which certainly is very complicated. This agglutination, 'Angliederung', is less obscured in the Selachians than in the Amphibia which, standing on a higher level, show manifold reductions. A secondary agglutination of vertebrae such as actually takes place in Holocephali, Ganoids, Teleosteans, Dipnoi and Amniota has therefore to be absolutely excluded from the Amphibians, so far as these have been investigated".

The above account is a syllogism of sophistry with safeguarding reservations and with annoying use of the word "Occipital-Bogen" which is certainly the occiput of the Amphibia, but by which he means the x, y, z lot, his primary "Occipital-Wirbel".

through which he unfortunately labelled XI, and when Watson discovered a similar, possibly double, hole in other Permian Stegocephalia, and recognised the respective nerve or nerves as hypoglossal, the whole problem has entered a new phase.

Instead of assuming that the recent Amphibia have stopped short of developing an intracranial hypoglossus, we have to conclude that they have lost it and that its function has been taken over by one or more of the first spinal or cervical nerves. The nucleus of this new hypoglossus remains the same in all Tetrapoda, and their atlas is also the same, despite the eventual existence of a proatlas.

This applies to the Anura, whose Stegocephalian ancestors still possessed, or were already in the stage of losing, the intracranial XII; and it applies to the recent Urodela because of their relationship to the Coeciliae.

Table I (p. 86) has been drawn up from Fuerbringer's ascertained facts, giving only the barest outlines, but arranged in order to bring out my conclusions. One zero line is possible for all known Tetrapoda, the occipital joint, behind which issues the first spinal nerve or nervus suboccipitalis. It is clear that this nervus suboccipitalis belongs to the proatlas, issuing behind the latter when present. As already explained, we assume that every spinal nerve issues through or behind the neural arch of that vertebra to which it belongs.[1]

Fuerbringer arranged the postvagal nerves in three groups, the division of which is arbitrary, as he confessed himself (Fig. 21):

(1) Nervi occipitales, issuing below and behind the vagus, consisting of ventral roots only and supplying somatic muscles belonging to the abdominal and rectus system; attached to the shoulder girdle, these muscles extend forwards along the ventri-lateral side of the branchial region. The nerves are the same as those which Gegenbaur originally

1 Where is the neural arch of this nervus suboccipitalis? Certainly not in the occiput, because this nerve belongs to the proatlas, through or behind which it must have issued. Complete reduction of the proatlas arch leaves a gap covered by what henceforth becomes the ligamentum occipito-atlanticum (strictly a compound of the ligamentum occipito-proatlanticum and ligamentum proatlanticum-atlanticum).

Theoretically there should be also a nerve leaving immediately behind the occiput, a condition which still prevails in cartilaginous Ganoids. But the occipital joint of the Tetrapoda dates from a stage in which all the dorsal arcualia were still of equal size, the basidorsals not yet larger than the interdorsals, the nerve issuing between them, so that when amalgamation into the formation of the occipital region did take place, all its nerves issued through this mass. In the Anamnia, even in those with an intracranial rest of a hypoglossus, no trace is left of the metameric origin of the occiput, but the Amniota, being of later date, still show traces of the skleral metamerism. In fact, they have done later what the Amphibia had done before they went in for reduction of their occipital complex.

described in Selachians as the ventral roots of the vagus. Later, after van Wijhe's researches, he considered them as representing a nervus prohypoglossus, the vestiges of an earlier now more or less vestigial hypoglossus. This newer term is preferable to Fuerbringer's nervi occipitales, which is misleading. Their respective metameres are certainly the oldest, completely amalgamated with the cranium, without a trace of segmentation of their skeletal, vertebral portions. Their reduction in numbers and strength proceeds clearly from the rostral to the caudal end. Fuerbringer labels them *z* to *v* in the opposite direction, *z* being the most posterior nerve without a dorsal root. According to him such occipital nerves exist in the adult stage only in Fishes; in the Tetrapoda they are present only in embryonic stages, and even then are already reduced to *x*, *y*, *z*, or less.

(2) Nervi occipito-spinales, so-called because they cannot be definitely allocated to either group. They differ mainly from the occipitales, because in the lower Fishes they are still complete spinal nerves in so far as they possess not only the ventral, but also dorsal roots with a ganglion spinale. Their respective metameres are those which eventually in some creatures are added as the latest and last instalment of the neocranium, and the nerves themselves then constitute the nervus hypoglossus. In this case they are labelled *a*, *b*, *c*, counting from *z* tailwards; they are supposed by Fuerbringer to be limited to three, rarely to two, the reduction being due to the loss of *a*. Reduction to *c* alone is disallowed.

(3) Nervi spinales are all those which follow behind *c*, between which and *d* lies the occipito-cervical joint; they are not added to the cranium. This difference he frequently expresses by the formula d^4 (1), etc., which means that *d*, the fourth postoccipital nerve, has become the first spinal; or rather is the first which remains spinal, instead of being converted into an occipito-spinal.

The above scheme seems plain enough, but, since in the vast majority of Elasmobranchs the occipital joint lies or will be formed behind *z*, Fuerbringer logically labels the next following segments 1, 2, 3, etc., spinals or cervicals.

Fuerbringer's scheme is invalidated by the following incontestable facts:

(1) The *v* to *z*, *a b c* and the postcranial spinal or cervical metameres form one continuous series, running into each other, an unbroken plexus extending from *v* to the brachial region. Proceeding from lower Fishes to the higher and highest Tetrapoda, this series undergoes a

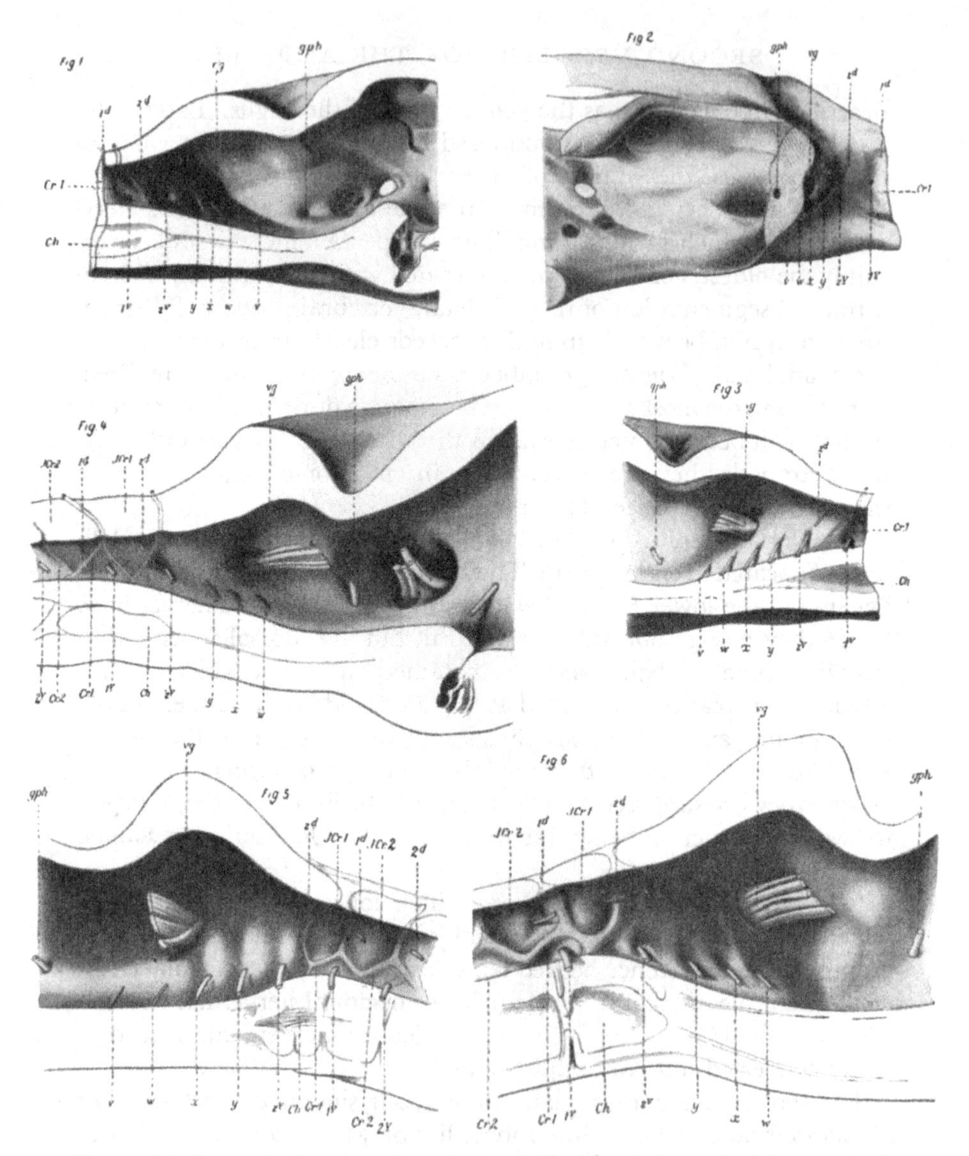

Fig. 21. Median sagittal sections through the skull and anterior end of the vertebral column of certain Selachian fish, after removal of the brain and spinal cord, to show the exits of the spino-occipital nerves.

Figs. 1–4. *Hexanchus griseus*. (Fig. 2 is a view of the posterior end of the skull from the lateral aspect.)

Figs. 5, 6. *Heptanchus cinereus*.

v, *w*, *x*, *y*, *z*, occipital nerves; z^d, dorsal root of last occipital nerve; z^v, ventral root of last occipital nerve; 1^d, dorsal root of first spinal nerve; 1^v, ventral root of first spinal nerve; *JCr*, intercalary; *Cr*, neural arch; *vg*, vagus nerve; *gph*, glossopharyngeal nerve; *Ch*, notochord. (Fuerbringer.)

steady reduction in numbers, beginning in front, and attacking so to speak each nerve, at first partially by suppression of the dorsal root, then completely by loss of the ventral root, the following nerve becoming in turn affected.

Nerve *z*, being the last occipitalis, should consist of a ventral root only, but Fuerbringer himself found and figured it with a ventral and dorsal root in *Heptanchus* and *Hexanchus* as a normal feature (Fig. 21); though in *Cestracion* the dorsal root is in a state of degeneration. Consequently the difference between occipitals and spino-occipitals is abolished. Certainly *z* cannot be used as the zero for a new count.

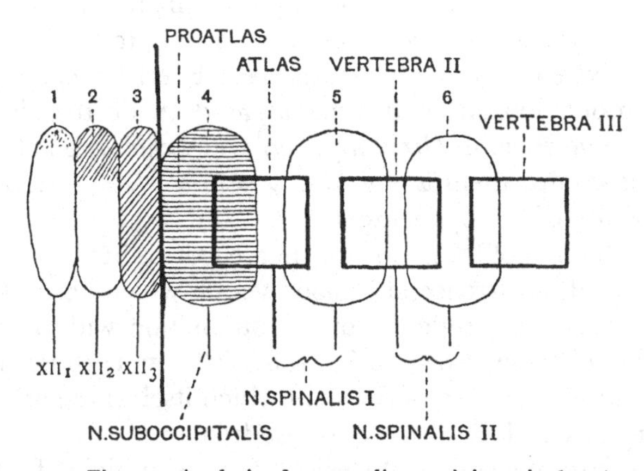

Fig. 22. Analysis of mammalian occipito-spinal region.

The heavy, vertical black line indicates the cranio-cervical boundary. The myotomes are numbered consecutively and are represented by the ovals with thin outline, the vertebrae by squares with thick outline. The hypoglossal nerve has three roots labelled XII_1, XII_2 and XII_3: XII_1 has a ventral but no dorsal root, XII_2 sometimes both and XII_3 always has both. The line labelled proatlas indicates the position the proatlas would occupy.

These reductions continue from the region of the vagus, which is excluded, to the brachial plexus, regardless of the level at which an occipito-cervical joint may be established. Thus, if we call *v* to *z* the prohypoglossus, the latter, on ceasing to exist, hands over its function to those nerves which thereby become the hypoglossus, and this in turn commandeers help from the several cervicals which always remain postcranial. Therefore, if the intracranial hypoglossus should be abolished, cervical spinals alone will take over the function of the tongue apparatus, which itself is the transformed branchial apparatus of the Fishes and Amphibians. It is at least certain that the skeletal framework is the same, the muscles themselves presenting the phenomenon of Fuerbringer's "imitatory homology".

(2) The second objection is that Fuerbringer assumed, and insisted, that the hypoglossus of the Amniota does not contain more than three metameres, *a b c*, and never less than the one intracranial *c*. But a few years later Howes found in embryos of *Sphenodon* four hypoglossal foramina and five roots, which by the time of hatching are reduced to three foramina with corresponding roots, the reduction being due to the ontogenetic disappearance of the anterior, foremost portion of the complex. If Fuerbringer had known of them, he would logically have designated them *z* and *y*, and thereby brought them into agreement with the embryos of other Sauropsida according to the discoveries by Chiarugi. Even in embryonic Mammals, Fuerbringer is obliged to concede a root z^v in front of the usual a^v, b^v, c^{d+v}. Further, he describes and figures five roots in *Ornithorhynchus* and *Echidna*, but labels them α to ϵ, and groups them $\alpha = a$, $\beta + \gamma = b$, $\delta + \epsilon = c$, apparently to save the trinity of the hypoglossus.

The ascertaining of these variable nerve exits is further complicated by the frequent occurrence of a blood vessel, which leaves the lateral occipitals either independently or in conjunction with a nerve root. The position of these vessels varies, sometimes in Tortoises lying very near the condylar portion of the skull which itself is remarkably long. Asymmetry of vessels and nerves is also frequent.

(3) The third objection is the existence in Amphibia of intracranial postvagal nerves which issue through the occipital region. There are probably two in Stegocephali (Watson). In the Coeciliae Waldschmidt discovered one in *Ichthyophis*, which leaves the skull by the same hole as the vagus, but separated from it by a septum. Similar conditions have been found in other species, notably in *Euthonerpeton*. The only instance in Urodela has been found by Fuerbringer in one adult *Cryptobranchus*, in which a small artery leaves the cranium well behind the vagus, nearer the end of the occiput, accompanied by an extremely fine nerve thread, which he identified histologically. Of course he labels it z^v in conformity with the view that in the Amphibia no such neocranial addition exists. In the Anura this last trace of such an intracranial nerve has disappeared.

In most cases, from Fishes to Mammalia, the occipital, spino-occipital and cervical nerves form a continuous chain by ansae or plexus connexions of the successive nerves. There are, however, exceptions, when a break occurs, as described and figured by Fuerbringer. Such a reduction turns the respective nerves into solitary strands and can proceed even beyond the skull, affecting the sub-occipital or proatlantic nerve. Thus the intracranial and the sub-occipital nerves of the Coeciliae have become solitary. The same applies to the suboccipital nerve of the Aglossa. Curiously enough in many Birds, although they have a well-developed hypoglossus, their first and second spinal nerves seem to be represented only by their rami dorsales.

The Stegocephali show at least with fair certainty that most of their postvagal region was undergoing a process of reduction similar to that which takes place in the living Amphibia to a still greater extent, reducing the basis cranii and leaving in its place a large gap, filled in the Urodela by the large parasphenoid and also by the ventral remnant of the proatlas.

There is thus justification for the grand idea, which Fuerbringer about 1878 announced to us prophetically, although only in jocular conversation: "The Anura are drawing their skull into their neck, the Amniota have added part of the neck to their skull"; this is the direct opposite of his later view concerning the Amphibia. The Amphibia are, however, not really drawing their skull into their vertebral column. What has happened is that, owing to the shortening reduction of the occipital region, the gap which would result is filled by the forward move of the column. The proatlas necessarily remained in contact with, or followed up, the withering portion of the cranium, so that the occipito-cervical joint remained the same.

At a later phyletic stage the proatlas gave way; first in its ventral half, a reduction which brought the basioccipital and lateral occipitals into direct contact with the atlas. The dorsal half or neural arch of the proatlas continued to link occiput and atlas, as it still does in Crocodiles and in *Sphenodon*, until the atlas alone took over the exclusive articulation with the occiput.

Fuerbringer has shown by his comprehensive and minute investigations that the reduction of the successive groups of postvagal nerves keeps step with, or rather is caused by, the reduction of their respective muscles. Thus the epibranchial muscles continue to exist only in the Elasmobranchs and are lost in all other Fishes and in the Tetrapoda. The reduction of the hypobranchial muscles (lateral trunk and rectus system) may, according to him, be due to the forward shift or concentration of the myomeres, so that the most anterior are crowded out, while the more posterior, being more favourably placed, have supplanted them. These hypobranchial muscles, not to be confounded with the branchial muscles proper which are supplied by the vagus, contain the material for the muscles of the tongue, an organ which does not exist in the Fishes, including the Dipnoi.

The muscularised tongue is a neomorph of the Tetrapoda, a new acquisition, supplied by the intracranial occipito-spinales.

Further elaboration of the tongue is due to an addition from cervical metameres, so that for instance in the mammalia the intracranial hypoglossus receives branches from at least the first and second cervical nerves, their corresponding myomeres becoming incorporated. The whole muscular and sensorial apparatus has undergone many important changes, culminating as usual in the perfect mammalian tongue, which itself is a mammalian neomorph, not homologous with that of the other Tetrapoda, whose tongue is what still exists as the "under-tongue" of the monotremes; vestigial in Man as the right and left plica fimbriata. The flat and thick tongue of the Crocodiles is movable in bulk only, contrasting with the agile tongue of many Lizards and Birds, which is protrusible, sometimes telescoped and prehensile. On the other hand, it and its framework are reduced to the utmost degree in Gannets, Cormorants and Pelicans.

It does not matter whether the Aglossa have lost the tongue; it can scarcely be doubted that it existed in Stegocephalia, as an organ typically supplied by some intracranial nerves, and it exists in the majority of the Anura as a highly specialised organ, though having an extracranial innervation, and so may be a neomorph of these newer Frogs and Toads. Fuerbringer seems to favour this possibility by his term Opisthoglossa for the usual Phaneroglossa, thus presupposing Proteroglossa. In this case there must have been continuity between the successive kinds of tongue, with a continuous sequence of nerve supply, in which a backward shift is taking place. Urodeles show this by their conjoint spinal nerves 1 and 2 commingling distally with a branch from 3 which, independent of 2, ramifies into the same branchial muscles. We can imagine how distal ansae between 2 and 3 may be formed, which then consolidate into one stronger ramus communicans, and that by analogy the same process of distal mixing has produced the junction of 1 and 2. Later reduction of this branch from 1 to 2 would be correlated with reduction of the foremost muscles of the dawning tongue while posterior myomeres took their place, the framework, the branchial arches, remaining fundamentally the same. During this critical time the communication between 1 and 2 weakened and vanished. But the hypoglossal nucleus remained where it was, though its ganglionic cells had already extended to the level of the issue of 2, which in Urodela is the main supply of the tongue. The threatened loss of the direct supply by 1 is compensated merely by its issuing one metamere farther back, conjointly with 2, so that the apparently missing 1 appears in the guise of a branch of 2, but supplies, as it should, its share of the tongue, and the musculus levator scapulae.[1]

Why then should the Stegocephali be in the stage of losing their intracranial hypoglossus, the Coeciliae practically, the Anura completely, having lost it? Palaeontology is not likely to help. As a matter of fact they have transferred the whole supply to nerves beyond the cranium. We do not even know whether the Permian Amphibia had a tongue proper, nor whether the tongue of the recent Amphibia has been muscularised from muscles farther back. To appeal as a cause to the enormous reduction of the cartilaginous posterior half of the cranium may be a *petitio principii*. In Coeciliae and Perenni-branchiate Urodeles the occipital arch arises like a pair of horns which join dorsally and then form the very narrow occipital ring. The huge deficiency in the cartilage is filled by a large parasphenoid, and dorso-laterally other membrane bones take the greater share in the formation of the occiput. The link between Fishes breathing by internal gills and the Tetrapods,

1 The musculus levator scapulae is the foremost more or less separate portion of the serratus system, which is supplied by the so-called nervi thoracici superiores sive dorsales. The levator is very variable in its extent. In the Amphibia, restricted to a muscle between the occiput and the top of the scapula, it is supplied only by the nervus spinalis 1. In the Amniota, conforming with the longer neck, it extends much farther back, frequently losing its anterior portion, but gaining new portions behind, until it arises only from the transverse processes and ribs of the lower cervical region, supplied by the corresponding spinal nerves.

The *musculus cucullaris sive sphincter colli* shows another case of transference, less easily understood. It belongs, with the musculus trapezius and musculus sterno-cleido-mastoideus, to the system of the nervus accessorius XI. It arises immediately behind the musculus levator and in the Amniota develops into a thin sheath of the neck with a tendency to cutaneous attachments. In the Mammals all these muscles are supplied extensively by the nervus accessorius, which itself increases with often considerable extension over the shoulder region. In the Sauropsids only the anterior, cephalic portion of the cucullaris is supplied by XI, thence down to the shoulder girdle the whole of the rest of this muscle is innervated from the cervical nerves. When, as in Crocodiles, the anterior portion is absent, having been lost, the whole muscle belongs to the typical spinal or trunk system.

This transference and change of nerve supply becomes less perplexing if we boldly consider the whole accessorius as representing the ventral branch or branches of the vagus, with which it is now usually combined as "vago-accessorius". As somatic motor of the vagus it was on the road to suppression during the change from Fishes to Tetrapoda, but took a new lease with the Amniota. Although certainly not belonging to XII it is intermediate between vagus and typical spinal nerves in more than one sense.

which ultimately breathe by lungs only, must have been formed by more or less Dipnoan-like creatures, Gano-Dipnoi, in which the apparatus of internal gills is so obviously being reduced, the critical transitional stage being assisted by the *ad hoc* development of external or cutaneous gills. The whole skeletal apparatus was preserved; but it stands to reason that the waning of the inner gills in its turn caused reduction of the moving muscles hitherto supplied by the vagus, and that more superficially placed trunk muscles caught on to the branchial arches and thus introduced a complete change of function.

These trunk muscles laid the foundation of the muscularisation of the tongue: as we know it now, thanks to Gegenbaur's scarcely enough appreciated "Zur Phylogenese der Zunge".

There still remains a difficulty. Why does the above speculation, that has just been elaborated, not apply to the Amniota? Why have they been able to retain their intracranial hypoglossus, although actually supplemented by postcranial, cervical, nerves? There can be no reasonable doubt that the early Amphibia were amphibious in the literal sense of the word. Aquatic, but adapting their whole constitution to terrestrial life. All of them had a disproportionately large, depressed head and no neck, the heavy head resting upon the ground. Those, however, which at the right time terrestrialised more emphatically than others, at a time when the process of reduction had not yet advanced too far although the limbs had reached a higher evolutionary stage, thereby became the founders of the Amniota. They acquired a neck, lifted the head by spino-occipital muscles which in turn by their attachment caused the whole occipital region to react by consolidation so as to form a higher rather than broader skull and one which was more roomy so as to adapt itself to the enlarged hind-brain; all characteristics typical of the Amniota.

Chapter XII

THE NERVE SUPPLY OF THE MUSCULATURE OF THE TONGUE APPARATUS

The essential agreement in Frog and Man, in that the tongue of both receives additional nerve supply from spinal 1, and from the loop connexion between 1 and 2, is not a mere coincidence. It shows that in both Amphibia and Amniota the second spinal nerve tends to assert its importance as a postcranial addition to XII. In the Mammalia part of the third is added similarly, and sometimes even more farther back. In the Amphibia, owing to the complete loss of XII since the Stegocephalian stage, the second spinal becomes the more important, often the only tongue nerve: further caudal supply of the tongue, such as occurs in Mammals and Sauropsids, is checked by the forward shift of the limb girdle, so that the third becomes the strongest element of the brachial plexus, almost its only nerve, the contributions from 2 and from 4 becoming insignificant: the third cannot therefore supply the tongue. Reduction of the Amniotic XII from five to three or even to two roots being known and compensation of the tongue's innervation by transference into the cervical region being easily demonstrated, completion of this process would be a striking convergent analogy to the Anuran condition. The limit of the nervus cervicalis descendens, formed by the ansae of the anterior spinal nerves, in such ultra-modern Amniota would be decided only by the backward migration of the fore-limbs; in other words by the length of the neck. This is not likely to happen, but is theoretically not impossible. It might be worth enquiring into the nerve supply of the enormous tongue apparatus of the African species of *Manis*, the Pangolins, the usual muscles of which seem to have lost contact with the hyoids and been converted into huge sternoglossi.

DEFINITION OF THE NERVUS HYPOGLOSSUS

The hypoglossus is that polymeric compound of nerves which supplies the muscles of the tongue; a definition which is incomplete from the physiological and morphological point of view. Leaving out of count the nervus glossopharyngeus, part of which supplies the specific sensorium of the tongue, the XIIth pair also contains sensory elements

TABLE I

The cranial and postcranial nerves lying behind the vagus which contribute to the innervation of the tongue apparatus.

	v	*w*	*x*	*y*	*z*								
						1							
Heptanchus	.	.	.	.	:	:							
Heptanchus	.	.	.	.	.	:							
							2						
Selachi, average				.	.	.	:						
								3					
Torpedo					.	.	.	:					
Torpedo						.	.	:					
									4				
Raja, average						.	.	.	:				
Raja, individual							.	.	:				
						a							
Carcharias, young				.	.	:							
							1						
Carcharias, adult				.	.	.	:						
							b	1					
Holocephali, average				.	.	.	.	:					
								c					
Teleostei, many							:	:					
									d	*e*	*f*	*g*	
Acipenser				.	.	:	:	:	:	:	:	:	
								1					
Ceratodus 1			.	.	:	:	:	:					
Ceratodus 2				.	:	:	:						
Ceratodus 3				.	.	:	:						
Ceratodus 4				.	.	:	:	:					
Protopterus				.	.	:	:						
						a	*b*	*c*	1	2	3		
Reptiles, various embryos				.	.	:	:	:	:	:	:		
Reptiles, most adult						.	.	.	+.	:	:		
Sphenodon							.	.	.	+:			
Birds, many embryos						.	.	:	:	:	:		
Birds, most adult							.	.	+.	+.	:		
Placentalia, embryos					.	.	:	:	:	:	:		
Placentalia, embryos and adult						.	.	:	+:	+:	+:		
												4	5
Placentalia, some adult, e.g. highest Anthropoids							.	.	+.	+:	+:	+:	+
Monotremes, adult			.	.	.	.	.	.	+:	+:	+:		
Macropidae							.	.	+.	+:	:		
Stegocephali							.	.	.	?	?		
Coeciliae								.	.	+.	:		
Cryptobranchus, individual								.	.	+:	+:		
Urodela and Aglossa									.	+:	+:		
Anura Phaneroglossa										.	+:		

Single dots indicate nerves with ventral roots only, double dots nerves with ventral and dorsal roots.

The thick lines show the level of the cranio-cervical joint, the primitive position of which is indicated throughout the table by the thin continuation line.

Nerves to the left of this continuous line are referred to by Fuerbringer's designation *v*, *w* *x*, *y*, *z*.

Nerves to the right of this line, if postcranial, are numbered 1, 2, 3, etc., if intracranial, *a*, *b*, *c*, etc. forming the XIIth cranial nerve. The first example of these have their numbering inserted over the top.

\+ indicates the presence of ansae between roots.

because of its frequent possession of dorsal roots with or without a ganglion. Moreover, it is not the only nerve which supplies muscles attached to the framework of the tongue apparatus. Lastly, it innervates muscles like the musculus levator scapulae and musculi thoracici superiores and thoracico-abdominales, which it would be pedantic to include in the tongue muscles, although all belong to the category of longitudinal somatic or trunk muscles, innervated from still typical, or no longer typical, spinal metameres.

Any terse morphological definition is equally precarious. The almost mediaeval notion of a round dozen of cranial nerves, the last thing XII, was upset by Gegenbaur, who however rightly retained the numerals and names as tokens in his new grouping. The hypoglossus as a whole may be composed of more than half a dozen, some of which are intracranial and therefore cranial nerves, while others issue behind the skull and are therefore cervical. They may all leave the skull (Fig. 22) or all lie behind it. They may form a continuous series, or this may be interrupted by the partial or complete reduction of some nerve.

Fuerbringer as a strict logician distinguishes between plexus cervicalis, composed of all the occipito-spinales plus those postcranials which contribute branches to the hypoglossus, and plexus brachialis, namely those cervicals, and eventually thoracics, which are not connected with the hypoglossus by ansae outside the vertebral column. The string resulting from the more or less straightened out ansae he calls the nervus cervicalis descendens. The extent of these connexions is indicated in Table I by +.

The Amphibia present a very different condition, the meaning of which has to be discussed in some detail.

Some of the Stegocephali show holes at least indicating an intracranial XII; about the nerves themselves we know nothing. In the Coeciliae one such nerve is known; also the tiny nerve discovered in *Cryptobranchus* by Fuerbringer, a nerve which seems to be reduced to a functionless vestige (Fig. 23*A*). With these exceptions the whole intracranial hypoglossus formed by the occipito-spinal nerves has disappeared in the recent Amphibia. The first postcranial, the suboccipital, is still present in Urodela (Fig. 23*B*) and Aglossa although reduced to a branch for the musculus levator scapulae in some Newts and in *Xenopus* (Fig. 23*D*). In *Pipa* (Fig. 23*C*) it is surprisingly long, sending a very slender thread to the combined stems of the second and third cervicals. But for this connexion with 2, which Fuerbringer found wanting only in one specimen which he considered as abnormal, the first nerve passes over the combined second and third and ends as a ramus abdominalis. There is in fact no tongue nerve, in conformity with the absence of a tongue. 2 and 3, both strong nerves, soon combine into the brachial plexus, which, besides supplying the limb, sends off two long abdominal branches and from the anterior side of 2 a smaller branch to the pectoral and supracoracoid muscle. Thus in this

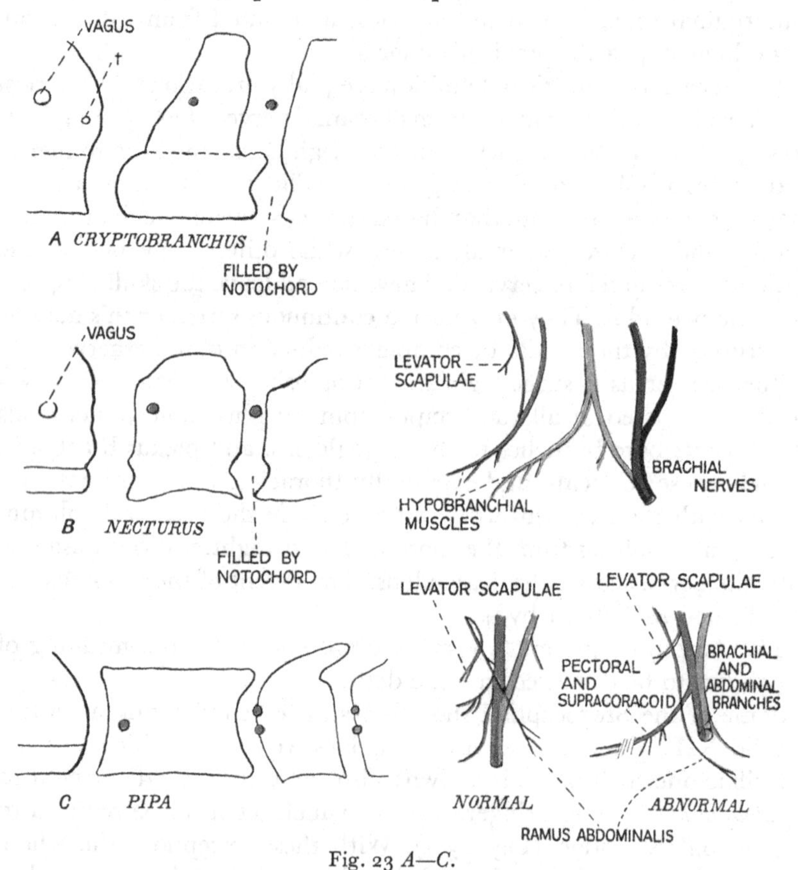

Fig. 23 *A—C.*

tongueless creature, the first, second and third spinal nerves each sends a long branch to supply the most superficial and latero-ventral muscles of the abdominal wall.

Xenopus marks a new departure. Nerve 1 is very short and weak and supplies the levator scapulae muscle; it sends, however, an extremely slender branch to 2. The second is fairly strong and goes to where the tongue should be, radiating out in the bottom of the mouth; it sends

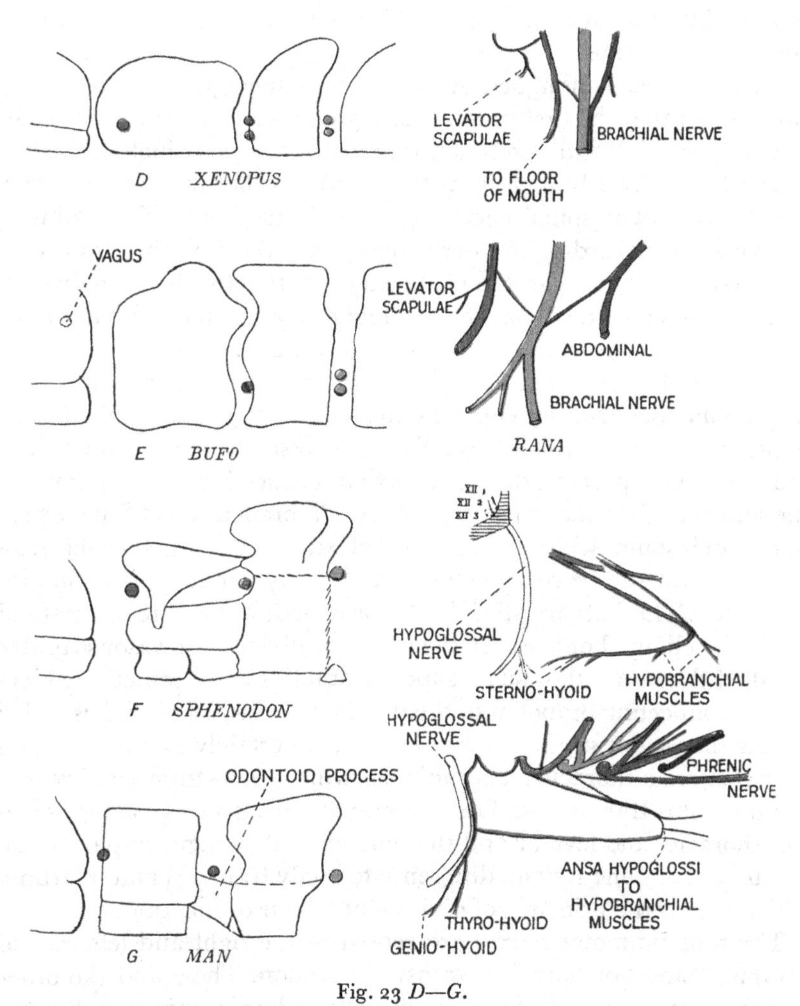

Fig. 23 *D—G*.

Fig. 23 *A–G*. Diagrams of the postcranial region of certain vertebrates to show the nerve plexuses and the nerve exits.

The following colour scheme holds throughout for the spinal nerves shown on the right, and for their foramina of exit shown on the left:

Nervus spinalis I in red.
Nervus spinalis II in green.
Nervus spinalis III in orange.
Nervus spinalis IV in blue.
Nervus spinalis V in black.

The posterior end of the skull is indicated on the left of each diagram with the successive vertebrae lying to the right. The † in *A* indicates the foramen in the skull for the exit of Fuerbringer's vestigial nerve. *F* and *G*, of *Sphenodon* and Man, have been put in for purposes of comparison though not directly referred to in the text.

a short thin branch to the third, which together with the fourth form the brachial plexus.

In the typically tongued Anura, the Phaneroglossa, it is universally stated that the first spinal is lost, that therefore the second is the sole hypoglossal and sends a branch to the 3rd which is further strengthened by a branch from the fourth, so that the third becomes the strongest of all spinal nerves. Frequently the branch from 2 into 3 is very weak; according to Fuerbringer, however, it is always present. It shows remarkable variations. It may run directly off the main stem of 2 into that of 3, or it may leave 2 higher up, together with a branch which supplies the levator scapulae. Since this muscle in *Xenopus* and *Pipa* is supplied by nerve 1, which in the abnormal specimen of *Pipa* is quite independent, we conclude that this same nerve 1 is the homologue of the branch which in the Phaneroglossa connects 2 with 3. The only, but all-important, difference is that it issues from the column one metamere farther back and in the guise of a branch of 2. We have here at last an instance which shows how a plexus can be formed which gross anatomy states to be composed of 2 and 3 only, but in reality contains 1, 2 and 3. The facts are all right, but are made difficult to understand by the labelling. For instance the nerve supplying the levator scapulae is called the ramus thoracicus superior anterior and "sometimes gives off the connecting branch into the brachial plexus". This link is called ramus thoracicus superior posterior and is certainly homodynamous with the ramus thoracicus abdominalis, which comes from 4 and goes to the musculus transversus. The abdominales transversus, rectus, serratus, thoracics and levator and the muscles of the tongue apparatus all belong to the same system, differentiated only by origin and insertion, which means the principle of division of labour or function.

The somatic motor nerves are rooted in the right and left ventral column, "anterior horn" in transverse section. These and the other columns are composed of ganglionic cells and are continuous, but the number of these cells varies metamerically in correlation with the thickness or importance of the issuing nerves. The resulting swellings in the horns may be likened to relay stations, while the nucleus, the central office of our simile, as the oldest and most important, lies and remains in the brain. The whole column of ganglia is the result of caudal growth which has kept step with the evolution of the trunk. It does not matter whether a whole nerve falls out, since this applies only to the no longer issuing nerve, whilst the columnar continuity is preserved.

These and kindred considerations have been made possible since W. H. Gaskell's investigations.

These investigations led him to the problem of the Origin of Vertebrates. He drew attention to the mode of formation of organs which he elaborated as traceable directly to those of Invertebrates. It is a question still unsolved, how far they are mere heterogeneous coincidences, or whether at least some of them are truly homologous. His "heresies" met with silence which, to the detriment of morphology, affected also his epoch-making neurological investigations. It was not suspected by the orthodox morphologist that one of our foremost physiologists could, during the hey-day of unrestrained embryology, become a well-equipped self-trained morphologist, with the advantage of approaching the new domain with an unbiased highly critical mind, which enabled him to discover and mercilessly to discuss not a few of the skeletons kept in our cupboard.

Our friendship of more than thirty years was enhanced by the discussion of many problems.

For the Anura the excellent Fig. 45 in Ecker's *Frog*, edited by Gaupp, should be consulted; and for individual changes of the brachial plexus Fig. 47, which is a partial adaptation from Adolphi's comprehensive series concerning *Bufo*, *Pelobates* and *Rana*. The tongue apparatus has shifted backwards, away from the head; the limb plexus has shifted headwards at the expense of a possible neck. The Urodela represent an earlier line of evolution which made a neck possible. In *Necturus* (Fig. 23*B*), nerves 1 and 2 behave exactly like the normal specimens of *Pipa*; but it is most suggestive, as Fuerbringer shows, that 2 and 3, instead of being directly connected with each other, each sends a nerve into the hypobranchial muscles, in which the terminal ramifications seem to mingle. These muscles present a close analogy to the anterior portion of the transversus of *Rana* as sketched by Gaupp; that part of the transversus which is supplied by the thoraco-abdominal branch of 4 being in the act of splitting into several bands, each of which is a potential separate muscle.

The problem of the supply of the tongue apparatus by addition from, or tapping of, nerves farther back, is, *caeteris paribus*, exactly the same as that of the shift of the whole brachial plexus of Birds, in the Pigeon composed of nerves 10–15, shifted back to the level of 22–26 in the Black Swan. The facts are these; how they have been brought about is still a much contested question. Fuerbringer's résumé "Nerv und Muskel" of fourteen full quarto pages left it unsolved.

Chapter XIII

THE FIRST THREE VERTEBRAE

THE PROATLAS, ATLAS AND EPISTROPHEUS

The analysis of the first three vertebrae of the Amphibia has been dealt with in Chap. x; that of the Amniota is in comparison easy, as two of these vertebrae are wellnigh complete.

The Atlas is by right of priority the carrier of the head, and by general consent has to retain the name of the first vertebra. The second is called Axis, preferably Epistropheus. But there is also a Proatlas which is to be indexed or numbered 0. To distinguish it as 1*a* would lead to confusion, since it is either incomplete or generally absent. Although these anterior vertebrae are, as discussed in the chapter on the cranio-cervical joint, the oldest of all vertebrae, contemporary with the formation of the first axial joint, they retain to a great extent their component parts in the separate temnospondylous condition, and in this respect are the youngest of all vertebrae and as such are still in the experimental stage, their constituent cartilages showing a surprising difference in combination and permutation; permutation, since one or more units may be reduced to vanishing point, while the remainder combine in various ways. Every reasonable arrangement is represented in some recent species or other, and often this phyletic variation is repeated by the embryo, or, better still, by some fossil form. This is a proof that the particular arrangement was actually in serviceable use, and although morphologically faulty, may have lasted a long time until an absolute terminus was reached incapable of further improvement. It is of the greatest interest that such specialised end results have been evolved by members of the most different groups. Instances of such convergence or parallel development, or so-called Isotely, if we waive the question of their strict homology, are the atlas plus epistropheus of *Diplocaulus* and Plesiosaurs, the trunk vertebrae of *Ichthyosaurus* and some stereospondylous Stegocephali.

THE FIRST THREE VERTEBRAE

The anterior vertebrae of the following forms will be described, discussed and figured in the systematic part:

Ichthyophis, p. 138.
Diplocaulus, p. 140.
Embolomeri, p. 160.
Anura, p. 164.
Seymouria, p. 193.
Cotylosauria (*Diadectes*), p. 196.
Chelonia, p. 201.
Pelycosauria, *Ophiacodon*, p. 215.
Dimetrodon, p. 215.
Moschops, p. 222.
Kannemeyria, p. 225.
Mammalia, p. 236.
Sphenodon, p. 262.
Geckones, p. 267.
Lacertae, p. 272.
Crocodilia, p. 296.
Dinosauria, p. 303.
Pterosauria, p. 307.
Aves, p. 310.
Ichthyosauria, p. 326.

The first three vertebrae can be analysed as follows:

PROATLAS

As has already been stated, its material is the spare half of the sklerotome A, the caudal half of which combines with the cranial or anterior half of sklerotome B to form the first complete skleromere or atlas.

At the most this spare half can be expected to form only the equivalent of half of a vertebra, serially homologous with a posterior intervertebral cartilage. The interdorsal portion produces the vestigial right and left arches of the proatlas, eventually articulating imperfectly with the lateral occipitals, and with the neural arch of the atlas. Such dorsal proatlantal pieces have been identified in *Ophiacodon*, *Sphenodon* and Crocodilia. The interventrals of the proatlas are represented by a small separate nodule in *Ophiacodon*. The centrum of the proatlas makes the odontoid-like process of the atlas of the Urodela (Fig. 33), sometimes of considerable size; in the Amniota it forms the tip of the odontoid process which is the centrum of the atlas, so to speak the dens of the true dens (Fig. 26); when containing osseous traces it is known as the ossiculum terminale, for example, in Man as an individual variation.

ATLAS

The Atlas consists normally of the right and left basidorsals, the halves of the neural arches; these may fuse, or in some fossils may remain separated from each other. In some ancient Reptiles, for instance, *Dimetrodon* (Fig. 30), the supradorsalia exist as a separate ossified piece. The basidorsals proper fuse generally with the first basiventral, making the atlas ring, while the centrum of the atlas usually fuses on to that of the second as the odontoid process; the name epistropheus refers to the fact that the atlas ring turns upon this process as upon a swivel. The odontoid can remain as an entirely separate ossification, articulating in front with the atlas ring, behind with centrum 2; this arrangement is rare; it is best developed in some Tortoises and may help in the interpretation of fossil Reptilia and Stegocephali. On the other hand, both centra may co-ossify completely leaving no trace of sutures, even abolishing the last traces of the axial portion of the second basiventrals, which are sometimes spoken of as

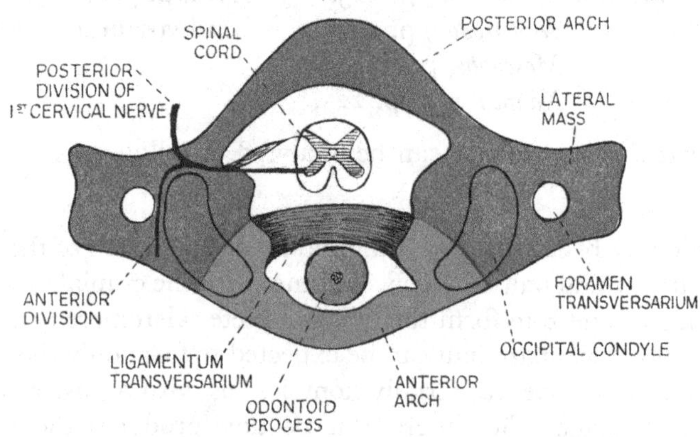

Fig. 24. Diagram of the human atlas, showing its morphological parts. The greater part is formed by the basidorsals: the basiventrals form a ring surrounding the odontoid process of the axis and composed of the bony anterior arch, including the anterior ends of the condyles, and the fibrous ligamentum transversarium. The odontoid process is the separated interventral of the atlas. The spinal cord with the first cervical nerve are shown in black.

the first intercentrum, a term scarcely fitting, especially if, as usually happens, there is no proatlas.

The first basiventrals vary much in size. If the first and second are large, they may be in contact. Rarely the first carries a rib, for example, in *Ophiacodon* (Fig. 31), or at least a facet which suggests a rib. The first basiventral has in *Dimetrodon* (Fig. 30) been called the proatlantal intercentrum, an unpardonable term which suggests either that it lies in front of the atlas or at least of its centrum as all basiventrals do, or that it is itself a piece of the proatlas, a terminal vagueness which has not failed to cause confusion. In some Marsupials the first basiventral is reduced to small size, in others it disappears with the result that the atlas ring is formed entirely of the co-ossified neural arches. Since the centrum has parted from it and is added to the second centrum, the first functioning vertebra has attained a unique condition. This is the absolute opposite to the large atlas of *Diplocaulus*, Coeciliae and Plesiosaurs, in which the first two vertebrae make one huge physiological unit. This principle is carried to the extreme in the Dinosaur *Triceratops* in which the first, second and third centra are co-ossified, obviously in correlation with the large and heavy head continued into the massive neck shield.

A very important feature of the atlas is that the axial and dorsal portion of the first basiventral ceases to be cartilaginous and forms the ligamentum transversarium atlantis,[1] grasping round the odontoid process (Fig. 24).

All these changes are brought about by the dislocation of the units. The ventral portion of the future bony mass of the odontoid process and the axial part of the first basiventral come to lie partly below centre 1, and this change seems to have induced Schauinsland (cf. *Sphenodon*, Fig. 34) to call this portion in the diagram the intervertebral zone, which if correct would of course contain the interventrals, i.e. the future centra, so that there would be two centra for the atlas, namely the true centrum 1 and his intervertebral zone. The not infrequent confusion is caused by the neglect of the fact that the basiventrals or intercentra, owing to reduction in the column, come in the final vertebra to lie in an intervertebral place, although they themselves are plainly of intravertebral origin as the homologues of the pseudocentra of the Urodela.

EPISTROPHEUS

The Epistropheus is a complete vertebra with large basidorsal, centre and basiventral. When this centre fuses with the next, its axial portion vanishes, is squeezed out of existence, exactly as happens to the menisci or intervertebral pads in the immovable region of the trunk.

The epistropheus of many Lizards, still more of Snakes, is remark-

1 Explanation of some terms:

Ligamentum suspensorium of the vertebral bodies (Jaeger) is the remnant of the reduced sheath of the notochord which passes through the central opening of the joint cavity. The degraded rest of the chorda itself is the *Nucleus pulposus*, in exactly intervertebral position. The first ligamentum suspensorium is the ligamentum suspensorium dentis.

Meniscus intervertebralis (Jaeger, 1858) = "Annulus fibrosus" of Man and other Mammals is the whole axial disk or pad, if fibro-cartilaginous and free, not passing imperceptibly into the opposed ends of the centra. The first meniscus is the Ligamentum transversarium atlantis.

The *ligamentum capsulare atlantis* and the *ligamentum atlanto-odontoideum* are the two halves of the first joint capsule.

able for its length and for the possession of two large ventral processes, an anterior and posterior of equal size, so that this vertebra looks as if it were the compound of the complete second and third. This erroneous notion is further enhanced by the very oblique position, slanting instead of horizontal, of the first neuro-central suture, see Fuerbringer's figure of Python (Fig. 25), which reminds one of the second and third suture of Seymouria with its deplorably erroneous interpretation. The fixing of the homology of the two ventral processes presented some difficulty. There was the possibility that the anterior is

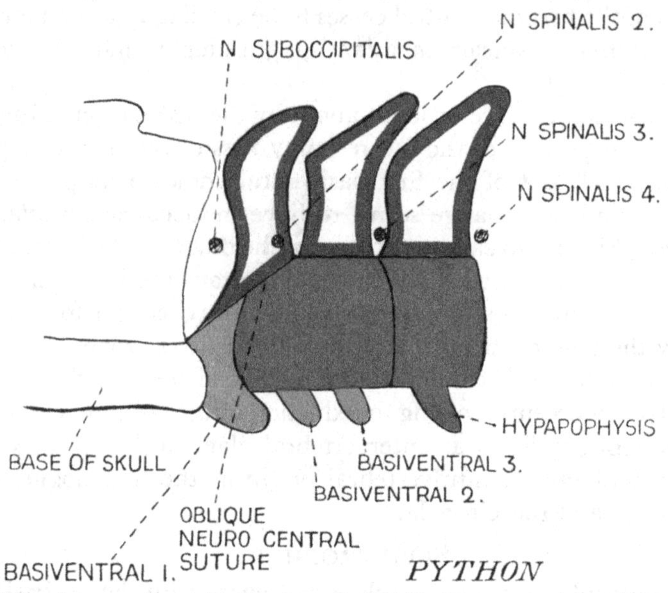

Fig. 25. First three vertebrae of Python from the left side. The position of exit of the nervus suboccipitalis and of the second, third and fourth nervi spinales is shown. The elongated body of the epistropheus carries two ventral processes which are formed by the second and third basiventral elements and are haemapophyses. The third vertebra on the other hand has a single ventral process formed from the centrum and therefore is an hypapophysis.

the haemapophysis or ventral prolongation of BV 2, while the posterior is the hypapophysis of the centrum 2, which occurs in many of the other vertebrae of Sauropsida. In 1896 I was able to show that in Lizards both processes belong to basiventrals, the anterior to BV 2, the posterior being the haemapophysis of BV 3, which fuses with the caudal end of the second centre. In this respect the posterior process behaves exactly like the chevrons of many Lizards. All these ventral processes, chevrons or intercentra, are now usually unpaired, but in

Heloderma they are sometimes still paired, as they should be, having been produced by the originally paired basiventrals. The same double, anterior and posterior, processes exist in the epistropheus of Snakes, with the apparent difficulty that the third vertebra (Fig. 25) has also one equally strong process; but this, arising from the middle of the centrum, is easily accounted for as the hypapophysis, these central ventral processes being well developed in the cervical region of Snakes, but absent in Lizards. In Snakes bony intercentra do not occur except on the atlas. Consequently in them, from the third vertebra backwards, the median ventral blades, the hypapophyses, have taken over

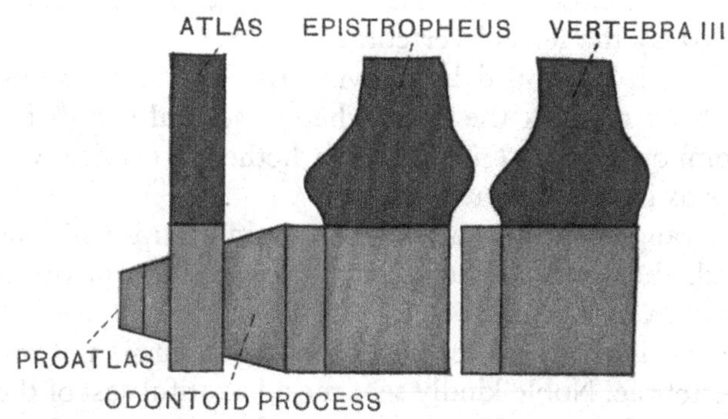

Fig. 26. Scheme of first three vertebrae. The first vertebra or atlas is composed of basidorsal and basiventral elements; the odontoid process of the epistropheus, which is seen passing through the first basiventral, is the interventral of the atlas. The tip of the odontoid process is formed by the interventral of the proatlas. The second vertebra or epistropheus consists of the fused and ossified basidorsal, interventral and basiventral, to the last of which the odontoid process is also fused. The third vertebra is a typical vertebra formed of the co-ossified basidorsal and interventral, with the basiventral forming the intervertebral fibro-cartilage between it and the epistropheus.

the function of attachment of the subaxial muscles which enable the long neck to strike forwards. My interpretation of the homologies of these processes had to wait a long time for embryological proof, which was finally given in 1899 by Maenner.

The complete skeleton of an adult Python in the Cambridge Museum affords further corroboration. It shows several abnormal features, helpful since they are instances of arrested development, relapses into an earlier stage. The atlas is quite flat on the ventral surface and the basiventral is divided with a median suture; there is also another suture between this basiventral and the right basidorsal. The first, second and third centra are quite separate from each other and each carries its own neural arch. The neural arch of the first vertebra has a long and slender process, but this vertebra carries nothing. The second vertebra carries a typical rib, two ventral processes and a sharp median centre-keel or blade. The third also carries a rib, only one, the anterior, ventral process, but no centre-keel. The fourth, fifth and sixth vertebrae carry ribs, no ventral processes, but each has a centre-keel.

A broad consideration of the foremost vertebra leads us irresistibly to the conclusion that what has happened to the proatlas is happening now to the atlas, so that it may hand its function over to the next following vertebra. Complete reduction of the odontoid process would reduce the originally complete atlas to an osseous ring-like pad analogous to another ligamentum transversarium and the occipital condyles would be carried by the second vertebra.

The principle implied in carrying the head by osseous structures farther back remains the same whether several centra fuse into one axial compound as in *Triceratops*, or whether all cervical vertebrae fuse together as in the present Whales.

The arrangement in examples of certain groups will now be briefly discussed, the general scheme of the composition of the atlas region being shown in Fig. 26.

ERYOPS (Fig. 27). After an interesting discussion of my models of fossil vertebrae, Noble kindly sent me a beautiful cast of the first three vertebrae of Eryops. The whole complex measures about 10 cm. in length and 11 cm. in height. The missing spine of the atlas and the first basiventral can be easily restored. On the dorsal surface of the neural arches proper, belonging to the atlas, is a clear indication that they were sutured to a supradorsal which carried a forked spine grasping the second. The anterior zygapophyses of II and the posterior zygapophyses of III, the diapophyses with their large and long facets for the first, second and third ribs, the very large basiventrals or second and third intercentra, and fortunately the interventral semi-ring between BV 2 and BV 3, are all clear, like a diagram. The dotted line in the figure indicates the suture between the first supradorsal and the rest of the first neural arch, the front of which contains a large and deep cavity, bordered ventrally by the second basiventral. Cope rightly drew attention to the absence of an odontoid. Case has cleverly restored the right and left interdorsals. They still lie behind their neural arches like the proper dorsal halves of the intervertebral cartilage. Therefore there cannot yet be an atlas centrum, although undoubtedly the cartilaginous intervertebral mass filled the cavity in the front of our cast. It is quite possible that the gap figured on the ventral side behind the atlas contained the missing first interventral pair, the potential Amniotic centrum, in which case *Eryops* would possess an incipient osseus gastrocentrum, or, if the still cartilaginous interdorsals, future notocentra, were present, *Eryops* would be in the interesting and most significant condition of the holocentrous Cricoti. This is one of the several indications that the tripartite type of the other relations of *Eryops* is a further specialisation of the quadripartite holocentrous or embolomerous type; not the reverse as has been suggested.

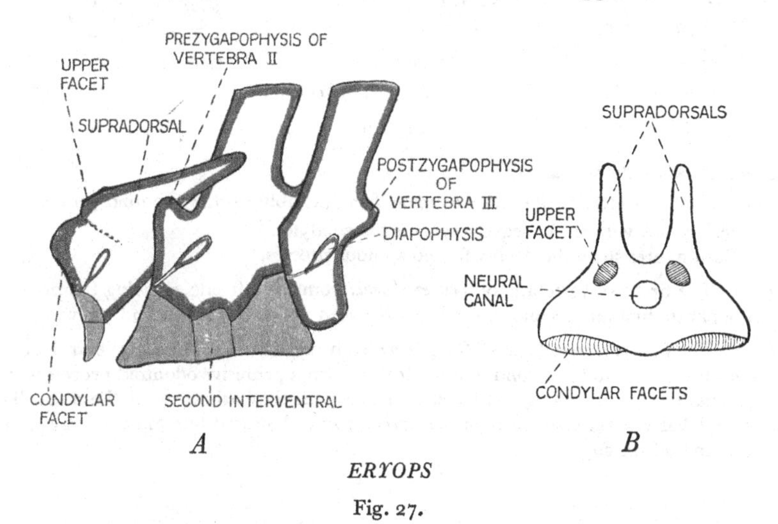

ERYOPS

Fig. 27.

A. The first three vertebrae of *Eryops* from the left side. The gap seen between the first and second basiventrals was presumably filled by the missing first interventral pair. The second interventral is shown between the second and third basiventrals. The dotted line indicates the suture between the first supradorsal and the rest of the first neural arch. The three diapophyses carry the ribs. The anterior zygapophysis of the second and the posterior zygapophysis of the third vertebra are prominent.

B. Anterior view of the atlas of *Eryops* showing the supradorsals which clasp the second neural arch, the two upper facets, and the two lower, condylar facets.

DISSOROPHUS (Fig. 28). Atlas tall like that of *Cacops*. The spinous process is deeply forked; BV 1 has a pair of clear concave facets for the condyles at its front end; the facets well below the notochord. On its back surface BV 1 has one large concave facet which has either to receive a large ventral odontoid or a centrum 1, unless this cavity lodged the still large chorda, which was not constricted. ID 1 is absent; the condyle therefore articulates only with BV 1.

CACOPS. Seems to have ID 1; ID 2 is doubtful.

In TREMATOPS both are absent.

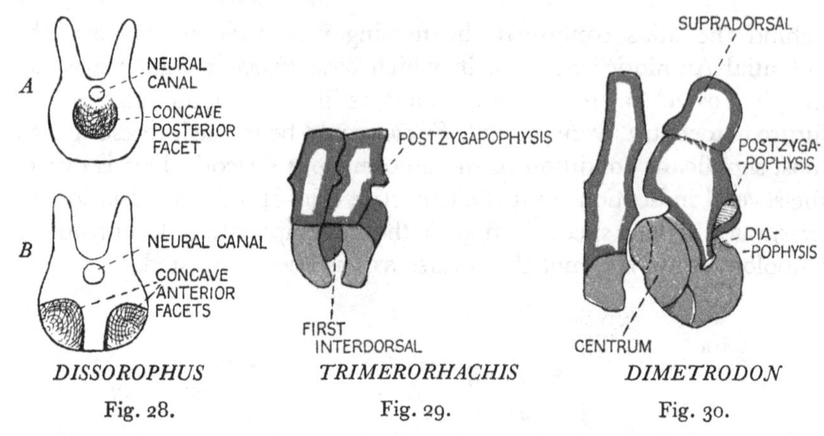

Fig. 28. Atlas of *Dissorophus*.

A. Posterior view showing the large concave facet possibly for the odontoid process.

B. Anterior view with the concave facets for the condyles.
Both figures show the deeply forked spinous process.

Fig. 29. The first two vertebrae of *Trimerorhachis* from the left side, showing the presence of separate first and second interdorsals between which lies the second basiventral.

Fig. 30. The first two vertebrae of *Dimetrodon* from the left side. The interventral of the atlas is fused on to the second interventral to form a primitive odontoid process which articulates with the elongated basidorsal and with the basiventral of the atlas. The second basiventral forms a typical intercentrum. A dotted line indicates the suture marking off the supradorsal.

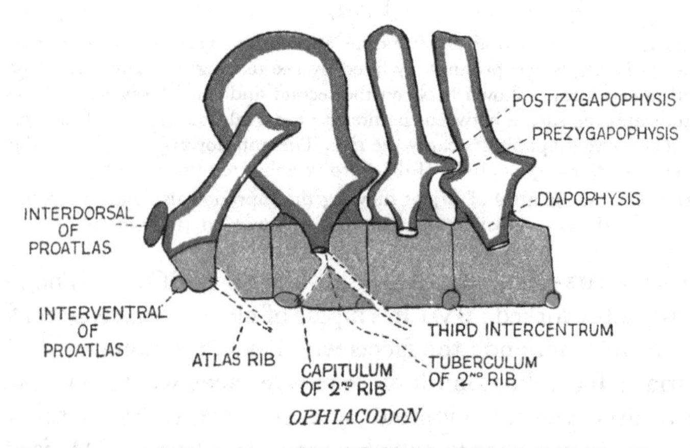

Fig. 31. First four vertebrae and proatlas of *Ophiacodon* from the left side. The proatlas is represented by both an interdorsal and an interventral portion separated from one another. The basiventrals from the second onwards are reduced to small intercentra which carry the capitula of the ribs, the first two of the latter being indicated by dotted lines. The strong diapophyses carry the tubercula.

TRIMERORHACHIS (Fig. 29) seems to possess both separate ID 1 and ID 2. The lateral occipital and the basioccipital parts of the condyle articulate with neural elements, and with a paired intercentrum (BV 1).

DIMETRODON (Fig. 30). From several rather discrepant figures we come to the following restoration. The atlas was but loosely attached to its basiventral, the neural arch being high. The second vertebra is much more bulky, with a suture for the supradorsal. The irregularly shaped centrum 1 articulated with the BD 1 and BV 1, the second basiventral forming a small typical intercentrum. A first atlas rib seems indicated by a facet on the centrum and the BV 1. The second rib must have been much stronger to judge from the large process for its tuberculum, the diapophysis, in one of Case's figures.

OPHIACODON (Fig. 31). One of the rare cases with a proatlas represented by an interdorsal, and a very small interventral piece. The neural arch of the atlas is large and seems to grasp the large second spinous process, and probably is fused with the first basiventral to which a rib is attached through its capitular portion perhaps by means of a ligament; the tuberculum is in contact with a knob belonging to the first centre, which is sutured to the second centre. A small second rib seems to be connected by a ligament with the slightly shifted small basiventral, while its tuberculum has formed a diapophysis upon the second neural arch. Intercentra seem to be present, although all are small throughout the neck at least.

COECILIAE (Fig. 32). Fig. 7 of Marcus and Blume. Atlas and second vertebra in sagittal section. The odontoid, which the authors declare to be absent, must be the anterior nozzle-like projection of the atlas which is attached to the occipital bone by a ligament in the figure. It included chordal cartilage, but there is no centrum. Fig. 11 (older larva than Fig. 12). The atlas has a pair of large flanges for articulation with the occiput; the chorda ends with the cartilage at the mid-level of the atlas, thence it is continued into the occiput by connective tissue only.

Fig. 12. Larva. Horizontal section through the atlas and occiput.

CRYPTOBRANCHUS (Fig. 33). First and second vertebrae with their ribs and the position of the exits of the nervus suboccipitalis and nervi spinales 2 and 3. The process of the atlas is the odontoid-like process or shelf, homologue of the proatlas centre. There seem to be indications of the pseudocentrum and centrum; the former as the basiventral mass agreeing in position with the first rib.

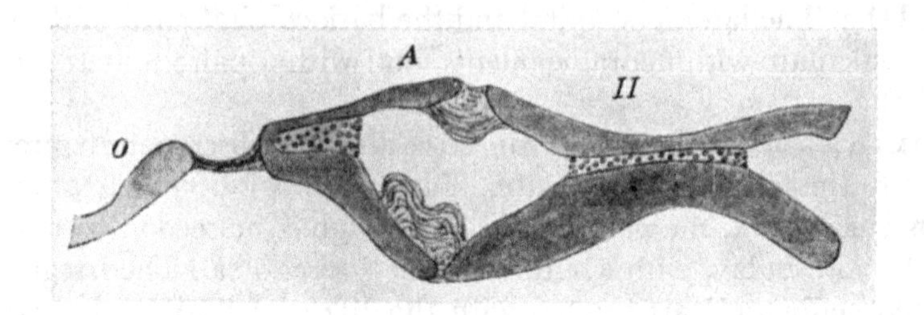

A

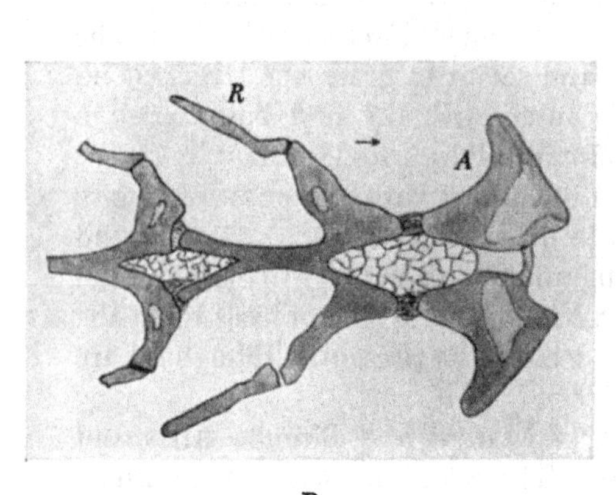

B

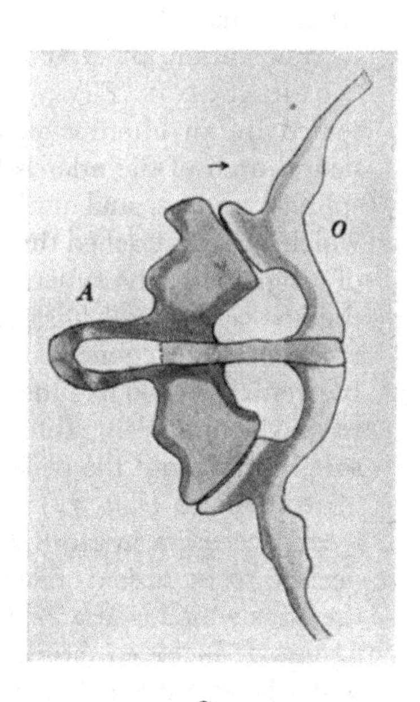

C

Fig. 32.

A. Fig. 7, p. 12. Sagittal section of *Hypogeophis*, 20 cm. long. Cartilage indicated by small rings. *A*, atlas; *II*, second cervical vertebra; *O*, occipital bone. Magnified 30 times.

B. Fig. 11, p. 14. Horizontal section through the first vertebra of a larva of *Hypogeophis*, 7·2 cm. long. Cartilage shown in lighter shading. *A*, atlas; *R*, rib of second vertebra. Magnified 37 times.

C. Fig. 12, p. 14. Horizontal section through the occipital joint of a larva of *Hypogeophis*, 5 cm. long. *A*, atlas; *O*, occipital. Magnified 50 times. (Marcus and Blume.)

SPHENODON (Fig. 34). Shows the small ID 0 of the proatlas lying in front of BD 1 with the suboccipital nerve issuing below it. The line of fusion of IV 0 with IV 1 is indicated by a dotted line. BV 2 lies in front of IV 2 and below IV 1.

CHELONIA (Fig. 35). These show great variation. In *Trionyx*, *Chitra*, *Clemmys*, *Testudo* and many others, the various pieces of atlas

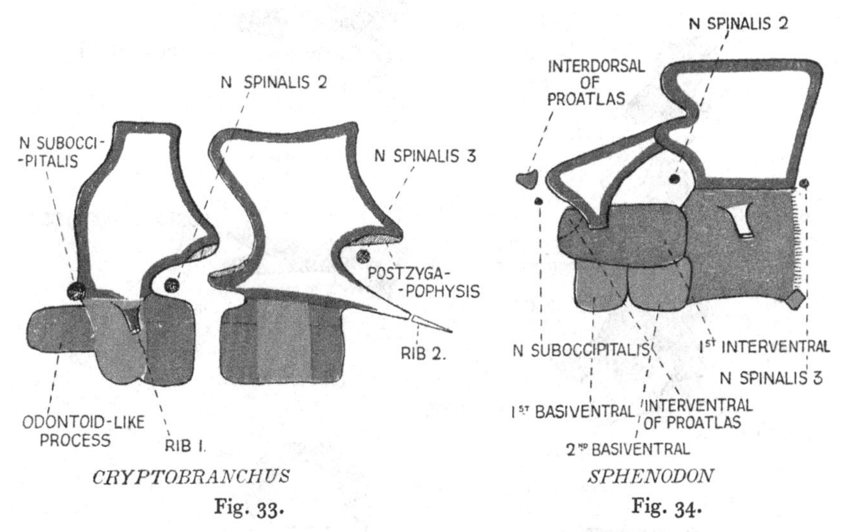

Fig. 33. Fig. 34.

Fig. 33. First two vertebrae of *Cryptobranchus* from the left side. The odontoid-like process of the atlas is the interventral of the proatlas. The rest of the basal part of the atlas is formed by its basiventral and the fused anterior halves of its interdorsal and interventral. The lower half of the second vertebra is formed by its basiventral on to which is fused anteriorly the posterior halves of the first interdorsal and interventral, and posteriorly the anterior halves of its own interdorsal and interventral. The first and second ribs are shown.

Fig. 34. First two vertebrae of *Sphenodon* from the left side. The proatlas is represented by a small interdorsal in front of the first basidorsal and by an interventral which is fused to the first interventral along the dotted line. Basiventrals 1 and 2 lie below the first interventral; basiventral 3 forms a small intercentrum.

do not anchylose, and the first centrum remains quite movable, although sometimes, for example, in *Trionyx hurum*, it is fused with the second basiventral. In others, for example, *Platemys* and *Chelys*, the pieces of the atlas fuse into a solid vertebra which freely articulates by a convex-concave joint with the second centrum. In *Chelys* a small BV 1 still exists; in *Platemys* it cannot be recognised. In *Chelydra* the right and left neural arches are quite separate, and each forms a long spinous process which bifurcates near its posterior end, one arm carrying

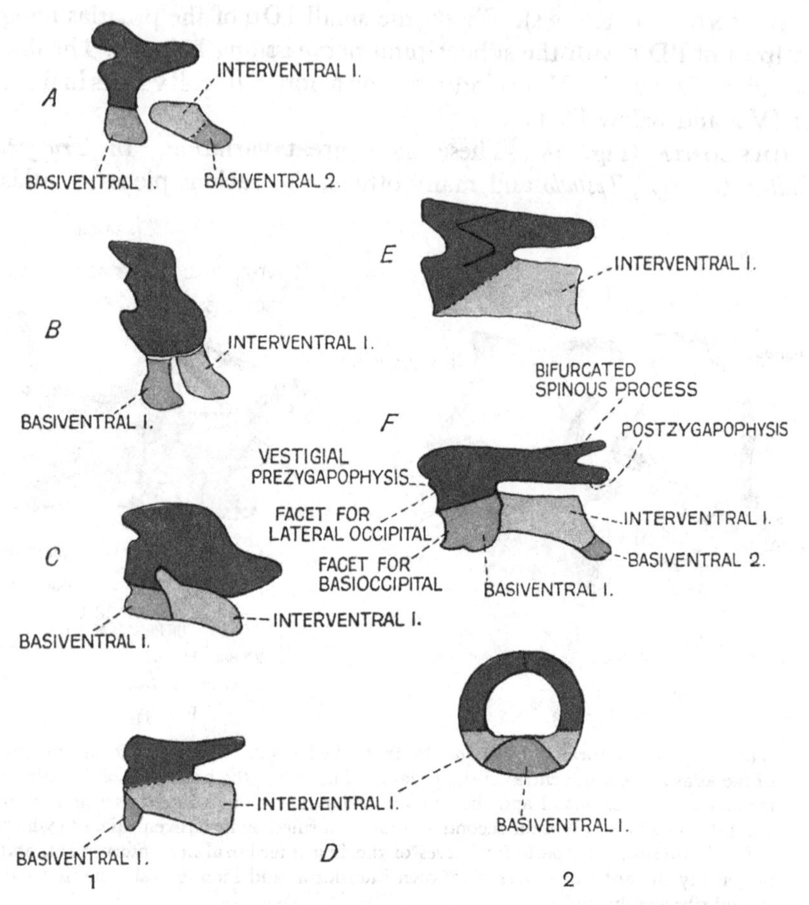

Fig. 35. The atlas of various Chelonia seen from the left side.

A. **Trionyx hurum.** **The first interventral is free from the atlas and is fused to the second basiventral.**

B. **Trionyx gangeticum.** **The three blocks composing the atlas are not fused.**

C. **Chitra indica.** **The blocks are still separate as in** ***Trionyx*** **but the interventral is larger.**

D. **Chelys matamata.** **1, lateral view; 2, anterior view. The interventral and basidorsal are fused: there is still a small basiventral element.**

E. **Platemys.** **The interventral and basidorsal are fused; there is no recognisable basiventral.**

F. **Chelydra.** **Shows the bifurcation of the non-fusing basidorsals. The second basiventral is fused to the first interventral. A postzygapophysial and a vestigial prezygapophysial facet are seen: above the latter emerges the suboccipital nerve. The condylar facets are shown.**

a facet for the zygapophysis of the second vertebra. The anterior end has a facet, a vestigial prezygapophysis; above it issues the first spinal nerve. The share of the lateral occipital and the basioccipital in the triple condyle varies considerably; sometimes the basioccipital is quite excluded from the foramen magnum, for example, in *Chelydra*, owing to the excessive elongation of the lateral occipitals, which have two or three hypoglossal foramina.

MAN (Fig. 36). The variations of the first and second vertebrae of this best studied of all vertebrates seem endless. Thanks to the morphological completeness of these vertebrae, they are still capable of variations which may possibly lead to new adaptive changes, so that this morphological chapter may not yet be absolutely closed. Most of these variations have of course been described; Macalister has written a special paper on them. The embryonic and phyletic meaning of such cases would often be enhanced if Anthropotomy did not preferably stick to its special jargon. However, we must not forget that the older Human Anatomy is the mother of Vertebrate Morphology.

The variations are shown in Fig. 36*A–G*.

Fig. 36*A*. A ventral view of a young epistropheus (Cunningham). The odontoid contains a right and a left piece, which unite in the seventh to eighth month; their paired nature proving the paired nature of the centre I. Where they diverge in front lies the "centre or summit of the dens", often dismissed as an epiphysis, though actually the remnant of the centre of the proatlas or ossiculum terminale. The second vertebra itself ossifies from five centres. One pair forms the articular and transverse processes, together with the right and left laminae, which combine to make the spinous process. The whole forms the first neural arch, being the basidorsals. The "superior and inferior articular processes" are of course the anterior and posterior zygapophyses, in conformity with the upright gait of Man, while anterior and posterior (human) means our ventral and dorsal. The normal vertebrate to the human anatomist stands on its tail. The centrum is the third centre of ossification; its originally paired nature is indicated by the occasional appearance of two nuclei within this centre.

Fig. 36*B*. A vertebra in horizontal section, after Keith. The body, composed of the interventral mass, is correlated in position with the rib, which in reality belongs to the hypochordals, i.e. the basiventrals. At this early stage represented by the figure, the paired units, which compress the chorda and carry the ribs, are in reality the basiventrals (hypocentra, intercentra or hypochordals). The pair of nodules, which do not yet constrict the chorda, are the halves of the future body or centrum, a valuable recapitulation by ontogeny of a truly phyletic process of evolution. Hypochordals are Froriep's term for the basi-

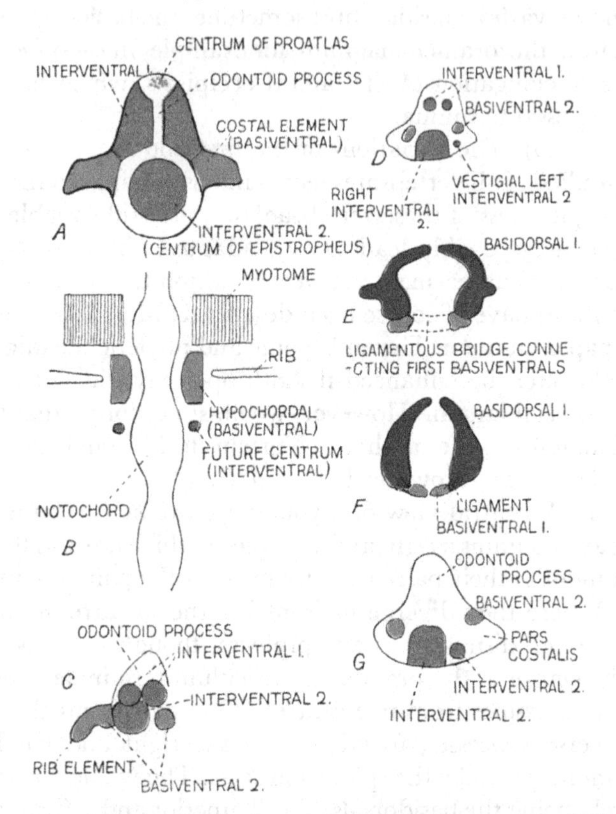

Fig. 36. Variations in the development of the first two vertebrae in Man.

A. Ventral view of young epistropheus. The stippling shows the ossiculum terminale or the centrum of the proatlas. The first interventrals, forming the odontoid process, are not yet fused.

B. Horizontal section through developing vertebra (after Keith). The basiventrals are shown compressing the chorda and carrying the ribs. At this stage the interventrals are much smaller and do not yet compress the chorda.

Figures *C–G* are various anomalous developmental conditions figured by Macalister and described in the text.

ventrals, this term being correctly applied by Macalister in the young atlas.

The following are from the Anatomical Museum, Cambridge; described and figured by Macalister.

Fig. 36*C*. Infant, 5 months. Delayed union of lateral dens centres. Right BV 2 containing rib.

Fig. 36*D*. Infant, 5 months. Dislocated and vestigial left half of second centre. Cf. Fig. 36*G*.

Fig. 36*E*. Aetat 4½ years. With unusually wide ligamentous bridge connecting the BV 1.

Fig. 36*F*. Aetat 4 years. Neural arches not yet united, ventrally connected by ligaments with the still separate right and left BV 1.

Fig. 36*G*. Aetat 5 months. A very anomalous case with two centres of cartilage in the left pars costalis; upper centre BV 2, cf. Fig. 36*A*; lower centre IV 2.[1]

1 Flower, *Osteology of the Mammalia*, p. 35, has the following: "If the axis is examined a year or two after birth, its body appears to be composed of two parts, one placed in front of the other, the first including the odontoid process and the anterior part of the body, the second all the remainder of the body. The arch is united to both. On the other hand, the atlas at the time of birth has nothing corresponding to the centrum of other vertebrae, its inferior arch being still cartilaginous". This is rather confusing; he looked upon the odontoid as part of the axis centrum instead of a centrum in its own right—an oversight which I ought to have corrected in the third edition of his work. That the neural arch "is united to both" shows that it extends its base upon both centra.

Chapter XIV

RIBS

The lowest living Craniata, the Cyclostomes, are still devoid of ribs, but cartilaginous dorsal and ventral arcualia evolve in the system of intermuscular septa, and by extending into them form, at least in the tail, paired, distally fused, dorsal and ventral arches. Strictly they cannot be called neurapophyses and haemapophyses, since each pair, or rather each arcuale, itself represents the whole outgrowth.

The first traces of lateral extension of some of the ventral arcualia appear in the Holocephali. They become further developed into horizontal outgrowths in the anterior trunk region of other Elasmobranchs, and are bordered by the peritoneum. Farther tailwards they become longer and their distal portion begins to separate itself off as a short rib (Fig. 37*A*), now carried by the basal stump, which is the basiventral proper in a restricted sense. As a further modification the basiventral stumps begin to approach each other ventrally (Fig. 37*B*), and, in adaptation to the dorsal aorta, send out a pair of ventral processes which tend to enclose the aorta; and in the tail, by distal fusion, to produce the caudal canal for protection of the caudal artery and vein (Fig. 37*C*). These most ventral outgrowths are the haemapophyses. The rest of the now forked basal stumps are the pleurapophyses, carrying the ribs (Fig. 37*D*).

Thus the originally simple mass has become more or less divided into a lower haemapophysis and a more latero-dorsal rib-bearing pleurapophysis, and this Selachian rib extends in a slightly dorsal curve into the horizontal septum between the hypaxial and the epaxial trunk muscles, at the crossing of the horizontal and the transverse septa. In many cases the ribs become vestigial in the posterior region of the trunk, and, owing to the eventual elongation of the whole basal stump before it bifurcates, the vanishing rib seems to have slid down distally upon the haemapophysis. In reality the original and unchanged place of insertion of the ribs is usually indicated by a conspicuous knob. The united Y-shaped pair of haemapophyses of the Dipnoi, various Selachians, and Chondrostei ends in one or more ventrispinalia, exactly analogous to the cartilaginous dorsospinalia. They are the infraspinalia of the general scheme (Fig. 37*E*).

Ossification of the ribs begins in the Dipnoi, is carried further in the Chondrostei, and becomes complete in the Teleostei.

The haemapophyses and pleurapophyses exhibit a great range of variation in size, completeness, position and presence or absence, not

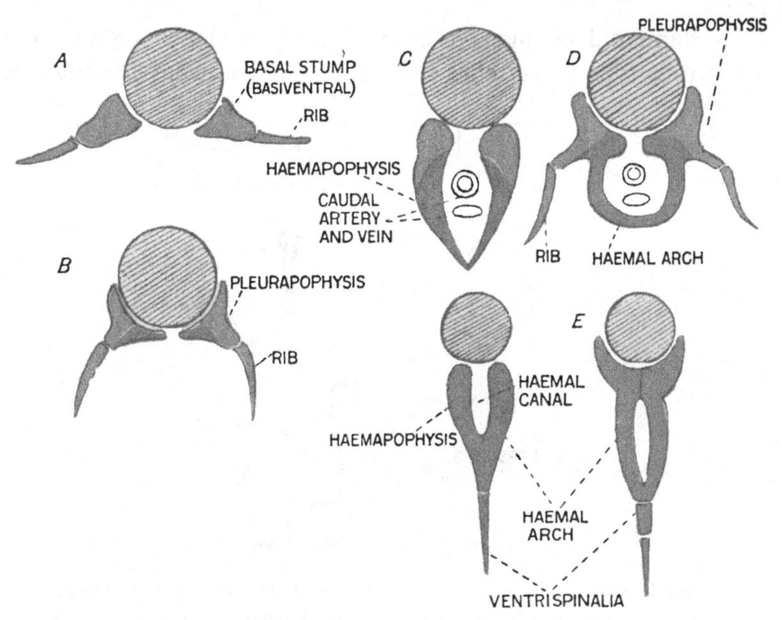

Fig. 37. The relation of ribs to haemapophyses in various fish, shown in transverse section.

A. Trunk region of Elasmobranch, showing short ribs separated from the basiventral stump.

B. A further development. The basiventrals begin to approach each other ventrally.

C. Tail region with haemapophysial outgrowths of the basiventrals, enclosing the caudal artery and vein.

D. Fully developed basiventrals forking into the pleurapophyses, carrying the ribs and the haemapophyses, fused distally to form the haemal arch.

E. The two forms of the haemapophyses of Dipnoi, various Selachians and Chondrostei, carrying the ventrispinalia at their fused extremities.

In this series the notochord is shaded, the pleurapophysis and rib are coloured orange, the haemapophysis blue.

only in the various vertebrates, but also in the different regions of the same individual. An important feature is that the basal stumps not only grow outwards and downwards, but grow to a considerable extent in an horizontal direction; but this growth does not keep step in the transverse plane at the anterior and posterior ends of the vertebral body. The haemapophyses, in particular, instead of remaining in the same transverse plane as the pleurapophyses, may be relegated to the posterior end, well separated from the more anterior and more dorsal pleurapophyses, a common feature in many Teleostei and in *Polypterus* (Fig. 38).

A fundamental change has been caused by the invention of the arcocentrous type of vertebra instead of the acentrous Dipnoan and

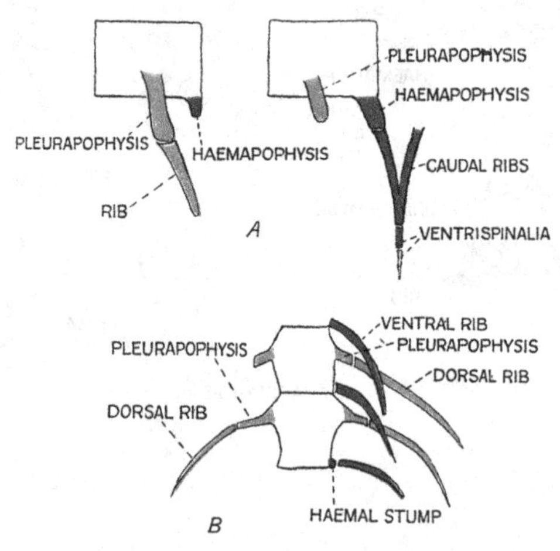

Fig. 38. The relationship of the dorsal and ventral ribs and haemapophyses of *Polypterus*.

A. Two vertebrae from the left side. That on the left is from the trunk region, showing a pleurapophysis carrying the rib, and a small haemapophysial stump. That on the right is from the tail region and shows a much smaller pleurapophysis with no rib and a haemapophysis carrying the caudal ribs forming a Y. These have been turned round through a right angle to come into the plane of the picture.

B. Two trunk vertebrae from the ventral aspect carrying both dorsal and ventral ribs. The anterior end is above. On the left side is shown a pleurapophysis, and below a rib articulating with it. On the right side the ventral, haemapophysial, ribs are also shown, crossing below the others and attached to the posterior end of the vertebra either directly or by means of a haemal stump.

The colour scheme of Fig. 37 is maintained, the pleurapophyses being orange, the haemapophyses blue.

the chordacentrous Selachian side departure. In many Teleostei, for example, the Codfish, the ribs of the anterior trunk region are directly inserted on to the side of the centra, there being no outstanding pleurapophyses. Farther back in the trunk the pleurapophyses are fully developed and carry the ribs attached to the caudal side, not at the distal end. In the tail the ribs fuse with the pleurapophyses and the right and left compounds produce a ventrispinal Y, strikingly like the haemapophysial Y of Selachians and Dipnoi and yet not homologous (Fig. 39).

The confusion between haemapophyses and pleurapophyses arose as follows. The haemapophyses, as the most ventral part of the lower arch (BV), were supposed to be the homologues of the ribs of all Teleostei, therefore ventral ribs; and the incipient most basal indications of the Selachian haemapophyses were supposed to be vestigial; therefore the ribs of the Selachians were declared to be dorsal elements. Moreover, this seemed to be corroborated by the position of these ribs

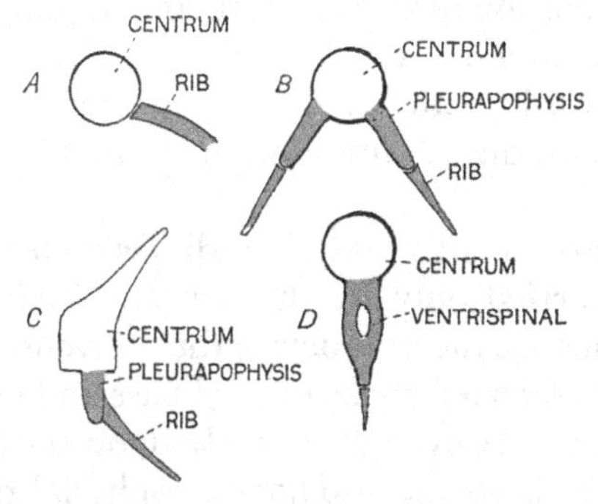

Fig. 39. Teleostean ribs as seen in the Cod.

A. Transverse view of the anterior trunk region. The rib is directly inserted on to the side of the centrum.

B. Transverse view of the posterior trunk region, with developed pleurapophyses to which the ribs are attached.

C. Lateral view of the same to show that the ribs are attached to the caudal side, not to the distal end of the pleurapophyses.

D. Transverse view of the tail region. The ventrispinal Y formed by fused ribs and pleurapophyses.

lying in a transverse septal level which is higher, more dorsal, and placed between the muscles. Further, since the ribs of Amphibia are ontogenetically exclusively connected with the neural arches, they are considered as dorsal ribs and therefore homologous with those of Selachians. The conclusion is therefore reached that the Amniotic ribs are homologous with the ribs of Frogs and Selachians in opposition to the ventral ribs of Dipnoi, Ganoids and Teleostei. As a proof of the former existence of double ribs stands *Polypterus* (Fig. 38), representing the original ancestral condition. The chain of argument is sound, but some of the links are misinterpreted.

Gegenbaur's comprehensive account in his *Vergleichende Anatomie der Wirbelthiere* makes difficult reading. On p. 278 he criticises the value of the upcurved distal portion of the Selachian ribs, and explains it most sagaciously as an adaptation to the new conditions required by the necessity of room for the adductor muscles of the pectoral fin. Therefore the Selachian ribs express no fundamental difference from other ventral ribs. On p. 280 he correlates the reduction or absence of the Selachian ribs of the trunk, with the reduction in volume of the ventro-lateral muscles. On the same page, in small type, he very reasonably makes little of the difference whether the ribs arise ontogenetically continuous with their basal supports, or at a distance, as it proves nothing for their phylogenetic standing. And yet there are so many amendments strewn into the whole account, and terse remarks which brush aside amendments accepted elsewhere, that almost any writer on this rib question may find his view recognised in part.

Supposing for argument's sake that all Tetrapods have only dorsal ribs, which are carried only by the neural arches. This could only mean that their ribs are the products of the basiventral or lower arches, and that they have then left their original base and been transferred on to the centra, sometimes even exclusively upon the diapophyses. But since many Tetrapods possess also haemapophysial ribs, it follows that there are, or were, two pairs of outgrowths of the basiventral, either or both of which can end as ribs. Consequently all ribs are in origin ventral elements. Because the only ribs of the Elasmobranchs in many Selachians grow into the horizontal septa, and the same applies to the "cephalic ribs" of Sturgeons, there is no need to stamp them as dorsal elements.

The ribs are in fact twins of the same basal stump which has bifurcated, and when this split affects the whole stump the resulting pleurapophyses and haemapophyses are separated. Either of these may become reduced so that their ribs may become directly attached to the body of their vertebra. This, however, could not yet happen in the acentrous and chordacentrous Fishes, but only in the arcocentrous creatures, in which the rib has gained support upon the parapophysis, which is an outgrowth of the interventral. It would be erroneous, on the same lines as the explanation of the ribs in Anura, to take this secondary attachment as a proof of the interventral origin of these "dorsal ribs".

The ribs of Amniota must be homologised with the upper ribs of *Polypterus* and with the ribs of the Anura, but if some authors declare the Amphibian ribs to be primitive exclusive growths from the neural arches, which clearly does not apply in the case of the Amniota, they are not justified in placing the Amphibia between the Selachians and the Amniota.

The following scheme represents my view:

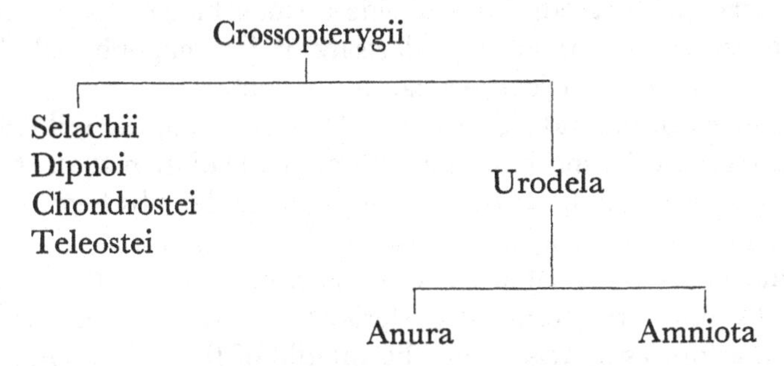

Since Teleosts are undoubted branches of more or less bony Ganoids, some may well retain preponderance of haemapophysial and others of pleurapophysial characters. The Selachii still suffer from the honour of supposed ancestry to all other Gnathostomes instead of being considered the chordacentrous side branch of not even a very ancient level.

The above scheme is based chiefly upon the ribs of *Polypterus* and the Newts.

(1) *Polypterus*.

The work of Budgett on the development of *Polypterus* was unfortunately limited to the description of the early larval stage owing to his untimely death. The later stages of this veritable key vertebrate are still uninvestigated. He discovered in the larva that the neural arches, which are by far the largest, form a continuous series all through the column. The lateral and ventral cartilaginous elements are much thinner and quite separate from each other, and all three series rest directly upon the notochordal sheath. The whole column, in this stage, looks like that of an adult acentrous fish, but there seem to be no interbasalia. The lateral elements may be the basalia in their still indifferent stage, although they are much more like the ribs themselves which, as in Teleosts, have lost their basal stumps. The ventral elements, which also look like baseless ribs and are rightly named haemal arches, are very small and delicate in the trunk; in the tail these ventral ribs are long, the first caudals still forming separate long haemal arches; the other caudals form by fusion a Y.

In the adult (Fig. 38) the upper ribs, which are movable, are carried by the large transverse horizontal processes, the pleurapophyses, which as in Teleosts are evolved later, and seem to consist mainly of directly ossified or calcified connective tissue. These large transverse processes and ribs are in the tail reduced to mere vestiges; in the trunk they carry on the medio-ventral side a small flange which Buetschli recognised as the "untere Querfortsatz or Haemapophysis". This is probably correct, provided we call it vestigial.

The lower or ventral ribs are best described from the tail forwards, as this seems to be more in line with actual evolution than the usual procedure from the head backwards. The caudal ribs form a Y with curved limbs convex outwards, attached to haemal stumps which are continuous with the ribless transverse processes. In the trunk the haemal stumps arise from the posterior end of the vertebra, while the transverse processes arise from the middle of the body. These lower ribs, likewise movable, now widely separated from each other, extend backwards in a very oblique direction, crossing the dorsal ribs and still arising from the haemapophysial stumps. Farther forwards in the trunk the lower ribs become much weaker than the strong upper ribs and lean against the under surface of the latter at the crossing point, and become ultimately fixed to them at this point by connective tissue. Thus it comes to pass that the proximal portion of the lower rib, together with its basal stump, is weakened and seems to be attached mainly to the powerful pleurapophysis. It can, however, be followed as a still more or less bony strand to its origin from the little process at the posterior end of its vertebra. This seems to explain the statement of Buetschli that in the trunk the lower rib is attached to the upper larger "Querfortsatz"; farther back in the trunk to the "untere" transverse process.

Polypterus shows that haemapophyses can sever their distal portions as movable ribs and that these, perhaps because they had become movable, can at least partly be reduced, while the upper ribs prevail as the functioning ribs. The deciding cause is the changed function from entirely protective and muscle serving service of the stiff united right and left haemapophysial mass, to rib service with its action in several planes. Why the dorsal ribs, or at least dorsally placed ribs, are superior in this respect is revealed to us by the Newts.

(2) The Urodela.

The most momentous step ever taken by Vertebrates was the step from water on to the land. The transverse section of the average fish is a long vertical oval, broadest above. The ribs, bordering the body cavity, are ventral hoops directed downwards. The free-swimming fish is buoyant, practically of the same specific gravity as the water. The early Eotetrapod lies with its whole ventral side upon the ground, literally flattened down, depressed by its entire weight. Hence the transverse section is a flat-bottomed transverse oval in both trunk and head. In correlation with the depressed head its arches follow suit; the mandibular joints point backwards and increase their distance from each other. The next arch remaining in a transverse plane is kinked into an articular "elbow", which coming into contact with the skin incidentally initiates the formation of a sound-conducting rod from the tympanum to the auditory capsule.

For the sake of better locomotion the legs, hitherto fin-like, raised the body above the ground and responded to the body weight by an elbow or knee and wrist or ankle bend. The raised portion made possible improvements of the body cavity and rearrangement of the viscera, so that the trunk assumed a more cylindrical shape. There is the clearly expressed tendency for the ribs to form more strongly curved hoops giving more room for the lungs, etc., and, since the ribs are ventral elements, the proximal portion of the hoops is transferred steadily farther dorsally, so that the bodies of the vertebrae, the axial portion of the column, seem to have sunk into the body cavity; in reality it is the latter which extends upwards on both sides of the bodies.

I interpret the meaning of the ontogenetic changes of the Urodele ribs as follows:

In the first stage the ribs arise from the basiventral with which alone they articulate (Fig. 40*A*).

In the second stage (Fig. 40*B*), since the basiventrals have become much reduced in the trunk, the strongly curved ribs are less dependent upon their original attachment, their ligamentous connections with the basidorsal become strengthened, resulting in the formation of a shoulder, while the basidorsal reacts by its diapophysial outgrowth. The rib, now double-headed, consists of a neck with capitulum, shoulder or tubercular process, and the shaft. Between the rib and the pedicle of the neural arch is a space, the true foramen transversarium through which runs the arteria vertebralis.

In the third stage (Fig. 40 *C*, *D*) the neck is reduced to a mere string, no longer cartilaginous, useless as a support for the rib, which reacts by improving its anchorings of connective tissue so as to form a new more dorsal shoulder or secondary diapophysis to which the

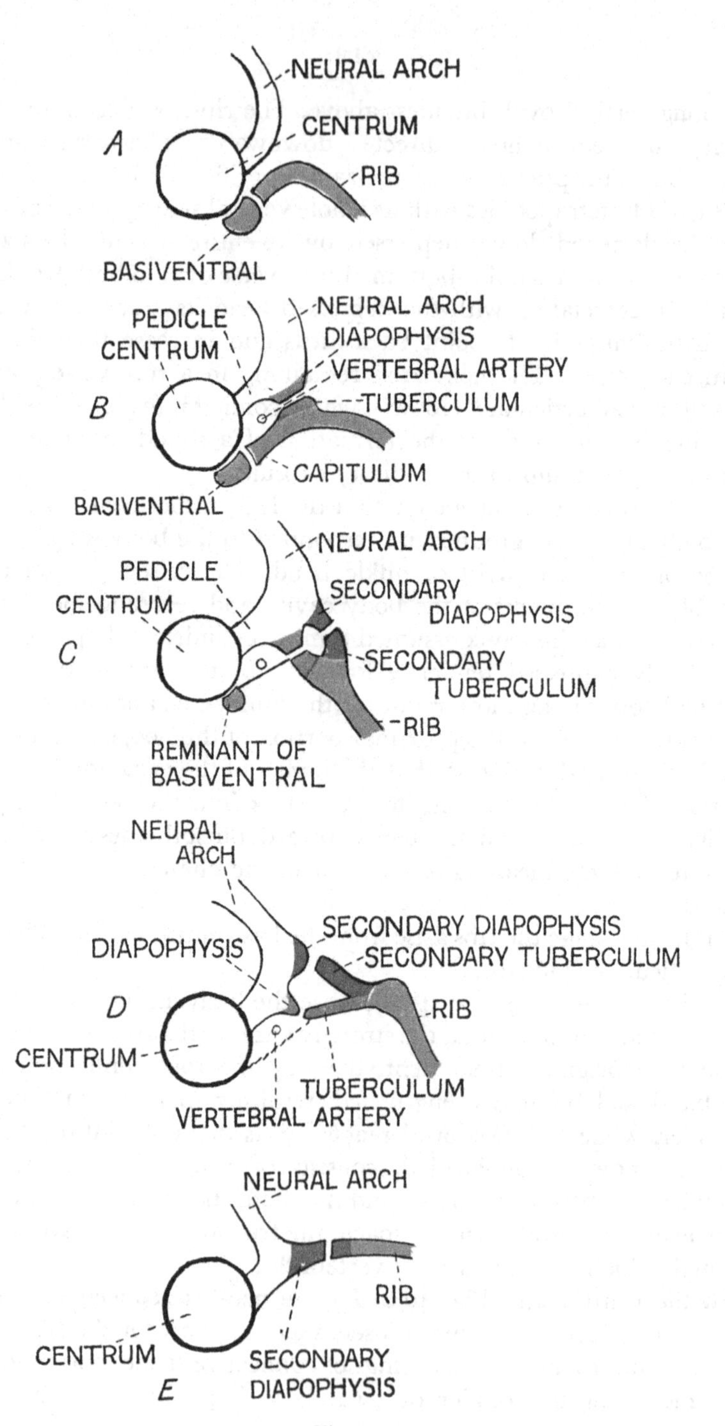
NEURAL ARCH
CENTRUM
A
RIB
BASIVENTRAL
PEDICLE
NEURAL ARCH
CENTRUM
DIAPOPHYSIS
VERTEBRAL ARTERY
B
TUBERCULUM
CAPITULUM
BASIVENTRAL
NEURAL ARCH
PEDICLE
SECONDARY
CENTRUM
DIAPOPHYSIS
C
SECONDARY
TUBERCULUM
RIB
REMNANT OF
BASIVENTRAL
NEURAL
ARCH
SECONDARY DIAPOPHYSIS
DIAPOPHYSIS
SECONDARY TUBERCULUM
D
RIB
CENTRUM
TUBERCULUM
VERTEBRAL ARTERY
NEURAL ARCH
RIB
CENTRUM
SECONDARY
DIAPOPHYSIS
E

Fig. 40.

neural arch also responds. The rib has regained the required double attachment, including a pseudo-transverse canal. If this also includes an artery, there is nothing against this being due to replacement of the true arteria vertebralis by collateral vessels; nor is it improbable that Goeppert's "Rippentraeger" has arisen out of the long tendinous anchorings which connect in various ways the rib shaft, or the secondary shoulder, with the neural arch.

Lastly, there is an apparent difficulty in that the ribs possess secondary, more distal joints between themselves and the "Traeger". The division seems explained by Wiedersheim's suggestion that in the sacral region of *Menobranchus* the two halves may be the pleurapophysial and haemapophysial processes; if the division extends to the base of the whole basal stump, the one may well be the rib of the Urodeles, the other the lower rib of *Polypterus*; but this would not militate against the secondary nature of the joints in question. *Menobranchus* is in many respects lower than the Salamanders and Tritons although it is a Pérennibranchiate. The whole upward trend of the rib attachment reaches its exaggerated terminus with Anura (Fig. 40*E*), in which the vestigial ribs are carried entirely by the enormously developed secondary diapophyses, developing ontogenetically without a repetition of the stages passed through by Urodela and Gymnophiona. Such dorsal ribs could easily be taken as proof that the ribs evolved from the neural arches. But this discovery of Goette dates from the time when we all hailed Ontogeny as the true revelation of Phylogeny, in spite of Haeckel's Cenogenesis and Gegenbaur's sharp criticism of the too sanguine faith of the embryographers.

While in the recent Amphibia the processes of the trunk vertebrae are preponderatingly formed by the neural arches, in the tail by the same and by the basiventrals with their hypapophysial preponderance, this pseudocentrous arrangement has, in the Amniota, given way to

---

Fig. 40. Five stages in the development of the ribs of Amphibia.

A. The rib is carried by the basiventral only (orange).

B. A secondary attachment (green) has been formed to the basidorsal which is throwing out its primary diapophysial process. The vertebral artery is shown lying in the foramen transversarium thus formed.

C and *D*. Progressive stages in the complete loss of connexion with the basiventral and the disappearance of the latter. A tertiary attachment (red), again to the neural arch, is being formed and strengthened at the expense of the earlier one (green). A secondary, false, foramen transversarium is thus formed.

E. The final stage in the Anura. The tertiary attachment (red) to the neural arch alone remains.

the autocentrous type, brought about by reduction of the basiventrals in favour of the interventrals, which ultimately, step by step, form the centrum exclusively. This results in the formation of chevrons in the tail, and articulation of the capitula of the ribs upon the centra of the trunk. A dorsal shift of the ribs then asserts itself in the Amniota. As there are no longer any axial bony portions of the basiventrals, the ribs seek support by a backward shift on to the parapophyses, the tubercular portion having previously reached the diapophyses, so that ultimately the whole rib may be carried by the latter. For an elaborate description of this shifting, compare Crocodiles, Chap. XXXIII.

The whole rib, or what remains of it, may thus be carried by the large horizontal "transverse process" in the Anura and in many Amniota. This is a beautiful instance of independent, parallel development in the rise and decline of the ribs, which are the distal portions of the same basiventral arcualia.

PART II

Chapter XV

A CLASSIFICATION OF TETRAPODA

In order to facilitate the detailed description of the vertebral apparatus of the various forms of Tetrapoda, the following table has been drawn up, so as to show the classification adopted and the order in which the different forms will be described. The terms in common use have been put first; they are in some instances followed by alternative terms, which will be most commonly used in the text, as for the present purpose they are often preferable.

PSEUDOCENTROUS TYPE

First Grade. PHYLLOSPONDYLOUS
- Branchiosaurus including Protriton.
- Pelosaurus including Melanerpeton.
- Leptorhophus and Micromelanerpeton.

Second Grade. LEPOSPONDYLOUS
1. *Lepocordyli.*
2. *Urodela.*
3. *Coeciliae.*
4. *Diplocauli.*
5. *Aistopoda.*

Third Grade. STEREOSPONDYLOUS
- Pre- and Post-Permian Labyrinthodontia.
- Loxomma and Capitosaurus.

AUTOCENTROUS TYPE

First Grade. HOLOCENTROUS.
- *Embolomeri*: here called *Cricoti*: Cricotus and Diplovertebron.

Second Grade. NOTOCENTROUS.
- *Rhachitomi*: or Permian Labyrinthodontia. Eryops, Cacops.
- *Anura.*

Third Grade. GASTROCENTROUS.
1. TOCOSAURIA.
   - *Microsauri*: Eosauravus, Hylonomus, Hyloplesion, Microscelis, Petrobates, Orthocosta, Seymouria (?).

2. THEROMORPHA.
   Cotylosauria: Diadectes, Limnoscelis, Pareiasaurus.
   Chelonia.
   Pelycosauria: Varanosaurus, Ophiacodon, Dimetrodon, Naosaurus.
   Theriodontia: Moschops, Kannemeyria, Cynognathus, Dicynodont theriodonts.
   (Incertae sedis: *Procolophonidae.*)
   MAMMALS.

3. SAUROMORPHA.
   (Insertae sedis: Araeoscelis, Kadaliosaurus, Protorosaurus.)
   Rhynchocephalia.
   AUTOSAURI. *Geckones*, *Lacertae*, (*Pythonomorpha*), *Ophidia.*
   ARCHOSAURI. *Eosuchia*, *Pseudosuchia*, *Parasuchia*, *Crocodilia*, *Dinosauria*, *Pterosauria.*
   AVES.
   (Incertae sedis: *Ichthyosauria*, *Sauropterygia.*)

Chapter XVI

PSEUDOCENTROUS TYPE

THE PHYLLOSPONDYLOUS AND LEPOSPONDYLOUS GROUPS

The osseous mantle or shell surrounding the chorda is formed by the principal arcualia (basidorsals and basiventrals) and not by the interarcualia. The latter, acting mainly as cartilaginous pads or conjoined blocks, are the intervertebral cartilages of Gegenbaur.

The term "Centrum" or "Body" of a vertebra belongs by priority to that axially placed bony mass which in Man and all other Amniota carries the neural arch and is formed by the interarcualia. Consequently if the axial body or bulk of a vertebra is formed by the other arcualia, namely the combination of the neural and ventral arches, such a "centre" can at best be but a structure analogous to our own or true centre. It is to be distinguished as a sham-centre, hence the term *pseudocentrous* (*Phil. Trans.* 1886).

FIRST GRADE. PHYLLOSPONDYLOUS

The shape of a vertebra is that of a barrel; there is cartilage around the persistent chorda, and the cartilage itself is surrounded by a very thin osseous mantle or shell, which consists of a pair of dorsal and a pair of ventral leaves which meet along the horizontal mid-level of the chorda without fusion. The dorsal leaves represent the dawning osseous neural arch with a low spinous process at their dorsal junction, with simple zygapophyses, and with thicker transverse apophyses. The latter are compounds, the dorsal share belonging to the dorsal leaf and the ventral share being contributed by the ventral leaf. Since these transverse apophyses carry the ribs, with their scarcely distinguishable capitular and tubercular portions, they are strictly speaking diapophyses and parapophyses (Fig. 41).

The bony mantle extends over the whole length of the vertebra and seems even to overlap a little the posterior end of the next vertebra in front. Owing to the extreme thinness of the ventral leaves, the only proof of the existence of separate ventralia in these fossils are the ribs which, behind the single but stout sacral vertebra, are continued as haemapophysial chevrons.

The intravertebral widening of the adult chorda is certainly restricted to these Phyllospondyli. It represents here an interesting persistent archaic stage; it occurs again as an embryonic stage in other Tetrapods at the period in development before the so-called chordal cartilage has appeared, but when the chorda is already being squeezed at the ends of the vertebral complex, by the intervertebral cartilage in Newts, or by the analogous growth of the basiventral in Lizards.

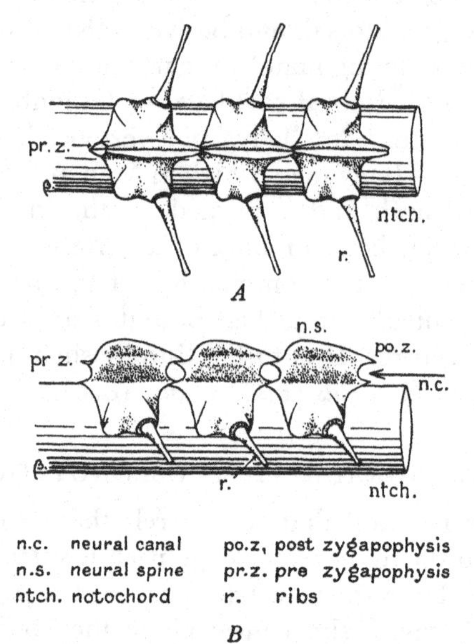

Fig. 41. Original drawing of three trunk vertebrae of a Branchiosaur. (O. F. Bulman.) *A*, from above. *B*, from left side.

Fritsch, the discoverer of *Branchiosaurus* in the Carbo-Permian of Bohemia, has devoted five plates of his invaluable monograph to the illustration of this type with non-metamorphosing gill arches. His large figure of the entire skeleton is a favourite textbook illustration. His account of the vertebral column is somewhat erroneous. According to him the vertebra consists of a lozenge-shaped, flat and very thin dorsal plate without indication of spinous and other processes; a corresponding extremely thin ventral plate; a pair of lateral ossifications carrying the ribs and a pair of posterior half-moon-shaped "Wuelste" (osseous lumps very conspicuous in his figures).

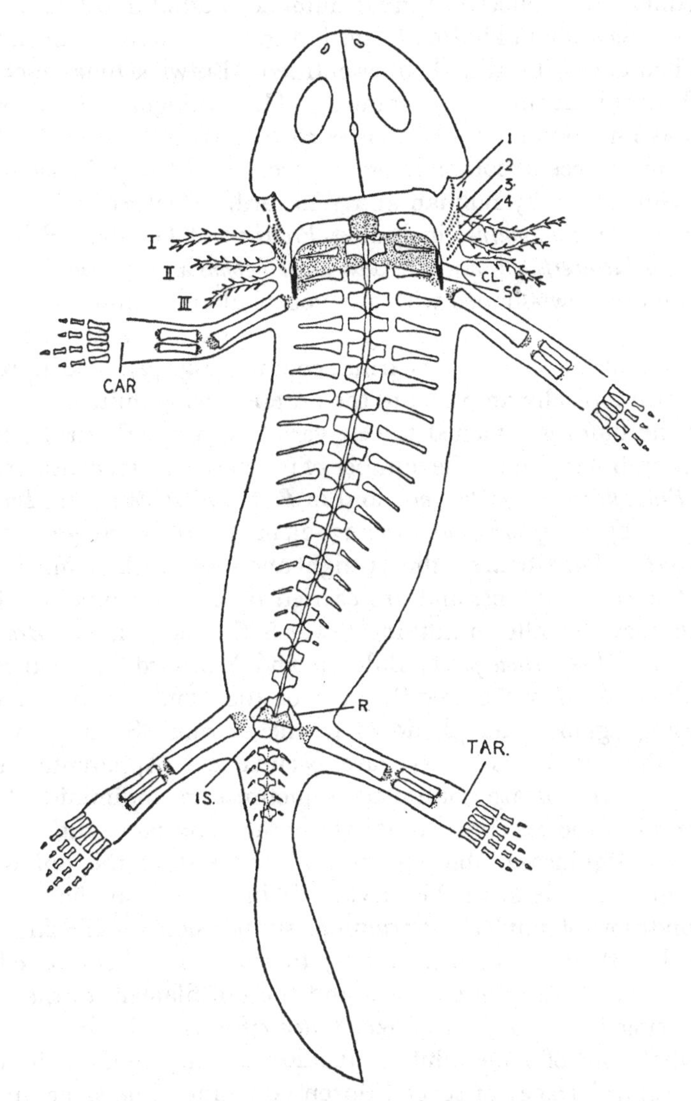

Fig. 42. Restoration of *Branchiosaurus amblystomus*, made from Odernheim specimens. $\times 2\frac{2}{3}$. (Bulman and Whittard.)

1st, 22nd and 23rd vertebrae omitted. Stippling indicates cartilage. I, II, III, external gills; 1, 2, 3, 4, loops of the internal gills; C, clavicle; CL, cleithrum; SC, scapula; CAR, carpus; TAR, tarsus; IS, ischium; R, ilium.

Credner, who collected a great number of adult and larval stages of *Branchiosaurus* and kindred forms in the Lower Red of Saxony, has published a well-illustrated paper entitled "Entwickelungsgeschichte von Branchiosaurus amblystomus". He distinguished the whole group as Phyllospondyli or "Blattwirbler".

The most recent comprehensive account of "*Branchiosaurus* and allied Genera" is by Bulman and Whittard. They recognise several species of *Branchiosaurus*, including *Protriton* of Gaudry; *Pelosaurus*, including *Melanerpeton*; and the new *Leptorhophus* and *Micromelanerpeton*. They all are Newt-shaped with a stegocephalous cranium; their total length is up to about 6 in. or 15 cm.; they have a rather short tail, less than half the length of the trunk column (Fig. 42). The limbs and girdles are typically amphibian, the anterior being quite close to the head; the ilium is attached to one vertebra, distinctly in the transverse acetabular plane. The number of presacrals varies considerably: 20 in *Branchiosaurus amblystomus* and in *Protriton petrolei*; 26 in *Protriton fayoli*; 21–23 in *Leptorhophus tener*; 23–25 in *Pelosaurus*; 29–30 in *Micromelanerpeton*. The ribs are rather straight and short, without any indication of sternal portions, and are carried by all the trunk vertebrae, though they dwindle to nothing towards the sacrum, e.g. *Branchiosaurus*. In *Micromelanerpeton*, Bulman and Whittard figure them as dwindling towards the middle part of the trunk, then gradually increasing again in length down to the sacrum. Such a "waist" within the trunk is of very rare occurrence. Apparently in all Branchiosauri ribs are continued, sometimes in undiminished size, on several of the anterior caudal vertebrae. The same authors have made the significant observation that the rest of the tail is still cartilaginous, at least in the larvae. It indicates that the tail is in the condition of phyletic shortening, suppression progressing from the end with delay of ossification; to this can be compared the whipcord tail ending of *Chimaera* and the tail filament of the metamorphosing larvae of Newts like *Triton cristatus* and *Triton vulgaris*. The whole tail of some adult Branchiosaurs may contain the more or less ossified traces of several dozen vertebrae. The same authors state that in *Branchiosaurus amblystomus*, the anterior caudals, besides bearing ribs, carry also a pair of flat blades, "the remnants of the haemal arches", i.e. chevrons not yet united.

PSEUDOCENTROUS TYPE

SECOND GRADE. LEPOSPONDYLOUS

The osseous vertebra is composed of (1) the basidorsals and basiventrals, which are fused together into a shell, for the most part sutureless, which surrounds the chorda and constricts it intravertebrally; (2) the intervertebral mass composed of the interdorsals and the interventrals, which co-ossifies with (1) and to a variable extent adds to the size of the axial portion of the body of the vertebra.

Because of the constricting shell, which extends fore and aft over the whole length of the "body" like a mantle but leaves a considerable portion of the chorda intact, Zittel called this kind of vertebra Lepospondylous,[1] i.e. Huelsen- or Hohlwirbel. However, he restricted it to the lowest group of Amphibia, the so-called Microsauri. But the four other groups which we include in this second grade are built upon essentially the same plan, although diverging and specialising, as will be explained in the text.

It is obvious that the lepospondyle structure represents a higher grade, one directly evolved from the Branchiosauri. The dorsal and ventral halves of the sheath, still imperfectly ossified in the phyllospondylous stage, are now not only fused together but have converted themselves into bone, which thereby prevents the intravertebral growth of the chorda. Finally a more or less solid central bony mass takes the place formerly occupied by the chorda. The result is the formation of an axial amphicoelous body; a pseudocentre, since it is partly formed by the basal portions of the principal arcualia. The condition of the interarcualia we can but surmise in the fossils. Their further behaviour and fate are discussed later on p. 128.

The jointing of the column has made great progress: there are now well-jointed prezygapophyses and postzygapophyses; frequently zygosphenes and zygantra, and in some groups even hyposphenes.

The lepospondylous Amphibia comprise:

(1) *Lepocordyli.* Small, tailed Tetrapods of the Carbo-Permian, with some ventral armour plates and abdominal mailed-chain-like ossifications.

(2) *Urodela.* Without any dermal armour; naked.

(3) *Gymnophiona* or *Coeciliae.* Recent. Snake-shaped; limbless; without girdles; extremely short tail. Apparently naked, but with

1 This very appropriate term has suffered confusion: *lepos*, husk; *leptos*, thin. The Phyllospondyli with their very thin shell have insinuated the amendment Leptospondyli, which is quite unsuitable for the thick husk of those creatures for which Zittel's term is intended.

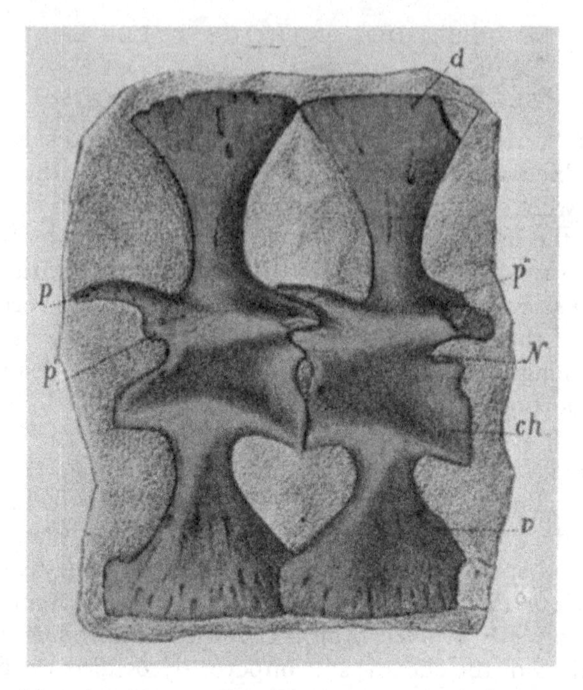

Fig. 43. *Urocordylus scalaris*. Two vertebrae from the anterior part of the tail. *d*, neural spine; *v*, haemal spine; *N*, spinal canal; *p*, postzygapophysis; *p″*, prezygapophysis. (After Fritsch.)

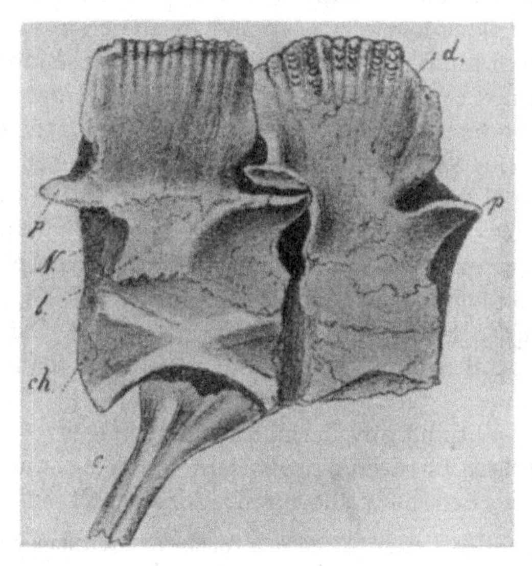

Fig. 44. *Ceraterpeton crassum*. Two vertebrae from the anterior half of the thoracic region. *d*, neural spine; *p*, zygapophysis; *N*, spinal canal; *l*, pedicle; *ch*, notochordal cone; *c*, rib. (After Fritsch.)

numerous vestiges of dermal armour hidden in their many imbricating cutaneous rings. Terrestrial; underground.

(4) *Diplocaulus* and *Crossotelos*. Permian. Transitional and highly specialised.

(5) *Aistopoda* (*aïstos*, unseen). Snake-shaped, limbless and without pectoral girdle. Aquatic.

(1) *Lepocordyli*.

Small Newt-like creatures of the Carbo-Permian of North America and Europe; notably *Urocordylus* (Fig. 43), *Ceraterpeton* (Fig. 44), or *Scincosaurus* (Fig. 45), *Tuditanus*, *Seeleya*, *Limnerpeton*.

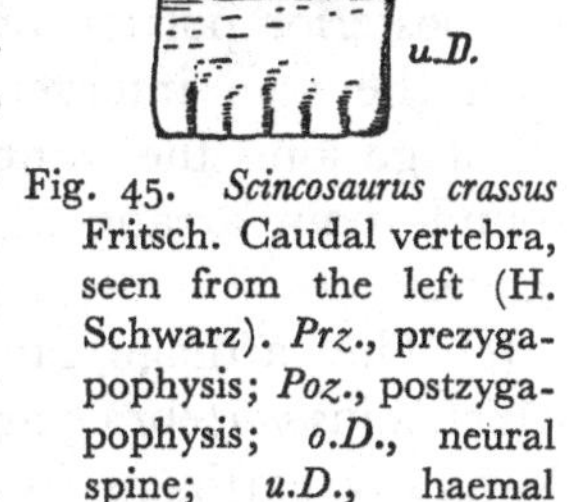

Fig. 45. *Scincosaurus crassus* Fritsch. Caudal vertebra, seen from the left (H. Schwarz). *Prz.*, prezygapophysis; *Poz.*, postzygapophysis; *o.D.*, neural spine; *u.D.*, haemal spine.

This group has essentially the vertebral characters just described and is of fundamental importance. Almost certainly directly descended from phyllospondylous creatures, it seems to have given origin not only to all the other lepospondylous groups, but also to contain the foundations of the Eucentrous Tetrapoda, via the Microsauri of Dawson as amended by Gadow, Baur, Williston and Abel. The amendment consists in the removal of *Hylonomus*, *Hyloplesion*, *Microbrachis*, *Petrobates* and *Sauravus* from the Stegocephali to the primitive Reptiles. No one single member of the Lepocordyli proves this by itself, but taken as a group they reveal themselves as in a state of flux, unstable, yet with clear tendencies in new directions, all of which have materialised in those Eucentrous Tetrapods, the lowest living members of which are *Sphenodon* and the Geckos.

(2) *Urodela* or *Caudata*.

Tailed, naked Amphibia without traces of dermal armour. Fore- and hind-limbs present, although sometimes very much reduced; the only exceptions are *Siren* and *Pseudobranchus*, in which the hind-limbs are altogether absent. Fossil "Salamanders and Newts" in the wider sense are still very rare; there exist a large specimen resembling *Cryptobranchus* from the Miocene and smaller specimens from the Oligocene of Europe. A single specimen, *Hylaeobatrachus croyi*, from the Wealden, i.e. lowest Cretaceous, of Belgium seems to be the earliest known

Urodele, still a long way off the Carbo-Permian, where they may meet the Lepocordyli, their possible ancestors.

One of the most interesting features is the formation of the intervertebral joint, since it throws a flood of light upon the ways and means of adaptation to necessary conditions. Every one of the theoretically possible arrangements has been tried and, what is more, kept in some creature or other.

(*a*) The whole intervertebral cartilage acts as a flexible mass between the bases or ends of the hollow cones of the successive vertebrae. There is a considerable amount of chorda, the cartilaginous ring being thin. This imperfect, morphologically very low, condition reminds us of the Lepocordyli, and it prevails in the recent Perennibranchiata.

(*b*) The whole cartilage is added to one of the neighbouring vertebrae, which is the usual condition of the autocentrous as distinct from the pseudocentrous type. The addition either forms the posterior portion of its own vertebra, when the joint coincides with the true intervertebral limit, e.g. most Tetrapoda; or the cartilage joins the front end of the vertebra following, a secondary round-about arrangement materialising in the odontoid-axis compound.

(*c*) The cartilage, say No. 9, is vertically split into a front portion which joins vertebra 9 and a posterior portion which joins vertebra 10. Naturally, cartilage No. 10 will be divided between vertebrae 10 and 11; now, therefore, vertebra 10 is a complex compound of the tenth basidorsal and basiventral with 9*b* added in front and 10*a* behind.

Some morphologists seem to be surprised at the number of primary and secondary units making up a single complete vertebra; but two and two, i.e. the interdorsal and interventral of each side, make four; double that number, as in the vertical splitting in type (*c*) above, and we get eight elements; adding the four pieces representing the paired basidorsal and basiventral, the total possible number of elements making up a complete vertebra is twelve. This is more than sufficient to produce almost endless variation in the shape of a vertebra, according to the relative growth of the various constituent units. All this is incontestable, although it may be a little confusing. Type (*c*) is that of the more terrestrial Urodela. Even in these the formation of the new joint does not always lead to the complete severance of the cartilage. As Gegenbaur discovered, only a differentiation of the cells may take place instead of a clean split. The whole process is faithfully repeated by ontogeny and the stages reached by the adults of the various Urodeles are good examples of epistasis.

The finished joint is of the cup-and-ball type, either procoelous or more usually opisthocoelous, with the ball in front and the cup at the hinder end of the vertebra. These differences depend upon which of the split pieces is the larger, upon the state of the calcification of the intervertebral cartilage, which in the pseudocentrous complex does not ossify, and upon the extent of the ossifying mantle of the whole vertebra. Strictly, these joints of the Urodela are intravertebral, since they divide a cartilage which by right of origin belongs to one vertebra.

To judge from the general resemblance of these features with those of the larvae of *Lepidosteus*, which are likewise opisthocoelous and form their joints by division of the intercartilage, we might assume them homologous. But in these fishes the spinal nerves issue at the level of the middle of the joint, i.e. between vertebrae. In the Urodela the nerves issue behind the neural arch and in front of the intercartilages, i.e. midvertebral. Consequently, although the principle of joint formation by the splitting of some cartilage is the same, the two cases are only analogous. The lepospondyle Amphibians are infinitely older than the osseous Ganoids, whence we may conclude that the Urodele joints are a feature acquired by these Amphibians alone and are unique.

We can guess how such splits were initiated as an incidental result of the spreading of the ossifying "leaves" of the Phyllospondyli. The resulting mantle of the chorda and of the skeletogenous layer extended forward and backward until it met the corresponding growths of the neighbouring vertebrae. In order to prevent the two growths from co-ossifying and so rendering the column stiff, the required centre of motion stepped into play and prevented this; the intercartilage was ultimately broken in two, because the shell mantle not only adhered to the cartilage but also colonised it.

It is worth while to follow up the behaviour of this cartilage. The process of ectochondral ossification attacks first the dorsal and ventral arches, which, whilst still growing in size, become honeycombed with osseous matter. But direct ossification of the connective tissue proceeds apace, notably at the base of the arches, so that their respective cartilages, the basidorsal and basiventral, are soon prevented from coming into contact with each other. Nevertheless, they are joined together by the ossifying connective tissue, which much more effectively than the cartilage causes the constricted waist of the body of the vertebra. The latter owes its further growth, first, to the steadily extending shell which is naturally thinnest at its free ends, and secondly, to the growth of the intercartilage. This cartilage, at first unfettered, extends and actually creeps in between the chorda and the osseous mantle. It is a question of degree when it, too, is converted into bone or calcified cartilage, until finally it is reduced in the adult to the thin, almost perichondral, lining of the cup-and-ball joint.

The above description explains also the paradoxical fact that in early stages the chorda is thickest in the middle of the vertebra, as in the ancient Phyllospondyli, and that later this level corresponds with the midvertebral constriction and with the position of the so-called chordal cartilage occurring in the tail and frequently in the trunk of Newts. The same midvertebral constriction and chordal cartilage appears in Reptiles, e.g. *Sphenodon* and Lizards, although the compressing agents are here the basidorsal and the true centrum.

Now we can understand how much the composition and configuration of the final vertebra must depend upon the activity of the intercartilage. If the anterior and posterior additions each amount to 20 per cent. of the whole vertebra, the share of the basidorsal and basiventral still preponderates with 60 per cent. If, on the other hand, the whole intercartilage with its 40 per cent. is added as an unsplit mass to the hind end of the vertebra, then 60 per cent. of the front portion of this vertebra is furnished entirely by the "arch". Intermediate conditions are met in the various pseudocentrous and temnospondylous vertebrae. The extreme case, an overwhelming preponderance of the original intervertebral mass and corresponding reduction of the basiventral, is the typical gastrocentrous vertebra of the Amniota.

When Miss Abbott and myself were trying to account for the much constricted waist of the Urodele trunk vertebrae, which clearly indicated some deficiency, Goeppert was studying the morphogeny of ribs and published a "Vorlaeufige Mittheilung" which contained an account of the gradual suppression of the basiventral cartilage in the trunk of *Salamandra* and *Menobranchus*. After this important discovery, it was easy to corroborate it with our own material of larvae and macerated adult specimens of various genera.

The following is a condensed account of Goeppert's work in my own diction. In the new-born larva of *Salamandra maculosa*, basiventral hyaline cartilage occurs in most vertebrae of the trunk. These arcualia are obviously in a vestigial, declining condition. They occur rarely in pairs, mostly asymmetrically, either on the right or on the left side of the chorda. They increase in size and in length towards the tail, where they gradually pass into the ventral arches which meet and fuse below the chorda. Still more primitive, normal conditions he found in a larva of *Menobranchus* of only 22 mm. in length (Fig. 46*A*). The ventral arcualia lie at the same transverse level as

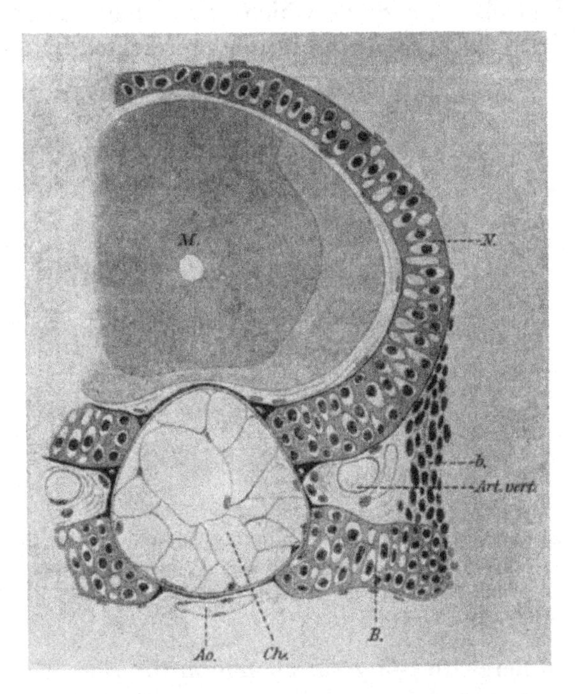

Fig. 46*A*. *Menobranchus lateralis*. Larva, 22 mm. Cross section of posterior trunk vertebra. Anlage of the dorsal process, *b*, of the basal stump *B*. *M*., spinal cord; *N*., neural arch; *Ch*., notochord; *Ao*., aorta; *Art.vert*., vertebral artery. (Goeppert.)

the neural arches, extend horizontally, and neither meet each other below nor meet the neural arches on the lateral side of the chorda. In the anterior trunk vertebrae of the same larva these ventral basalia do, however, meet the neural arches. The anterior vertebrae, notably the first and second, are therefore more complete than the more and more constricted vertebrae, especially those of the middle of the trunk. Towards the tail they recover or rather retain their completeness. Lastly, in the tail the right and left basiventrals are bifurcated, sending out a lateral process which eventually carries the capitular or primary basal portion of the rib and therefore may conveniently be called the pleurapophysis; and a vertical downward process, the haemapophysis, which, with its fellow of the other side, tends to enclose the caudal artery and vein (Fig. 46*B*).

In the metamorphosing larva the little vestigial basiventral cartilages disappear, becoming engulfed by the rapidly ensheathing osseous mantle of the vertebra, which in its middle third is reduced to the basidorsals. The space which would have been filled by a full-grown basiventral semi-ring is now represented by the waist.

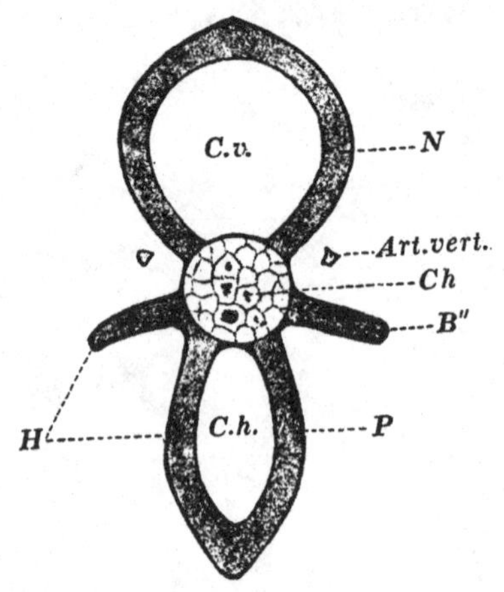

Fig. 46*B*. *Menobranchus lateralis*. Larva, 43 mm. Section of a vertebra from the anterior part of the tail region. *B''*, pleurapophysis; *P*, haemapophysis; *H*, basiventral; *C.v.*, neural canal; *C.h.*, haemal canal; *N*, neural arch; *Ch*, chorda; *Art.vert.*, vertebral artery. (Goeppert.)

Correlated herewith is a great change in the ribs. Here is a first-class illustrative case of cenogenesis, i.e. acquisition of a new character, combined with tenacious inheritance. The rib, having lost contact with its progenitor, no longer develops its capitulum, and therefore the rib-cartilage appears occasionally at some distance from the chorda; but a tubercular-like knob meets an oblique transverse outgrowth of the dorsal arch, Goeppert's "Rippentraeger" or rib-bearer (Fig. 40). This osseous transverse process of the adult, the rib-bearer being a kind of diapophysis, is further complicated by a broad string of connective tissue, which extends out of the lateral mass of the skeletogenous layer to the capitular portion of the rib and thus represents a part of the lost basiventral. The vertebral artery lies in the triangular space enclosed by this sometimes ossifying string, the chorda and the rib-bearer. It is the remnant of the foramen transversarium still enclosing the artery. A false foramen transversarium is formed higher up and more lateral, between the dorsal arch, rib-bearer and the basal portion of the rib itself.

The phyletic stages in the changes of the rib attachment may be summed up as follows:

(1) Only by capitular attachment, or rather jointing off from its basiventral.

(2) A tuberculum is added from the shoulder for more dorsal attachment, either to the same basiventral, or to the dorsal arch with corresponding elongation.

(3) The capitulum vanishes, the tubercular portion taking its physiological place.

(4) A still more dorso-lateral connexion is formed with the arch; a new, more distal shoulder connexion, the necessary cartilage being drawn out as rib-bearer by the connective tissue.

The rib has once more become two-headed, and is now carried entirely by the dorsal arch. Co-ossification eventually produces a very broad oblique process, to which the similarly very broad-based rib is attached. Resemblance to the Aistopods is unmistakable.

The transverse processes of the tail are composed entirely of ossified connective tissue, no longer preformed in cartilage. If they carry ribs, these are only vestigial.

Remnants of cartilaginous interdorsals in Urodela. Goette stated that the neural arch of the first vertebra of *Salamandra* is composed of two "Knorpelspangen" (loops or arches).

Schauinsland examined young *Siredon* and discovered such double cartilaginous upper arches; a large anterior arch, which later becomes surrounded by the typical bony mantle and fused with the body of the vertebra; the much more slender posterior arch the author rightly homologises with the intercalaria (our interdorsals) of Cyclostomes, Selachians and Ganoids. The intervertebral cartilage is much reduced, particularly, as the author remarks, in Coeciliae, next in the Perennibranchiates and least in the terrestrial Newts.

It is not difficult to explain this intervertebral cartilage, which is independent of the interdorsal, as the homologue of that half of the split (originally entire) intervertebral disk which is added to the preceding vertebra, while the remaining half forms the jointing tissue. The inducement for a downward tendency in these future interdorsal wedges seems to be the fact that they are overlaid by the large basidorsal; in any case their dorsal portion is reduced by pressure. *Eryops* (Fig. 57) is at a stage in which the interdorsals are being pushed down by the overlying and large basidorsals. In the figure the space marked *S* contains the traces of the interventral half ring of cartilage and also the notochord. If the interdorsal was pushed ventrally so as to occupy this space and replace these remnants, a notocentrous condition could arise by suppression of the basiventral. Such a condition would favour the assumption that the notocentrous vertebra is a side departure superimposed upon the original quadripartite stage.

(3) *Coeciliae; Gymnophiona or Apoda.*

This Amphibian order, whose larvae have three pairs of external gills, suffers from too many names. Mueller, in 1831, called them Gymnophiona, i.e. naked, snakish creatures. Bell, in 1836, distinguished them as Amphibia Apoda. Haeckel, in 1866, named them Peromela, maimed limbs, and joined them with the Labyrinthodonts as Phractamphibia, he being the first to recognise the phyletic importance of the Coecilian's hidden dermal armour, in opposition to the Urodela and Anura which are Lissamphibia. Cope, in 1869, invented the objectionable term Gymnophidia, naked snakes. Coeciliae may also be objected to, because not a few kinds are not blind; their eyes vary from being normal to being concealed by their skin, or they may even be buried beneath the cranial bones; but they are considered as forming one family only, Coeciliidae, *Coecilia* being a fairly representative genus. There are about 40 recent species in the palaeotropical and neotropical regions, all leading an underground life, burying with their head in decaying vegetable soil. Fossils are still unknown.

They are to be looked upon as snake-shaped, limbless, "naked" Amphibia, with about 200 trunk vertebrae and a vestigial tail. The vertebrae are typically pseudocentrous and deeply biconcave, the chorda is long persistent and the intervertebral joints are very imperfect, being in as low a stage as those of the Geckos. A peculiarity is the hyposphene apparatus. Another peculiar feature concerns the skin; it is apparently naked, because the whole body is covered by a continuous smooth, slimy epidermis, but the thick cutis forms metameric ring folds which are divided into an anterior and a posterior pocket, the former containing large glands; from the bottom of the recess between the adjoining halves arise several cycloid scales, which are themselves covered by smaller scales ossifying and partly calcareous. These form an unmistakable remnant of a dermal armour greatly resembling the scales which covered the body of the Lepocordyli. The osteodermal scales do not reach the surface; to see their edges, the epidermis has to be removed and the slightly imbricating cutaneous ring folds to be lifted asunder. The scales are therefore not assisting locomotion, which is effected by peristaltic waves passing over the body. The whole cutaneous apparatus must have a long history. First the scales, each with its own epidermal scale lying on the surface; then the thickened cutis thrown by metameric contraction of the muscles into rings which, owing to the ever-repeated forward motion, imbricated and ultimately buried the scales. The next stage would be a complete loss of the scales but retention of the rings; this has been reached by about half the number of recent genera. Their other organs cannot be similarly correlated but vary considerably, for example, the eyes, rows of mandibular teeth, and the bones of the stegocephalous skull more carefully called pseudostegal by Jaeckel and Abel. It stands to reason that one should refer most of the peculiar features of the Coeciliae to their burrowing life; for instance, the completely roofed and extremely solidly built skull was connected by Wiedersheim directly with the burrowing. But Marcus and Blume, the latest monographers of these creatures, find fault with this reasonable explanation and, as strict predestinationists, point out that such a skull is a necessary antecedent to burrowing! At some larval stage the whole dawning skeleton of the trunk is outlined by two continuous tubes of cartilage, the lower of which surrounds the chorda, while the upper forms the walls of the much larger spinal canal; meeting above the latter, they make up the neural arches. It is significant that the mass dorsal to the chorda is much larger than that ventral to it, a foreshadowing of the notocentrous tendency of Amphibia. The bases of the principal arches, basidorsal and basiventral, are confluent, forming metameric pillars which are connected with each other by cartilaginous masses representing the intervertebral cartilage containing the future interdorsal and interventral units. At first, these intermediate masses and the pillars are of equal length. Later, they become very unequal in size and structure. While the pillars or arches grow in length and undergo ossification from the surface inwards, the intermediate mass remains cartilaginous and is reduced until it becomes more and more restricted to the region of the intervertebral joints. The latter, however, remain in a very low stage, resembling in this respect those of the Geckos. How much this mass, which calcifies by preference, contributes to the building of the arches it is impossible to say, because the transition is so gradual. Suffice it to state that the ossifying mantle of the arches spreads forwards and backwards, covering the remainder of the intervertebral cartilage, but leaving on its outside a non-ossifying intervertebral ligament which passes on to the next vertebra. Thus the main bulk of the axial part of each vertebra becomes an elongated, hourglass-shaped, osseous cylinder, biconcave, with the ends more or less calcified. And since this cylinder acts as the centrum of the vertebra and is made up of neither interdorsals nor interventrals as in the Amniota, it is undoubtedly a pseudocentrum in contradistinction to an autocentrum, be this noto-, gastro- or holo-centrous.

In the middle of this pseudocentrum the chorda is interrupted by a rather thick cartilaginous septum. It was first described and figured by Gegenbaur, who concluded that this cartilage is derived entirely from chorda cells with a greater or lesser amount of calcification. Such metaplasy is of course not accepted by histologists, for example, Luoff and Zykoff. I described and figured this septum in the case of a Gecko (Fig. 47) and of *Lacerta*, explaining that the cartilage is derived from that of the arcualia, which thicken and constrict the chorda, then burst the sheath of the latter and ultimately form the complete septum. Howes accepted this for *Lacerta*, but not for the Gecko. I now have to point out that constrictions leading eventually to the formation of such a transverse cartilaginous septum, or to partial destruction or degeneration of the chorda, are not homologous in Amphibia and Reptiles. The chorda can be constricted in three ways:

(1) By the basidorsal and basiventral; Coeciliae and Urodela; this, of course, cannot happen in the Anura, because of the early reduction of the basiventrals.

(2) By the basidorsal and the true centrum; *Sphenodon* and Lizards.

(3) The intervertebral ring slightly constricts the chorda and causes its partial degeneration; Urodela.

Quite recently Marcus and Blume have come to the rescue, much against their wish. The line of argument is so curiously twisted that it is here translated. They started by thinking that metaplasia of chordal cells into cartilage cells had been ascertained, partly as a result of their own observations that the elastica remained intact, but especially because "the opposite and rather occasional statements by Luoff and Gadow were not supported by figures, and those of Zykoff not convincing...but as this question is of the highest importance we did not rest until we had found pictures which showed

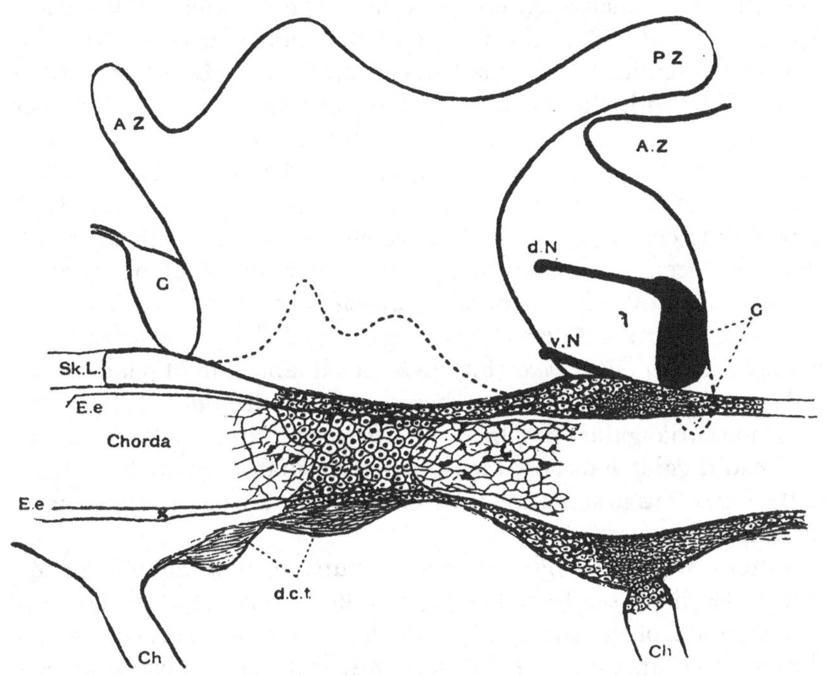

Fig. 47. Sagittal section through the tail of an adult Gecko (*Hoplodactylus pacificus*). *d.c.t.*, dense connective tissue; *d.N.*, *v.N.*, dorsal and ventral roots of spinal nerve; *E.e.*, elastica externa; *G.*, spinal ganglion; *Sk.L.*, skeletogenous layer. An × indicates the space which is due to shrinkage of the chorda and its sheath away from the skeletogenous layer. *A.Z.*, prezygapophysis; *P.Z.*, postzygapophysis; *Ch*, chevron. (Gadow.)

beyond doubt the tearing and eventually also the absence of the chorda membrane, and the penetration of perichordal cells into the interior of the chorda". They fail entirely to mention the following evidence given in my paper of 1896: (1) the clear figure of a Gecko, Fig. 15; (2) p. 26, "chorda is rent asunder"; (3) Fig. 29 (*Lacerta*); (4) p. 33, "chorda interrupted by cartilaginous septa"; (5) p. 35, "obvious that the chorda must be broken or destroyed somewhere, when the septum is complete". They also ignore Howes on *Sphenodon*, who discusses the question at some length, and while accepting my statements concerning Lizards, reserves judgment in the case of the Gecko and *Sphenodon*, but unfortunately omits the possibility that the difference may represent two evolutionary stages.

In the Coeciliae the atlas lies in a slanting position (Fig. 48*A*). The ventral half articulates with the condyles, the dorsal half touches in front the foramen magnum. On the dorsal midline of this flanged hood is a weak ridge, the vestige of the spinous process, which is practically reduced to a small dorsal projection above the surface articulating with the second vertebra. The spinal canal, or rather its wall, is continued forward as a complete projecting tube, which fits into and against a very short portion of the inside of the borders of the foramen magnum. It is not obvious why this connecting, articulating central tube should have escaped notice, with the result that the Coecilian atlas is usually stated to be devoid of an odontoid process. This tube in *Ichthyophis glutinosa* is more than a dens, it is a ventral, lateral and dorsal process combined. The same remarkable feature exists in *Diplocaulus* (Fig. 48*B*), a valuable hint of relationship, unless the homology is false, being due on the one hand to the enormous triangular, flat head of *Diplocaulus*, and on the other to the head-digging habit of at least some Coecilians, although the head of the latter fails to show any trace of the broad flattening so common in Amphibia.

Marcus and Blume give excellent figures of the odontoid region of the Coeciliae; see their Figs. 7, 11 and 12 (Fig. 32). In the horizontal section of the younger larva (Fig. 12) the chorda is continuous through the atlas into the basioccipital; in the older larva already hatched (Fig. 11) it is interrupted by a cartilaginous septum at the level of the middle of the atlas ring; in Fig. 7 such chordal cartilage is seen at the level of the middle of the second vertebra and in the atlas, and the dotted portion pointing to the occiput must represent the odontoid.

The neural arches of all the vertebrae produce low spinous processes; the roof of the arch extends shelf-like over the next following vertebra to form postzygapophyses which articulate with the prezygapophyses. The overlap is greatest between the atlas and the epistropheus. The articulation likewise prevents torsion. In the succeeding vertebrae Marcus and Blume discovered a longitudinal ridge arising from the midline of the ventral side of the spinal canal; it is highest in the midvertebra, declining towards the ends. According to them it helps to level the bottom of the spinal canal. They call this process, which also occurs in Snakes, processus ventralis superior, which is misleading, because processes receive their names from their direction of extension, further qualified by their origin; therefore processus dorsales superiores would be better, to distinguish them from the processus ventrales inferiores sent out from the same basiventrals. In the first few vertebrae these true processus ventrales are indicated by paired rugosities for the attachment of cervical muscles; farther back, they gradually combine into unpaired hypapophyses, which elongate backwards beyond the end of their vertebrae, forming a long spur (Wiedersheim's hyposphenes) which is grasped by a long pair of processes arising from the anterior parapophysial region of

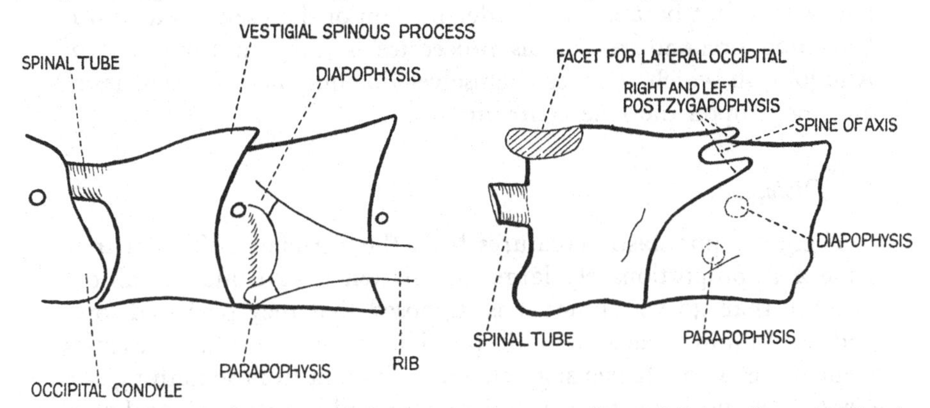

Fig. 48*A*. Atlas and Epistropheus of *Ichthyophis*. Fig. 48*B*. Atlas and Epistropheus of *Diplocaulus*.

the next following vertebra. The resulting articulation is rather syndesmotic, but also prevents torsion. The respective grasping processes, recognisable also in the Aistopods, e.g., *Dolichosoma*, are called processus inferiores anteriores by Marcus and Blume, infrazygapophyses by Abel. On p. 15, Marcus and Blume speculate that if the dorsal zygantra, the postzygapophyses of the Stegocephali, joined to form a median unpaired thorn, the result would be the zygosphene apparatus of the Coeciliae, and that therefore a sharp division of the two groups by their vertebrae would be impossible. They forget, however, that the zygantra are as clearly outgrowths of the neural arches as the zygosphene is a hypapophysis of the basiventral elements. It is to be remembered that intervertebral articulations are made of homotypes, zygapophyses by neuralia, the infrazygapophyses, or hyposphene apparatus, by ventral units.

In the trunk the parapophyses arise with broad bases from below the diapophyses. A hole close behind and between them, for a branch of the vertebral artery, serves as a landmark. If these strictly metameric branches are homologous with the primordial sklerotomic arteries, their holes would indicate the respective share of the re-combined sklerotome halves in the finished vertebra. The position of the diapophyses and parapophyses varies according to the spines and still more according to the region of the vertebral column. On an average the ribs nearer the head articulate with the anterior third or even quarter of the vertebra; farther back often upon the middle third. The ribs, and of course their processes, behave in the same way as they do in the Lepocordyli, not because they are shifting backwards, but because the hinder portion of the pseudocentrum is becoming preponderant. This procedure is different from that of Amniota, in which the ribs themselves do shift their position more and more upon the true centrum.

(4) *Diplocauli.*

These are phantastic creatures from the Permian of Texas; they have a disproportionately large and flattened skull and a column of more than 20 vertebrae; it is supposed that they possessed fore- and hind-limbs. The systematic position of these peculiar creatures is much debated. Jaekel suggested a derivation from something like *Ceraterpeton*, whereby they would remain within reach of the Lepocordyli. The vertebrae, however, bear a very great resemblance in their proportions and structure to the Coecilian vertebrae.

Diplocaulus seems to find its place with the lower groups of Lepospondyli; the Aistopoda and the Coeciliae would then be radiating relations, while the Urodela continue as the main stem. Every one of these groups went in for adaptation to a highly specialised life: the digging, terrestrial Coeciliae; the rapacious, swift, snakish, probably aquatic Aistopoda; and *Diplocaulus*, slithering or crawling "through the slime and mud at the bottom of some shallow body of water".

The largest of the several species described is *Diplocaulus magnicornis* (Fig. 49). The enormous head is very much depressed and triangular; it is crescent-shaped in outline, the horns of the crescent being formed by the supraoccipitals and by the much elongated epiotics or tabulars. The ends of the crescent are about 14 in. apart;

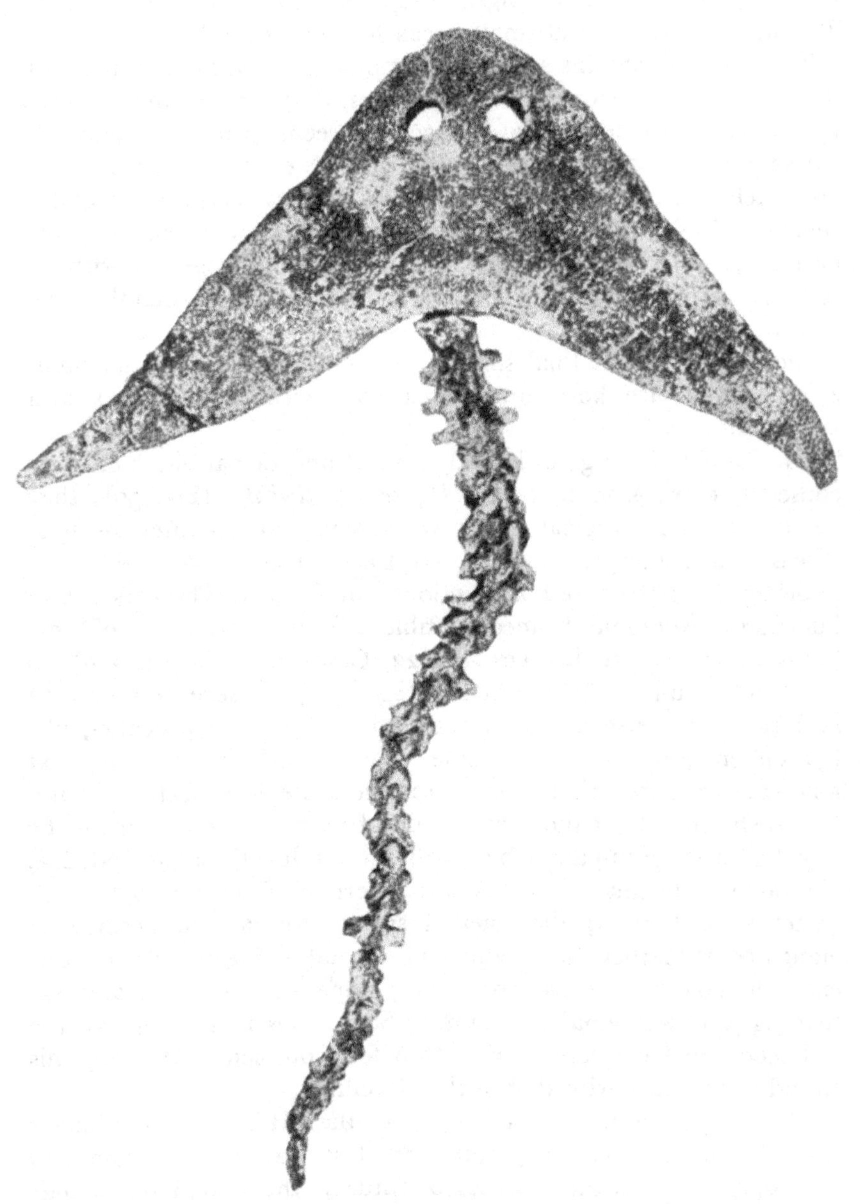

Fig. 49. *Diplocaulus magnicornis*. No. 4472, Am. Mus. (Case.)

this is equal to the whole length of the apparently complete column. The shoulder girdle and small, weak fore-limbs are known.

The column contains at least 20 amphicoelous vertebrae, those of the trunk having strong lateral processes, diapophyses and parapophyses, with which the two-headed ribs seem to have articulated. These processes, arising midway between the ends of a vertebra, are absolutely intravertebral and the whole vertebra is typically pseudocentrous; Case's remark that intercentra are absent is therefore unnecessary. All the constituent elements are fused together without sutures, perhaps by a calcified or a directly ossified mantle. The neural arches are depressed and broadened out so that the spine is reduced to a longitudinal sharp ridge and the facets of the small zygapophyses are horizontal. There are weak zygosphenes and zygantra.

On Plate 1, Case gives lateral, ventral and dorsal views of three consecutive trunk vertebrae of *Diplocaulus limbatus* (Fig. 50); they seem to show interdorsal pieces, wedged in and separately ossified. Whether this interpretation is correct cannot be ascertained on the phototype and they are not mentioned in the text. The presence of interdorsals would not be incompatible with the relationships of these funny creatures. In his Text-fig. 22, Case shows the sixteenth to nineteenth trunk vertebrae in situ (Fig. 51); the seventeenth, with its larger transverse processes and with the peculiar processes, like hyposphenes, on its ventral side which interlock with the next following vertebra, is supposed to represent the degenerated sacrum. On to the middle of the ventral side of its body there seems to be attached a large unpaired bone, almost the length of the vertebra. Neural and haemal elements are described from the eighteenth vertebra backwards; the haemal spines are said to become so elongated that they touch those of the adjoining vertebrae. The exact homology of these apparently unpaired, longitudinally extending processes remains unknown. So far as is actually figured the tail seems to have contained only a few complete vertebrae; this would form a link with that of the Coeciliae.

The most remarkable vertebra is the atlas. It is larger and longer than the second vertebra; both are close fitting, quite compact, stereospondylous without traces of sutures; the second has a pair of typical facets for the two-headed ribs. On the anterior dorsal end of the atlas is a pair of large facets in the position of prezygapophyses, which can have articulated only with the cranium (Fig. 52). From below the anterior end of these facets extends another pair of facets downwards over the anterior or cranial face of the atlas; this pair is borne by right and left flange-like extensions or broadenings of the lateral and ventral portion of the vertebra. The facets, slightly con-

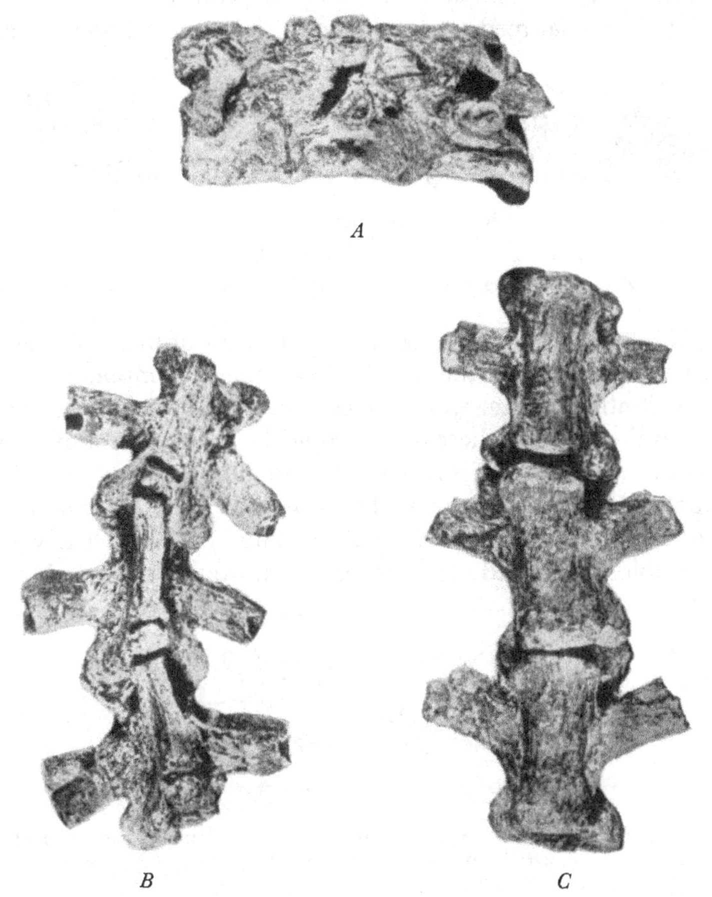

Fig. 50. *Diplocaulus limbatus*. Trunk vertebrae. *A*, from side; *B*, from above; *C*, from below. No. 4470, Am. Mus. (Case.)

cave, look forwards and somewhat upwards, and are divided from each other by a median swelling: they receive the occipital condyle, which, although it is certainly basioccipital, must itself be rather kidney-shaped.

The resemblance of this atlas of *Diplocaulus* to that of the Coeciliae is further enhanced by a stout osseous tube, which looks like the continuation of the walls of the spinal canal of the atlas: this tube "fits into the foramen of the foramen magnum" (Case). Its dorsal half is longer than its ventral half and ends like a spout, which perhaps touches the inner walls of the lateral occipital portions of the kidney-shaped condyle. Certainly it protects the medulla by bridging the dorsal median gap between the occipital and the atlas.

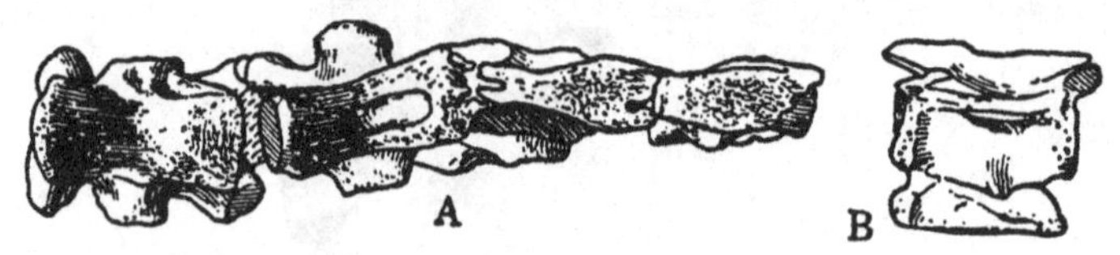

Fig. 51. *Diplocaulus magnicornis.*

A. Sixteenth to nineteenth dorsal vertebrae from below.

B. Lateral view of seventeenth vertebra, shown in *A.* No. 4472, Am. Mus. (Case.)

The whole tube recalls the "odontoid" process of the Coeciliae and the Urodela, but it contains in addition a dorsal component, so that this proatlantic remnant would be composed of both interdorsal and interventral elements; therefore it would be the most complete proatlas known, especially if it should contain also the largely developed ligamentum transversarium of the atlas, the BV 1, in conjunction with other atlanto-occipital ligaments. The enormous heavy head, scarcely able to be lifted above the ground, would require additional

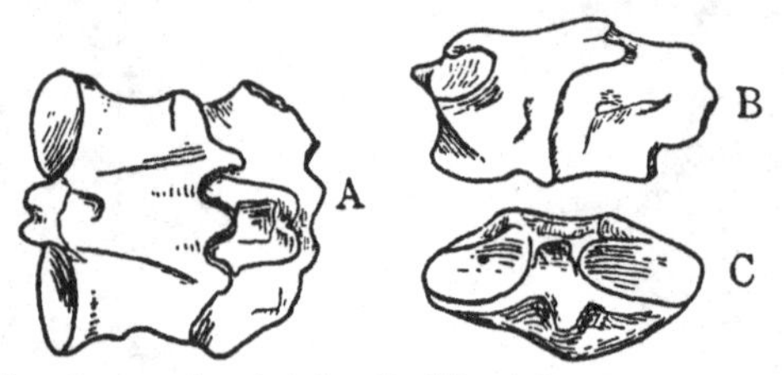

Fig. 52. *Diplocaulus magnicornis* (after Broili). Atlas and axis. *A*, from above; *B*, from side; *C*, anterior view of atlas.

fastening and strengthening of the joint just as much as that of the head-boring Coeciliae.

(5) *Aistopoda.*

Snake-shaped, long-tailed, limbless creatures with numerous small traces of dermal bones on the ventral side of the body; Upper Carboniferous of Ireland, Bohemia and the U.S.A. Notable members are *Dolichosoma* (Fig. 53), and *Ophiderpeton,* whose total length was about 2 ft. The giant of this group is *Palaeosiren*, from the Bohemian Permian; only three vertebrae are known and these greatly resemble

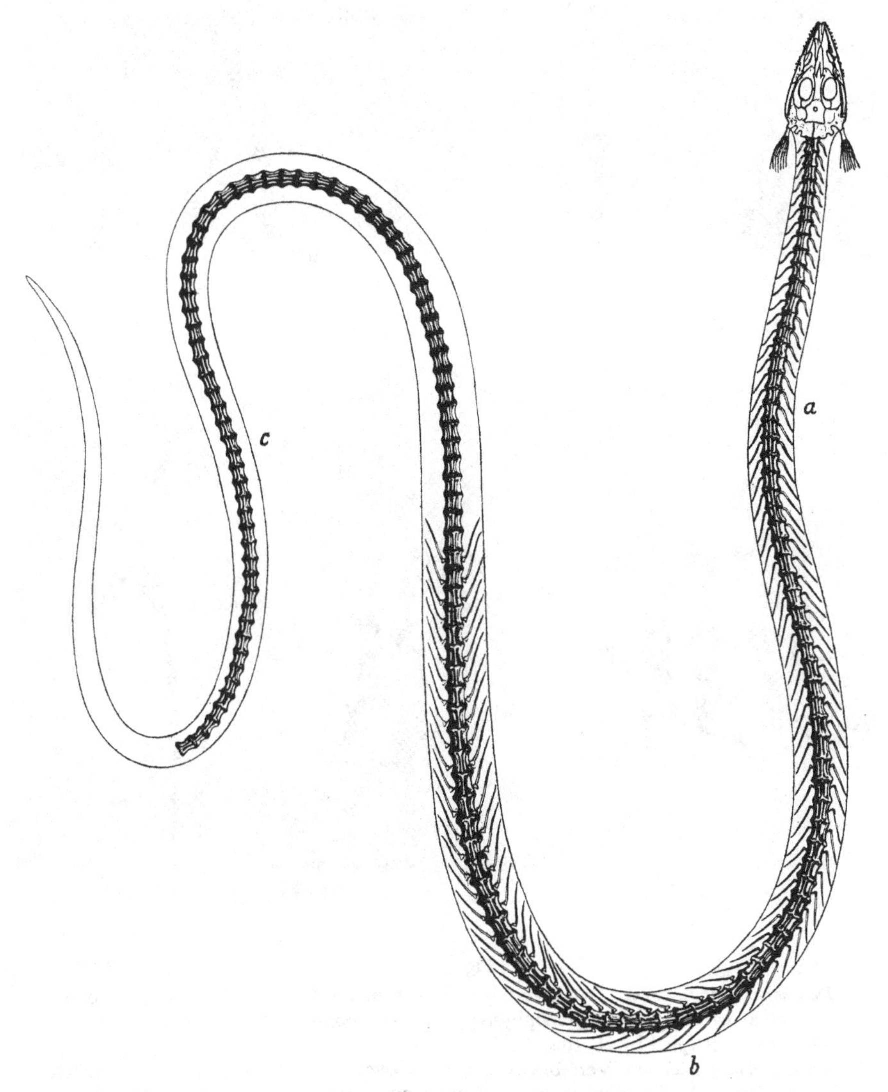

Fig. 53. *Dolichosoma longissimum*. Restored. About ⅔rds natural size. *a*, cervical vertebrae; *b*, trunk vertebrae; *c*, tail vertebrae. (Fritsch.)

those of *Ophiderpeton*, but are 10 cm. long and 8 cm. broad. Fritsch calculated that they belonged to a regular sea-snake about 15 m. or

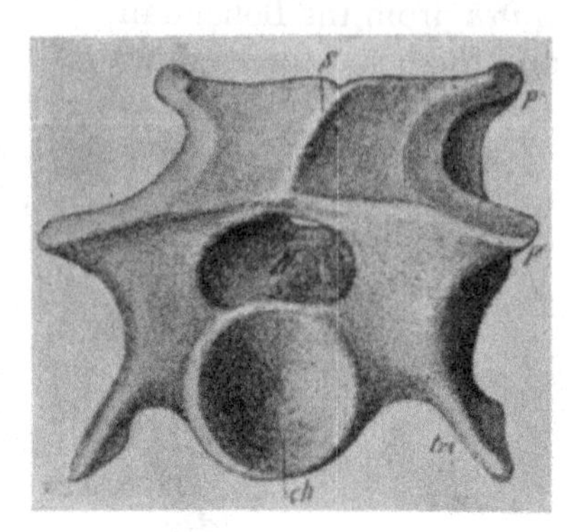

A

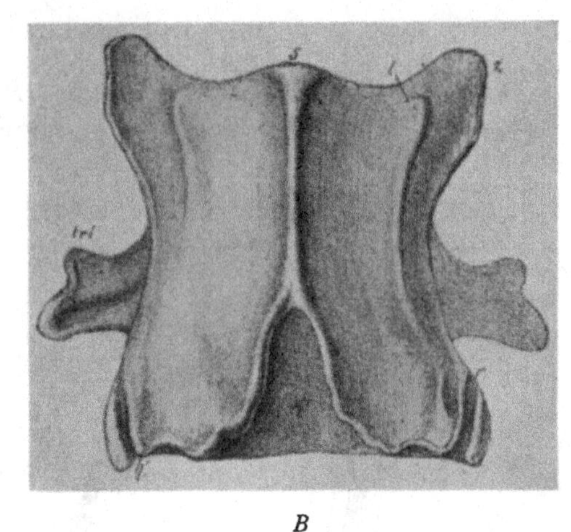

B

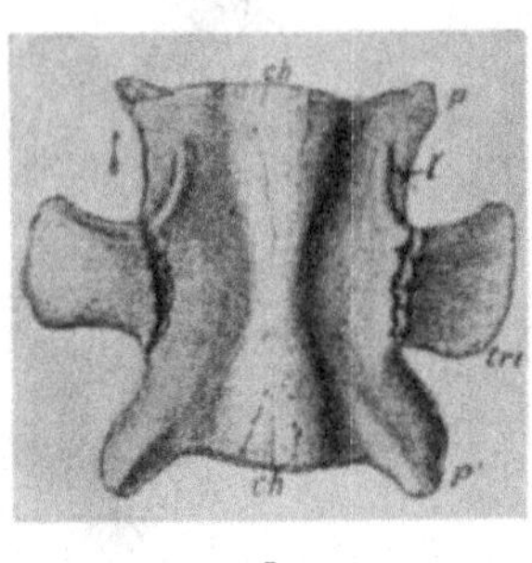

C

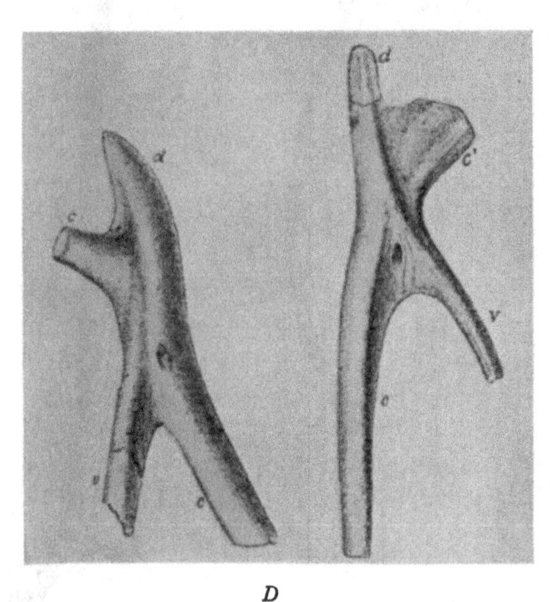

D

Fig. 54.

A. Dolichosoma longissimum. Trunk vertebra seen from the front, × 6. *S*, neural spine; *N*, spinal canal; *p*, postzygapophysis; *p'*, prezygapophysis; *tri*, transverse process; *ch*, hollow occupied by chorda.

B. Ophiderpeton granulosum. Vertebra seen from above, × 12. *S*, neural spine; *i*, notch in hinder end of vertebra; *tri*, transverse process; *l*, ridge of upper "transverse process"; *l'*, ridge on the postzygapophysis; *r*, groove; *z*, prezygapophysis.

C. Ophiderpeton granulosum. Vertebra of a small specimen seen from below, × 12. *ch*, notochordal cone; *p*, prezygapophysis; *p'*, postzygapophysis; *l'*, ridge continuous with transverse process; *tri*, transverse process.

D. Ophiderpeton granulosum. Rib, × 12. *c'*, *c*, rib; *d*, dorsal process; *v*, ventral process. (Fritsch.)

50 ft. long. The vertebrae of this group (Fig. 54) are of the typical pseudocentrous type; they are sharply constricted but deeply biconcave; they number more than 150 and bear numerous ribs. The dorsal spinous processes are represented by low but long ridges. The ribs have a long shaft, which carries three diverging prongs proximally, in which we can recognise the capitulum and the "Rippentraeger" of Goeppert's figures of embryonic Urodela, and especially Coeciliae. Fritsch suggests, with fair probability, that only the middle prong is, as capitulum, attached to the larger parapophysis; the other prongs form a bone laid obliquely across the rib and fused with it, connected with the vertebra by tendon, a condition which might explain the origin of the still mysterious rib-carrier.

Of further special interest are the processes of the vertebrae. There are anterior and posterior zygapophyses, with indications of apparently vestigial zygosphenes. There is a lateral ridge connecting the zygapophyses, from which ridge arise the diapophyses for the ribs; below these "upper processes" of Fritsch is a vascular hole both in the trunk and tail vertebrae. The parapophyses, sometimes very broad, arise from the anterior portion or from the middle third of the pseudocentra, extending downwards, forwards and sideways.

All these features reoccur in the Coeciliae, whose ancestors Fritsch rightly recognises the Aistopoda to be, warning us, however, not to see in them the ancestors of all the other Stegocephali.

Chapter XVII

AUTOCENTROUS TYPE

ITS EARLY ORIGINS IN THE PALAEOZOIC AMPHIBIA

So many Palaeozoic temnospondylous creatures are now known that we are able to discern their phyletic lines with their stages or grades, but we can no longer classify them and put them into compartments adorned with group names based upon some prominent morphological character. If such morpho-taxonomic terms permit an understandable diagnosis peculiar to the groups, well and good, but when these groups are found to run into each other, then even the best conceived terms and diagnoses are not elastic enough to follow suit, and they thus become insufficient, vague and contradictory. This is exactly as it should be if there is evolution.

Group diagnoses should contain not one but several characters, to permit of eventual alternative combination. This, however, is precisely what our well-meant morphological terms cannot do. It almost comes to Linnaeus' humorous dictum "optima nomina quae nihil significant"; we need not go so far as that, but vernacular names would sometimes be better since they are vague, or perhaps the official name of some representative member in the plural, or ending in "oides", providing it could be agreed that such a term purposely means neither order, suborder nor family, but nothing more nor less than ...oides and allies, or ...& Co.!

There are few greater group terms so well established as Mammals and Birds, since it will still take some time before the longed for Sauro-mammal shows cause why it should not be grouped with creatures which suckled their young; or when the Mesornithes, or even the comically pictured Proavis is discovered, he may still be called a Bird, even if it could be shown that his scales were not yet split into feathers. Elasmobranchi is a term we cannot do without, although the Chimaeras are Tectobranch. We get over this difficulty by the fact that cartilaginous Ganoids, bony Ganoids and Teleostei run into each other and can be referred to as Actinopterygians; lower down the stem, they are referred to as Gano-Dipnoi, which group it will still take some time to diagnose properly. "Amphibia" and "Reptilia" are good, being truly vernacular elastic terms, well understood, and we leave the little items of amnion and allantois in abeyance when dealing with fossils. And now, studying these fossils in their abundance, we find that they can no longer be sharply separated. About their respiratory organs we know nothing; the carpals and tarsals are no criterion, whether still cartilaginous or already ossifying. But *Cricotus* has developed centra which a strict diagnosis shows to agree with those of typical reptiles, as Cope was the first to emphasise. If the posterior disk of *Cricotus* could be shown to be the co-ossified interdorsal and interventral, the whole vertebra would be quadripartite like that of *Chelydosaurus*; at the same time it is, so far as one can see, glaringly tripartite, consisting of three independent blocks, certainly as much temnospondylous as the Rhachitomi, the Triassic descendants of which are supposed to have turned into the Stereospondyli by loss of, first the ancient interventrals, then of interdorsals and co-ossification of the remainder. Where in this case is the line to be drawn between the two types which morphologically are so categorically diagnosed? An attempt to consider *Loxomma* and allies as directly ancestral to the Rhachitomi would be wrong, due to the same error of judgment which tries to dispose of the interventrals as secondary, stopgaps of no importance. The result of all this is that Temnospondyli and Stereospondyli as taxonomic names break down if hard pressed; so do notocentrous and gastrocentrous if *Cricotus* is still really quadripartite. The systematist deplores this state of things and composes other labels, the morphologist should delight in it since verily the fossils have come to life again and his series are becoming as fascinating as the solar spectrum, which radiates in all directions and of which we see only a little part between the ultra at either end. We cannot conceive what future forms our present biota will produce, nor do we know what the ancestors of the oldest known fossils were like. There are plenty of footspoors, reputedly even from Devonian deposits, unmistakably belonging to Tetrapoda of considerable size.

The following is a working morphological key of Palaeozoic Amphibia based upon the constitution of the vertebrae.

The intervertebral, incipient autocentra-forming mass is represented as osseous units by:

(1) IV and ID:

- (*a*) if in separate units: *Chelydosaurus*, *Sphenosaurus*, *Archegosaurus*;
- (*b*) if co-ossified into one disk: certain *Embolomeri*, which in this case would represent the Holocentrous type.

(2) Only the fused IV:
Certain *Embolomeri* and *Amniota*. GASTROCENTROUS type.

(3) Only the paired ID:
Rhachitomi or *Permian Labyrinthodontia*, representing an imperfect NOTOCENTROUS type.

(4) Both IV and ID are absent as osseous units because

either (*a*) the intervertebral mass never got beyond the cartilaginous stage: *Loxomma* and allies (i.e. the *Pre-Permian Labyrintodontia*).

or (*b*) the osseous units have been lost: *Post-Permian Labyrinthodontia*.

In either case both groups have *Stereospondylous* PSEUDOCENTRA.

There must have been a very potent cause for the notocentrous departure by which most of the available material was thrown on to the dorsal side of the chorda, thus building up the main axis of the column, a process terminating orthogenetically in the epichordal type of the Anura. Perhaps we may correlate this with the marked, often extreme depression and broadening of the head and trunk of these creatures, compared with the normal terrestrial type of Newts and Amniota. The most flattened of all are *Pipa*, *Pelobates* and *Bombinator*, as representatives of the epichordal type.

In the Amniota the intercartilage does not split and their basiventrals are reduced to an anterior disk, so that these units have physiologically changed duty and position. Their interventrals are therefore not impeded in their growth, and henceforth are able to form a ventrally placed axial column. This gastrocentrous type is beyond cavil the highest.

When, on the other hand, the interdorsals had the better chance of expansion or growth, thus leading to the notocentrous arrangement of the Rhachitomi, we discern clearly the result of this departure. It looks like a struggle to make up by various devices for the incidental loss of their interventrals. This can be done either by the downgrowth of the interdorsals, so that these occasionally almost reach the ventral level; or, the only available material being the basiventrals, these are so much enlarged that they can form, as in *Cacops*, a continuous ventral series.

In the Triassic Labyrinthodonts the interdorsals were also discarded and the stereospondylous type was produced, superficially resembling in elegant simplicity the vertebrae of many Amniota, even Mammals. The working axis was made entirely of ventral elements, the joints were almost plane, and yet these large creatures did not survive the Trias, nor did the *Loxomma* group, which had tried the same solution ages before them, leave any descendants.

We may now discuss the group in which the intervertebral, incipient autocentra forming elements are represented by separate interdorsals and interventrals.

ARCHEGOSAURI

These are the only known Tetrapoda with osseous quadripartite vertebrae. They were large-headed creatures, up to 5 ft. in length, from the Lower Red Sandstone of Europe. The bulk of the vertebrae is formed by the basidorsals and basiventrals. The interdorsals or pleurocentra are considerably smaller, consisting of a separate right and left half, wedged in and more or less slanting; in *Sparagmites*, which is known from two caudal vertebrae only, the interdorsals seem to be in the process of fusing with their interventrals, indicating thereby the possible incipient formation of a posterior central disk. The most interesting units are the distinctly small interventrals which, owing to right and left fusion, are represented by unpaired ventral semi-rings, best preserved in the tail and trunk of *Chelydosaurus*; they were discovered by Fritsch, and described by him as hypocentralia pleuralia (Fig. 55). They are likewise well developed in the trunk of *Sphenosaurus* (tail unknown); in the tail of *Sparagmites* (trunk unknown); restricted to the tail, but absent in the trunk of *Archegosaurus*. Plainly these ventral elements are on the wane. The Archegosauri may fairly be considered as directly ancestral to the Rhachitomi.

RHACHITOMI, ERYOPES OR PERMIAN LABYRINTHODONTIA

These are mostly heavily built Tetrapods from the Lower Red Sandstone, through the Middle and Upper Permian, perhaps to the New Red Sandstone or Lower Trias. The vertebrae are tripartite, consisting of basidorsal, basiventral and interdorsal. If *Actinodon* or *Euchirosaurus* from the Lower Red Sandstone of France should belong to the Archegosauri, the large widely spread assembly of Rhachitomi would, with few exceptions, be restricted to the U.S.A. and to South

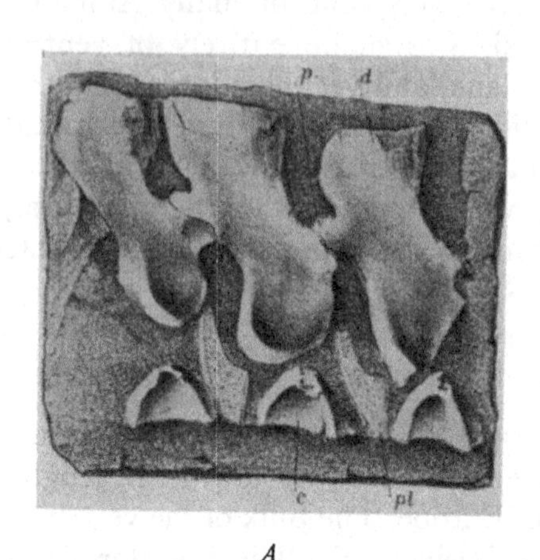

A

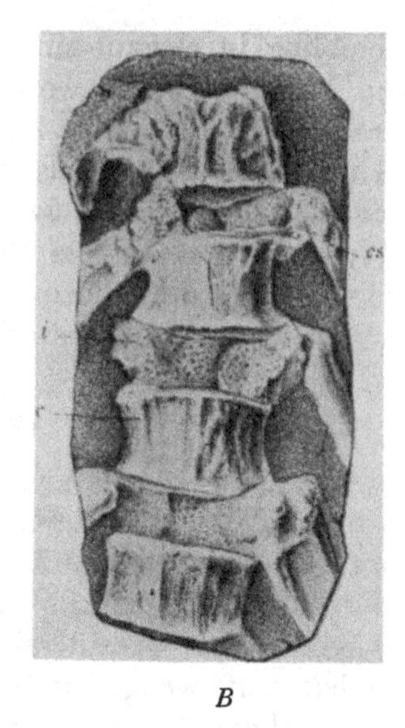

B

C

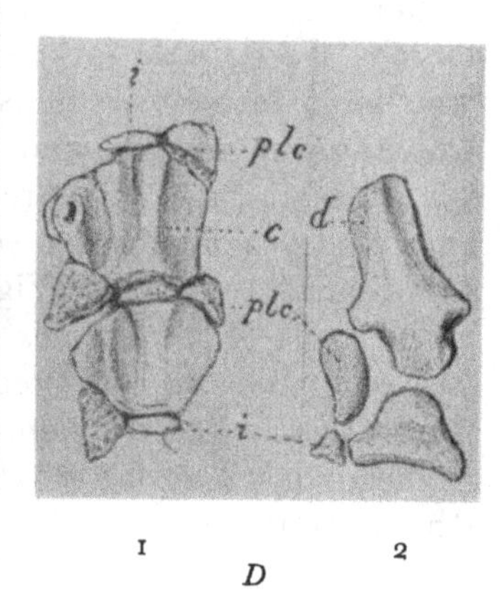

1 2

D

Fig. 55.

A. *Archegosaurus Decheni*. Three trunk vertebrae from a composite cast. Viewed from the right side. *d*, neural arch; *pl*, pleurocentrum; *p*, postzygapophysis; *c*, centrum or basiventral.

B. *Archegosaurus Decheni*. Four caudal vertebrae from a cast. Seen from below. *i*, interventral; *c*, centrum or basiventral; *cs*, rib.

C. *Chelydosaurus Vrani*. Pelvis with sacral and presacral vertebrae. Seen from the right side. *d, d′, d″, d‴*, neural arches; *pl*, plurocentrum or interdorsal; *ic*, basiventral; *sp*, interventral; *f*, femur.

D. *Chelydosaurus Vrani*. 1. Two vertebrae of a young specimen. Seen from below. 2. A restored vertebra, seen from the right side. *c*, basiventral; *i*, interventral; *plc*, pleurocentrum or interdorsal; *d*, neural arch. (After Fritsch.)

Africa. Altogether about 20 genera have been described; of many the vertebral column is still unknown.

North American: e.g. *Eryops*, Cope's type of Rhachitomi (Fig. 56); *Acheloma, Trematops, Trimerorhachis, Cacops, Dissorophus.*

South African: e.g. *Rhinosuchus, Luccocephalus, Micropholis, Lydekkerina.*

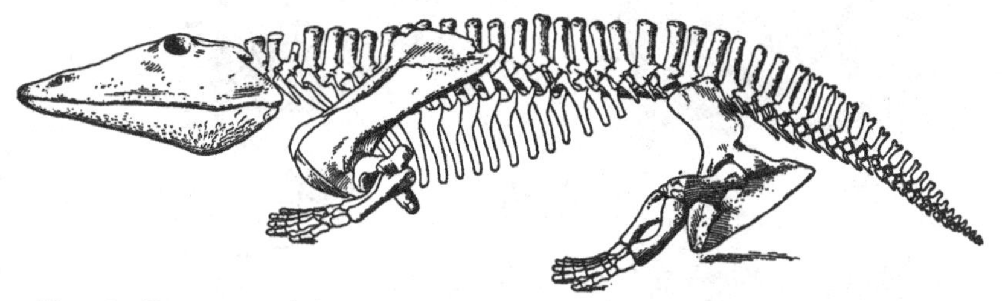

Fig. 56. *Eryops megacephalus*. Restoration of specimen. No. 4893, Am. Mus. (Case.)

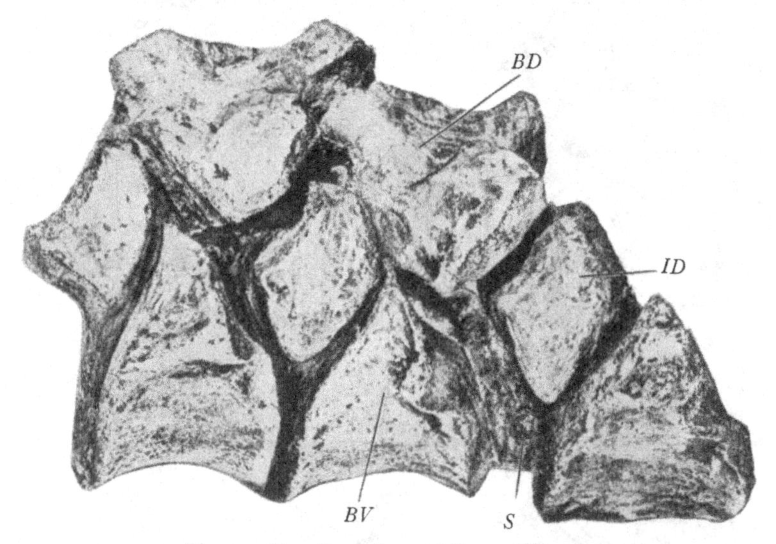

Fig. 57. Dorsal vertebrae of *Eryops*. (Case.)
BD, basidorsal; *BV*, basiventral; *ID*, interdorsal; *S*, space containing notochord and traces of interventral cartilage.

Whilst the Americans belong to the Lower Permian and have, according to Watson, still three occipital condyles, the Africans have only two condyles and range from Mid-Permian into the Lower Trias.

Interventrals are absent, with the exception of a pair of small nodules attached to the posterior ventral corner of the first ventral semi-ring of *Eryops*. The same may be discovered in other genera.

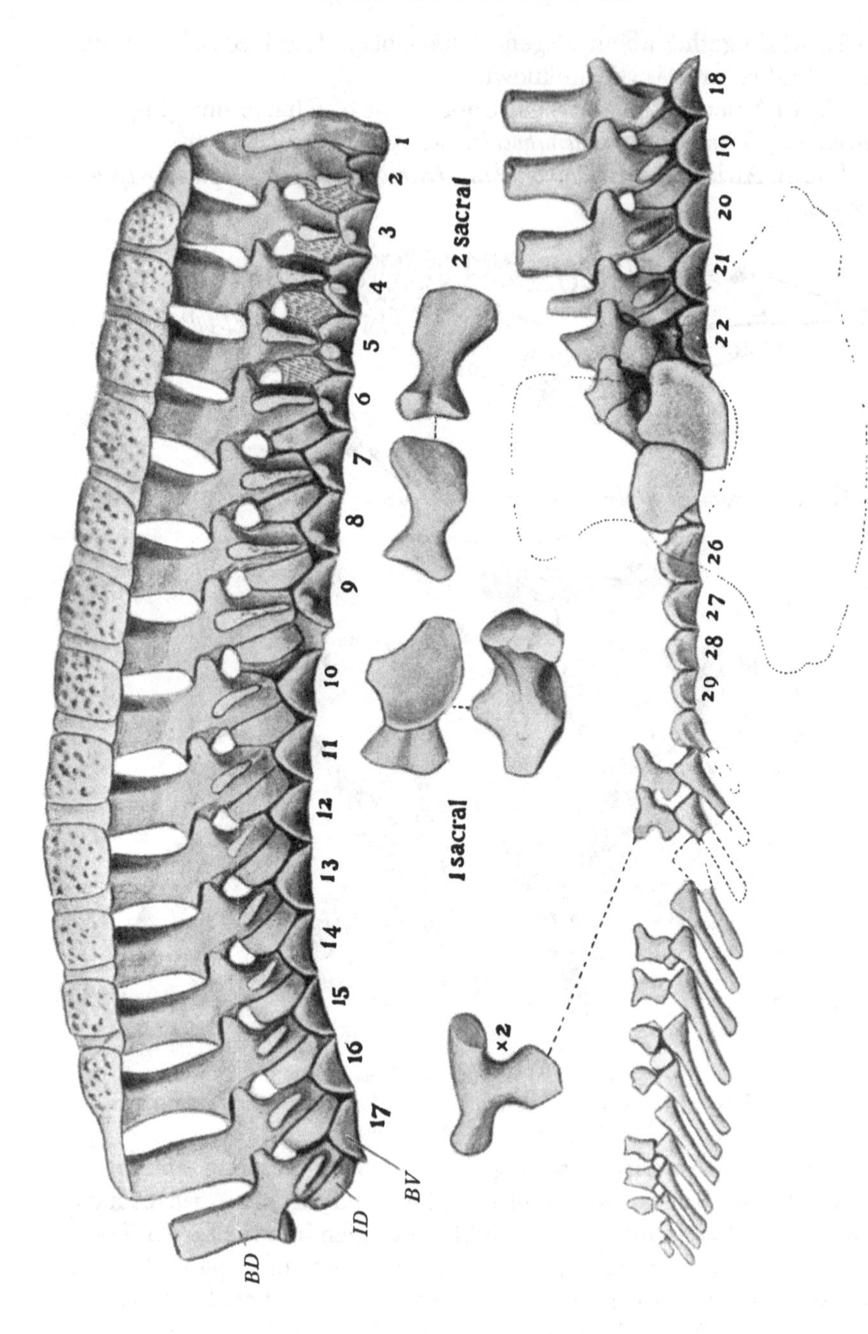

Fig. 58. *Cacops aspidephorus* (after Williston). No. 647, Univ. of Chicago. *BD*, basidorsal; *ID*, interdorsal; *BV*, basiventral.

Cope recognised them as the hypocentralia pleuralia discovered by Fritsch in *Chelydosaurus*. Branson corroborated them in *Eryops*. The bulk of the other vertebrae is formed by the BD and BV, the ID being much smaller wedges, intercalated between the successive vertebrae, but belonging to the vertebra in front of them (Fig. 57). The vertebrae are therefore still mainly pseudocentrous. The basidorsals are generally the largest elements, forming a large united spinous process, with distinctly small, rather unfinished zygapophyses. The rib-bearing diapophyses are powerful, sometimes, e.g. in *Cacops* (Fig. 58), drawn out into long, oblique processes. The most remarkable feature of this genus is the great height of the spinous processes, the tops of which are fused with a single or double series of neatly sculptured dermal bony plates, a truly homologous anticipation of the neural scutes or plates of the Tortoises!

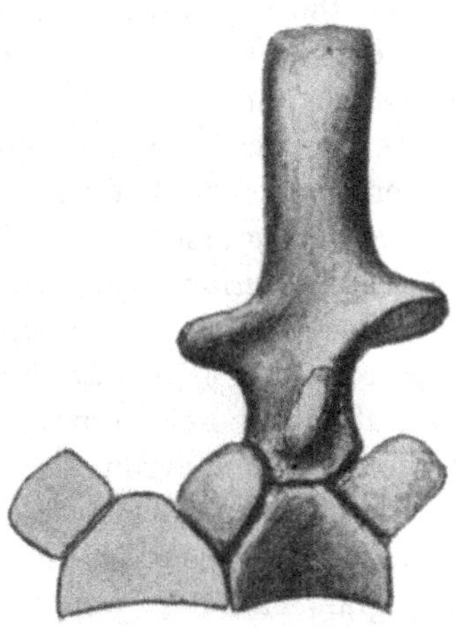

Fig. 59. *Trematops milleri* (after Williston). Ninth vertebra, from the left side. No. 640, Univ. of Chicago.

The connexion of the basidorsal with the basiventral varies considerably. Either their respective bases meet broadly with a more or less horizontal suture, e.g. *Acheloma*, *Trematops* (Fig. 59), and some thoracic vertebrae of *Eryops*; or the basidorsal and basiventral whilst grasping the chorda do not meet on end, but pass each other in such a way that the pointed apex of the basiventral *comes to lie in front of* the downgrowing apex of its basidorsal. This is an infallible guide applying to all Tetrapoda. The only, not infrequent, exceptions are those committed by the artist who sketches what he does not understand, or the erroneous view that the interdorsal lies in front of the basidorsal to which it belongs.[1]

1 *Examples*: Case, op. cit. Pl. 16, fig. 11. The ID on the left side should not be shaded since it belongs to the tenth vertebra; cf. the clearly drawn zygapophyses. The text-figure 34, four vertebrae of *Acheloma*, is quite correct. Pl. 21, fig. 5, twelfth vertebra of *Cacops*, "anterior view" is in reality *posterior* view, cf. the zygapophyses, and therefore correctly presenting ID behind its BD. Text-figure 7; *Trimerorhachis*, four dorsal vertebrae from left side; in reality the figure comprises three vertebrae complete; the first on the left wants its BV, its pair of ID is dislodged by pressure; the fifth vertebra on the right wants its BD. Text-figure 39*B*; *Trimerorhachis*, "4 dorsal vertebrae from the right side showing intercentra and pleurocentra"; but there are only three complete, with one BV to spare on the left end of the sketch.

The basiventrals are usually fused into a semi-ring, rarely remaining as separate halves. The first and second pair are sometimes more or less suppressed. Those of the trunk may be so largely developed as to form a continuous series, best shown in *Cacops* and *Trematops*. If they are small,[1] as is especially the case in the slanting type, they are kept somewhat asunder by the correspondingly better developed interdorsals, which clearly tend by downgrowth to wedge themselves between their basidorsals and the basiventral next following. The better they are developed and the closer they meet right and left above the chorda, although they but rarely seem to form dorsal semi-rings, the more they tend to offer a foothold or additional support for their basidorsals, until these come to rest mainly upon the interdorsals, which thereby have assumed the function of bearers of the neural arches. In fact the interdorsals are the incipient true centres, as Cope was the first to divine.

These interdorsals, as builders of the centra, are the most interesting elements. They are said to be absent, at least as osseous pieces, in the first two vertebrae, unless they are hidden as odontoids or centres, perhaps still partly cartilaginous. They are present throughout the trunk; in the tail they diminish and then vanish (cf. *Eryops*, Fig. 10*a*, Case), so that these caudal vertebrae are reduced to the neural arch loosely combined with the basiventral, which is reduced almost to the chevron.

Taking a broad view of the interdorsals as osseous entities, they seem to be rather in an experimental, initial stage than merely vestigial. This may well be the case and does not prevent the end of the tail from being in an arrested condition. A short tail is no doubt due to reduction, but it is an antiquated idea that even the longest tail was derived from one with still more segments. Of course the units were preformed in cartilage which had a fair share in the composition of the whole vertebra, but not as osseous units. This stage came later and seems to have been reserved chiefly for the trunk. It served for a while for the interdorsals to increase sufficiently to keep

1 In *Lydekkerina* (Lowest Trias, Africa), Watson, the discoverer of this latest and highly specialised form, describes the "intercentra" (basiventrals) as very thin. If thereby is meant that they are reduced to a kind of narrow, ventral semi-ring, or imperfect meniscus, what would in this case form the axial portion of such a vertebra? It would bear a striking resemblance to a vertebra with a large centrum and a much reduced intercentrum, reminding one of the original interpretation and figure of *Desmospondylus*; or has the pair of interdorsals at last succeeded in producing a notocentrum which by its ventral extension and broadness has come to resemble a gastrocentrum?

asunder the basiventrals and even to extend to the ventral side of the chorda. It was, however, at best a half-hearted attempt to transfer the axial centre more and more from above the chorda to the underside. The same half-hearted attempt applies to them as the bearers of the arches, a new function which did not come to more than a lean-to. Moreover, as remarked by Cope, in *Eryops*, and occasionally in others, appears the tendency towards a more intimate attachment between the interdorsal and the anterior portion of the neural arch next following, instead of its own in front. This anomaly amounts to a morphological indecision, resulting in no benefit to the neural arch, which is literally finding itself between two stools. This affords an instance of a contingency being met by adaptation, but without leading to an established feature.

STEREOSPONDYLI OR PRE-PERMIAN AND POST-PERMIAN LABYRINTHODONTIA

These Labyrinthodonts have bipartite vertebrae composed of co-ossified basidorsals and basiventrals which form the typical pseudo-centra. As here defined, these animals form two groups containing some of the largest fossil Amphibians known.

Pre-Permian: *Loxomma* and *Anthracosaurus*, from the Upper Carboniferous of England; *Eosaurus*, from Nova Scotia.

Post-Permian: *Trematosaurus*, Lower and Middle Trias; *Capitosaurus*, *Mastodonsaurus* with a skull of 3 ft. in length, from the New Red Sandstone to the Kemper; *Metoposaurus* and *Labyrinthodon*, the latest of all the "large Palaeozoic and Mesozoic Amphibia", from the Kemper; all these are European. *Anaschioma*, Trias of the U.S.A.

Zittel, the author of the term Stereospondyli, "Vollwirbler", i.e. with solid, not hollow, vertebrae, was struck by the fact that this, which was, according to his view, the highest stage of Stegocephalian vertebral development, should occur already in the Carboniferous era. This requires reconsideration. The whole vertebra is usually described and figured as consisting of the neural arch completely fused on to the "body", i.e. the hypocentrum, intercentrum or basiventrals. This "body" is formed entirely by the basiventrals, which make up a distinctly short but high disk, slightly amphicoelous, perforated in the middle by a small hole for the constricted chorda, the rest of the chorda causing of course the biconcave feature. In *Mastodonsaurus* the whole disk is so clearly a ventral mass that the chorda has left a furrow running closely beneath the medullary canal.

How much of the original intervertebral cartilage existed to act, as in the Phyllospondyli and Lepospondyli, as a connecting pad is not known. The vertebrae of *Anaschioma* are said to be "platycoelous", i.e. plane, in which case the intercartilage must either have been lost or reduced to a very narrow meniscus-like ring. This is interesting because of its analogy with the reduction of the basiventral or intercentral mass in many Amniota.

If the presence of intervertebral cartilage is not be conceded to *Loxomma*, we can but conclude that it had been lost by these contemporaries of Branchiosauri. They all had ancestors, even very large ones, which can be traced by their footspoors to the bottom of the Carboniferous era if Marsh's *Thinopus* is reliable. These unknown Tetrapods had vertebrae composed of the primordial four units, but still unossified; therefore, according to our definition, not yet temnospondylous. To small creatures this did not matter much, cf. the Phyllospondyli; increased weight induced ossification which proceeded apace, and this in turn made still larger, even gigantic, size possible; this temnospondylous stage in turn was superseded by the simple but effective process of co-ossification, restricted, however, to the BD and BV in the manner of Phyllospondyli and Lepospondyli. Other contemporaries, like the Archegosauri, with osseous centres to their ID and IV, who neglected the IV and enlarged the ID, thereby founded the Rhachitomi; this was an offshoot which seems to have died out with the Permian, except for the Anura, whose ancestry has still to be traced from Mid-Jurassic all the way back to the Permian. Of course, if some typically tripartite Rhachitomi reduced and ultimately lost their interdorsals, and then co-ossified the remaining units, they too would thereby turn into typical Stereospondyli. This suggestion by Fraas would satisfactorily account for the awkward absence of Stereospondyli in the Permian, and is therefore favoured by various palaeontologists, notably by Watson, whose *Lydekkerina*, however, would go the wrong way with its reduced "intercentrum".

All this would amount to a more perplexing state of phylogenesis.

1. The Pre-Permian Stereospondyli cannot be directly ancestral to the quadripartite Temnospondyli; they must be assumed to have died out, and are therefore not represented in the Permian era except for the *Rhinosuchus* of Broom.

2. The quadripartite Stegocephali flourished in the Lower Permian as typical Rhachitomi in America, and with the Trias they changed

into a close counterfeit of the meanwhile long extinct Pre-Permian Stereospondyli.

This would indeed be the most astounding, although quite possible, case of repetition by similar means. The term parallel development would scarcely be applicable, since the two groups were never contemporaries; "convergence" might be preferable, although it seems awkward to converge, by loss, with an ancient anticipation or precedence in its own rights. Watson may have been troubled by the above dilemma; he therefore dislikes *Loxomma* as of Carboniferous age: that he does not admit the often reproduced figure of von Mayer of the exquisitely tripartite vertebrae, which he referred to a very young *Mastodonsaurus*, is quite reasonable. He was not the first to doubt that identification.

Watson remarks that in the Upper Trias the Stereospondyli distinctly approach the modern Amphibia, and that certainly these large Amphibia have acquired these characters independently. This parallel evolution between them and modern Amphibia "is the most striking case that palaeontology has revealed to us". I suppose he means that the Stereospondyli have become what, since 1896, is known as pseudocentrous. But this is not a case of parallelism. Certainly we must refer them all back to the lepospondylous condition with potentially quadripartite vertebrae; henceforth we note a parting of the ways of further evolution.

1. The Urodela divided the intervertebral cartilage fore and aft; the bulk of the vertebra remained as basidorsal and basiventral with considerable fore and aft cartilage, and then in the trunk most of the basiventral was suppressed. This is a departure into a cul-de-sac, reserved for most of the modern Newts.

2. Quadripartite Lepospondyli were also the immediate ancestors of the tripartite Rhachitomi. If these also became pseudocentrous by gradual diminution of the interdorsals or pleurocentra, they still differed from the Urodela by enlarging the basiventrals and turning them into a practical centrum instead of losing it.

Consequently the "large Triassic Amphibia" and the Urodela agree only in so far as neither of them produces a true centrum. The Stereospondyli succeeded actually in making a surprisingly efficient centrum out of the ventral units, the basiventrals. The Urodela are actually losing these and their joints are formed across the intervertebral mass, i.e. the very elements that the Stereospondyli lost, having first suppressed the interventral and then also the interdorsal;

this latter element shows a steady decrease in the long pedigree of the Labyrinthodonts, a sign that the interdorsal proved unsuitable for the formation of a centrum.

EMBOLOMERI

Holocentrous, with interdorsal and interventral co-ossified into one disk.

These were four-footed, terrestrial, long-tailed creatures from the Bohemian Upper Carboniferous and from the North American Permian.

The axial portion of the vertebrae is composed of two nearly equally strong osseous disks, of which the anterior carries the neural arch with the ribs, and the chevrons of the tail, while the posterior disk acts mainly as a completely independent bony link between the

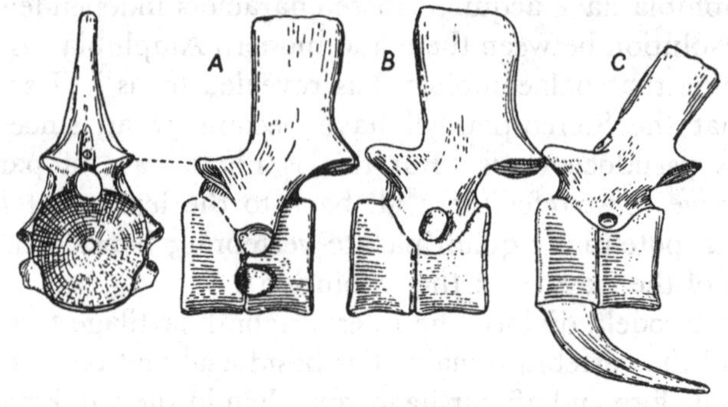

Fig. 60. Vertebrae of *Cricotus*.
A. Anterior and left-side views of dorsal (thoracic) vertebra.
B and *C*. Left-side views of anterior and middle caudals. (Goodrich.)

vertebrae. The vertebrae are therefore double, or rather consist of two jointed halves (Fig. 60). Considered as one whole, the vertebra is slightly biconcave, suggesting traces of a non-constricted chorda; the intravertebral joint seems to be nearly plane; the much constricted chorda runs through the middle of both disks.

The neural arch has simple anterior and posterior zygapophyses, and, at least in *Cricotus*, structures like zygosphenes and zygantra, to which the name of "Hypantrapophyses" has been given! Strong diapophyses carry the single-headed ribs. The arches themselves, apparently according to age, are imperfectly sutured or co-ossified with the anterior disk. This disk is of course formed by the completely united basiventrals, the mass of which has extended upwards around the chorda so as to reach the level of the spinal cord. It bears no capitular rib facet. In the tail the disk is continued to carry the paired chevrons.

The posterior disk is of course the homologue of the intervertebral cartilage ossified. The important question is whether it contains both interdorsals and interventrals, as seems to be assumed by some writers. This is quite possible, cf. the occasionally fused pleurocentrum and hypocentrum pleurale of *Chelydosaurus*. The vertebra of *Cricotus* would in this case still be quadripartite. American authorities prefer to homologise the whole posterior disk with the pleurocentra, i.e. interdorsals, enlarged ventralwards, and so reciprocating the process applying to the formation of the anterior disk. There is a third possibility, that the disk is formed entirely by the interventrals. There is a direct indication favourable to this view; the disk is generally broader ventrally than dorsally, where it narrows usually to a rounded-off apex. The reverse is not observed. Another indication is more theoretical: whilst in the trunk the neural arches are carried exclusively by the anterior disks, especially when they are fused to it, they are loose in the tail and have extended their base upon the posterior following disk to a considerable extent. The latter is, therefore, now in the incipient condition of becoming the bearer of the arch, in short, of becoming the arch-bearing centrum, as is the case with the Amniota, whose centrum is composed of the interventral elements.

It may, for argument's sake, be a case of mere convergence, but there are other weighty supports that these embolomerous creatures show the way to the Reptiles. Cope was already struck with the build of the pelvis; Fritsch has figured and commented on the long and slender ischia of *Diplovertebron*. Cope and Case mention that "the Ilium is much more reptilian in appearance than in most of the Amphibia; the crest has a considerable posterior prolongation to the rear".

The pelvis was attached by one pair of sacral ribs, which were very stout and left a broad, almost circular, impression right across both the anterior and posterior disk, inducing thereby their partial co-ossification. The ilium has left a curved impression on the neural arch.

The first, second and third cervical vertebrae are remarkably alike. The large neural arch, apparently fused with the basiventral, with separate diapophyses and parapophyses for the tubercula and capitula of the ribs, form the bulk of the vertebrae. These are

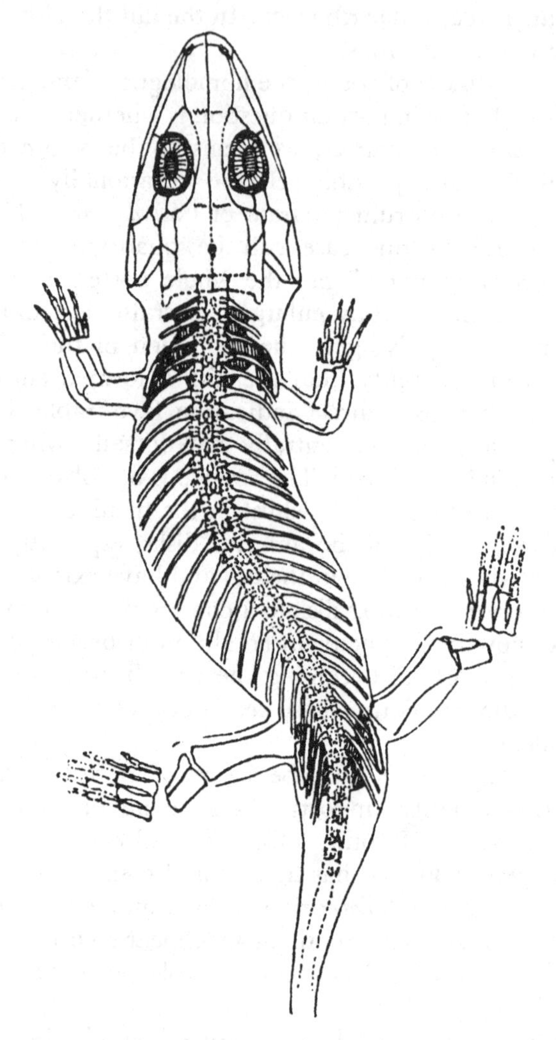

Fig. 61. *Diplovertebron punctatum.* Restoration of the skeleton, founded entirely on a single articulated individual. (Watson.)

separated, or rather joined, by the famous "embolomeres", wedge-shaped disks with a broad base ventrally and narrowing to an apex dorsally. They are, of course, the posterior disks, the incipient "true centra" of the Amniota, the still independent, osseous "hypocentralia pleuralia" of the temnospondylous Amphibia. Naturally there is one disk to spare, a conical mass attached to the atlas; Cope correctly recognised it as the homologue of the dens or odontoid-like process of the Newts, and his genius divined it as a remnant of the proatlas. Unfortunately, he spoiled it by labelling all these wedges *i*, homologising them with his "intercentra" of the Amniota, which he was the first to recognise as the remnants of the hypocentra, i.e. basiventrals, reappearing as the caudal chevrons. Thus he caused, by temporary aberrance, confusion for many years; it is also responsible for the still lingering tendency to reconstruct the temnospondyle with the pleurocentrum as lying in front of its vertebra.

Diplovertebron punctatum, Nyran, Bohemia; cf. Watson's Fig. 31 of a young restored specimen (Fig. 61). The ribs are similar throughout the column, all long, slender and considerably curved; they are mostly capitular only, but with indications of a tubercular shoulder; no reduction in the lumbar region. There are at least three cervical vertebrae, indicating a distinct neck. About 25 presacral, at least one sacral and several postsacral vertebrae. Five fingers and toes. In the restoration appear 27 ribs, then follows a *bona fide* tail apparently without ribs. The sacral region is difficult to analyse; the twenty-third rib is quite free but the twenty-fourth is in contact with it, the ilium being laid against the rib at a considerable distance from its proximal end. This sacral attachment is about acetabular, the vertebra itself is distinctly at a preacetabular level, but the ilium has, as in other Embolomeri, a rather long horizontal backward extension which, although not touching the twenty-fifth and certainly not the twenty-sixth and twenty-seventh, shows that the last three ribs are in the condition of potential sacrals, either incipient, preparatory to the conversion of the twenty-fifth into a sacral, or preserving them as useful for ligamentous attachment. These ribs although still long are doomed, an indication of forward shift of the whole posterior limb and girdle.

The above arguments tend to show that the evidence for the existence of a true holocentrous vertebra in the Tetrapoda is at present incomplete. The Embolomeri, though grouped as the chief members of the holocentrous type, may really have incipient gastrocentrous vertebrae and so lead to the Amniota.

Chapter XVIII

NOTOCENTROUS TYPE

ITS FURTHER DEVELOPMENT IN THE ANURA

The tailless Frogs and Toads are the most modern and highly specialised branch of the Amphibia, and by the irony of fate the almost cosmopolitan and cheap Frog has become the paradigm and martyr of Biology. It contains scarcely a feature that is not difficult to bring into line with other vertebrates. It is notocentrous, with the first ten postsacral vertebrae fused into a long os coccygeum, which serves as a backbone to the abdomen; the tail proper is completely absorbed during the metamorphosis of the Tadpole larva into the anurous Amphibian. The whole group comprises about one thousand recent species arranged upon slender grounds into numerous families and subfamilies, which as a whole find their geographical limit on the permanently frozen subsoil of the world. Fossils are rare, the oldest dating no farther back than the Mid-Jurassic.

The vertebrae of the whole order are in a decidedly transitional stage. They are slightly amphicoelous, verging to procoelous in the Hemiphractidae, a subfamily of the Cystignathidae, which are procoelous like the Bufonidae, Hylidae, Ranidae and some of the toothed Toads or Pelobatidae. Opisthocoelous are the Aglossa, Discoglossidae and some of the Pelobatidae. Similar unsettled conditions apply to the diapophyses of the sacral vertebrae; either horizontally dilated for the attachment of the ilium as in the Ranidae, or cylindrical as in the majority. In the Cystignathidae these processes vary. Ribs are present only in the Aglossa and the Discoglossidae, and are carried entirely by the diapophyses of the neural arches. Goette, who studied the ontogeny of *Bombinator*, the Firetoad, one of the Discoglossidae common near Strassburg, promptly falls into the deplorable error of saying that the ribs of the Anura, as "dorsal ribs", are not homologous with those of the Amniota! A condensed account of the ontogeny of the vertebral column of the Anura reads like a summing up of the phylogenetic, comparative study of Palaeozoic Amphibia. First there appears the material of the dorsalia, which forms a continuous right and left string of cartilage from the atlas into the tail. Then appear metameric swellings which enclose the spinal cord and also grasp the dorsal half of the chorda dorsalis, thereby producing in cartilage what, in the Amniota from *Sphenodon* upwards, appears already in the precartilaginous stage as that continuous mantle in which appear the nuclei of the future arches and interdorsals. On the ventral side of the chorda appears only an unpaired band, thicker and broader than the pair on the dorsal side. This "hypochordal band" is equivalent to the basiventrals, interventrals being no longer discernible. This band undergoes considerable reduction. Only in the atlas and perhaps the second vertebra (Ridewood, *Xenopus*) is the basiventral in broad contact with its basidorsal; in the rest of the trunk the ventral and dorsal bands are held together by non-cartilaginous connective tissue only, but there is attached to the diapophysis a small cartilage which produces a short transverse rib. Thus it looks as if the lateral portion of the basiventrals has been shifted over that connective tissue into the basidorsal. This change explains also the otherwise puzzling condition shown in *Sphenodon* (Howes, Text-figures 2 and 3), and it is responsible for the conclusion of Goette and others that the ribs of the Anura are parts of the dorsal "Bogen". This affords a demonstration of a not uncommon phenomenon, that parts long ago separated from their original home appear now as independent growths from the newly acquired basis; cenogenetic shortened recapitulation.

Next appear the interdorsals, hitherto not recognisable in the cartilaginous string. They begin to grow and completely to surround the chorda, reducing it either to a narrow vertical band or, in transverse section, to a more or less triangular mass with its long and slender apex pointing dorsally. This is important, showing that the squeezing of the chorda is caused by the paired interdorsal blocks growing with their apices downwards. We have here an exact ontogenetic recapitulation of the morphogeny of the rhachitomous column as revealed by comparison of the various adult fossil forms.

The interdorsal cartilage is of course homologous with the dorsal half of the intervertebral cartilage, and as such arises out of what Schauinsland called the "cranial sklerotom Abschnitt"; this is rather confusing, since it forms the caudal half of the final vertebra. It is important, because it shows beyond doubt that the interdorsals or future notocentra of the finished vertebra complex lie *behind* their basidorsals and basiventrals, not in front.

The intervertebral joint of Anura is usually of the cup-and-ball type, very rarely biconcave as it was in nearly all the infinitely older Labyrinthodonts and Cricoti.

As I pointed out long ago, and as has since been accepted by Ridewood, Schauinsland and Schwan, the procoelous and opisthocoelous joint depends upon which vertebra the interdorsal fuses with; procoelous when it fuses with its own basidorsal, the cup then belonging to the next following vertebra; opisthocoelous when the ball fuses on to the vertebra to which it does not belong. The portion with which it fuses belongs to a basidorsal, the base of which extends ventrally and is moreover strengthened by the ossifying mantle of connective tissue which enwraps the Amphibian vertebrae, so that the bases of basidorsals and interdorsals come to lie in the same horizontal line. Noble has added the interesting information that the "ball", interdorsal, sometimes remains unattached, e.g. in *Megalophrys*, *Heleiporus* (a Cystignathid) and partly in *Hymenochirus*, the small equatorial African member of the Aglossa. Boulenger has discovered that sometimes in adult specimens of the Spadefoot *Pelobates*, the knob ossifies and calcifies from a centre of its own and does not fuse with the bony investing shell of the whole vertebra, which includes the facets and joints themselves. Schauinsland suggests that by the existence of this separate entity "je zwei Halbwirbel angedeutet sind", a somewhat far-fetched reminiscence of embolomerous conditions. It is quite sufficient to homologise this knob with the right and left interdorsals of the Labyrinthodonts.

Stannis had previously described a *Pelobates* in which the second, fourth and ninth vertebrae were biconvex, the third, sixth and eighth biconcave. Moreover, since the sacral vertebra, generally the ninth, is invariably biconvex in the Anura, the eighth being biconcave in the procoelous families, opisthocoelous, like the remaining seven vertebrae, in the Aglossae and the Discoglossidae, it is not difficult to imagine that in the Anura the production of procoelous or opisthocoelous vertebrae is simply determined by a mechanical problem of motion. In any case, the shape of these joints cannot be of fundamental importance, although naturally beloved by the systematist as affording an easily expressed character. Even this undecided condition, as to whether the interdorsal mass is to fuse with its own or with the wrong vertebra, occurs already in Palaeozoic fossils; at least Cope mentions that in the trunk of *Eryops* some of the "pleurocentra" tend to lean more firmly on to the neural arch following.

As regards the fate of the hypochordal band and the chorda dorsalis during metamorphosis, Dugés found that the adult bony vertebra of the Anura was formed mainly by the dorsal elements, on the dorsal side of the chorda. He therefore distinguished this type as "epichordal". Gegenbaur accepted this term in opposition to the normal "perichordal" development. The chief studies of it have been made by Hasse, Ridewood and Schauinsland. The hypochordal cartilage degenerates from cartilaginous to ordinary connective tissue which, throughout the trunk, shrinks and gets mixed

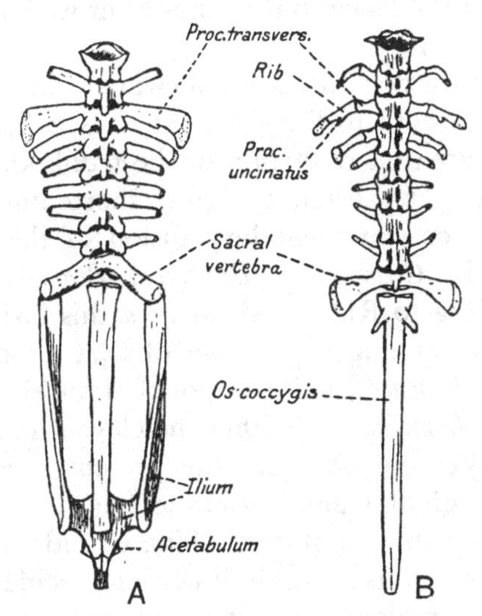

Fig. 62. Vertebral column of Anura seen from above.

A. Rana esculenta (after Gaupp).

B. Discoglossus pictus, with free ribs (after Wedersheim).

with the remnants of the chorda dorsalis as this is squeezed and destroyed by the interdorsals. The extent and severity of this change varies considerably in the various genera; it reaches its height in *Bombinator*, *Discoglossus*, *Pelobates*, *Alytes* and the Aglossa, when this compound of the chorda and the degenerated hypochordal cartilage forms a flat band running along the ventral side of the bony vertebral column, whence it can be stripped off; the column of the trunk is therefore reduced to the neural arches and the centra, absolutely Notocentrous.

We may now consider the sacral vertebrae and the os coccygeum, coccyx or urostyle. The term "urostyle" is scarcely applicable, since the composing bony units never were caudal. In the majority of the Anura, e.g. *Rana* (Fig. 62), and *Bufo*, the ninth is the only sacral vertebra, connecting the ilium by means of the transverse diapophysis. *Pelobates*, *Pipa* and *Hymenochirus* have normally two sacrals, whose diapophyses are fused into double-sized horizontal broad blades and their original duplicity is indicated by the exits of the spinal nerves. *Bombinator* is still in a transitional condition: normally the ninth is the only sacral, but in certain individuals the tenth has also diapophyses, weak but long enough to reach the ilium. The eleventh is free in the larva, but it fuses later with the twelfth, which already belongs to the coccyx.

The recent Anura demonstrate a forward shift of the iliac attachment from the eleventh to the sixth vertebra, whereby *Hymenochirus* has reached the greatest reduction of the trunk known amongst the vertebrates. Another, although slight, shortening of the trunk is effected by the frequent secondary fusion of the first and second vertebrae, e.g. in *Ceratophrys*, *Breviceps*, *Brachycephalus*, *Pipa* and *Xenopus*. According to Ridewood, in *Pipa* this fusion takes place so early that there is no longer any trace of separation in the cartilage.

The coccyx of *Hymenochirus* is composed of the eighth and following vertebrae, but in *Bombinator* of either the eleventh or even the twelfth and following. We conclude that the present coccyx is formed by the adding of original trunk vertebrae one after another. This is clearly enough shown by ontogeny. The considerable weakening of the neural vertebral units makes it difficult to decide upon the number of vertebrae thus amalgamated, it is probably at least ten, so that the twentieth was the first or original of the sacral vertebrae. This creeping headwards of the ilium, which is always satisfied with being attached to two or only one vertebra, accounts sufficiently also for the fusion. In fact the coccyx represents a syn-sacrum. It is the same process or principle which has produced our own sacrum, our ilium having been once attached to the twenty-ninth vertebra, now to only two of the anchylosed complex, but occasionally drawing upon the twenty-fourth, indicating by this individual, truly orthogenetic variation that the trunk shortening may yet continue.

Concerning the coccyx of the Anura, the hypochordal cartilage reappears together with the chorda, the cartilage farther back in the coccyx becoming predominant so as apparently to squeeze the chorda out of existence; the coccyx is thus composed of all three co-ossifying vertebral units. Beyond the coccyx comes of course the true tail, which is absorbed during metamorphosis. But curiously enough in this temporary tail the chorda dorsalis reappears, because there is nothing to squeeze it out, the tail containing no hypochordal cartilage. Thus the Tadpole's tail seems to represent or to recapitulate the stage when or where this axial prolongation of the whole creature consisted only of the elastic chorda dorsalis surrounded by contractile, though perhaps not yet metameric matter.

TABLE II. ANURA

Table showing the orthogenetic shortening of the trunk, owing to progressive forward attachment of the ilium to the vertebral column. The names of the Amphibia are put against the serial number of their sacral vertebrae.

Vertebra No.		
5		
6	...*Hymenochirus*. *Palaeobatrachus*?	
7		
8	...*Pipa*.	
9	...*Xenopus*. *Pelobates*;	...normal majority, e.g. *Rana*, *Bufo*, *Hyla*.
10	.... *Bombinator* (general)	
11	...............*Bombinator*	
12	(individual).	
13		
14	Trunk vertebrae which have been	
15	....	*Triton palmatus*: 22–25 caudals.
16	transformed into the os coccygeum.	
17		
18	....	*Triton cristatus*: 36 caudals.
19		*Salamandra maculosa*: 27 caudals.
20	Original first sacral of Anura...	BRANCHIOSAURUS...*Necturus*, with (caudals 25 ±) 29 caudal vertebrae.
21	....	Cryptobranchus: sacral either 21st or 22nd; about 24 caudals.
22	Tail of the Tadpole.	
23	..................................ERYOPS and CACOPS, about 24 caudals.	
24 to 28 +		
....		
60	....	*Amphiuma*: 20–35 caudals.
150 ±	Anguilliform, secondary increase.	Dolichosoma.
200 ±	....	Gymnophiona: very few caudals.

We have satisfactorily bridged the gulf which seems to exist between the Anura, Urodela and the equally well-tailed Stegocephalia. The difference between the Anura and *Eryops* with *Cacops* has become small beyond expectation.

The idea that the Anura are related to the Labyrinthodonts can be traced back to Quenstedt, who used the term Batrachii in Joh. Mueller's restricted sense of the Anura. Owen followed the same line, this being sufficient for Huxley to attack it: "No Labyrinthodont presents the slightest approximation towards the Anura". Cope published a phyletic diagram in which the Stegocephali appear as ancestral to the Anura on the one hand, and by another line to the Urodela, but he spoilt it by adding the Gymnophiona to the latter as a terminus. Moody, 1909, established a subclass "Euamphibia" with the four orders of Branchiosauri, Apoda, Caudata and Salientia (Anura). His other two subclasses contained all the extinct orders of the Amphibia, minus the Branchiosauria. Broili is equally unsatisfactory; he placed Amphibia into the four orders: (1) Stegocephali, containing Phyllo-, Lepo-, Temno- and Stereospondyli, (2) Coeciliae, (3) Urodela, and (4) Anura. Haeckel in his phyletic tables committed the error of deriving the Anura directly from the Urodela as a terminal and most flourishing branch, chiefly upon the strength of the popular statement that "the developing and metamorphosing frog climbs up its own tree".

Clearly the Anura are highly specialised; notably the jumping apparatus of the hind-limbs is intimately correlated with the elongated ilium, the exaggerated preacetabular position of the ilio-sacral connexion, and the formation of the coccyx. There are many other specialisations, but there are also features which are not yet sufficiently appreciated as to their ancestral value, in addition to the general shape and the large depressed skull. Certainly the Anura belong to the *Lissamphibia* (Haeckel's term for the "naked Amphibia" in opposition to his *Phractamphibia* or armoured forms, the Labyrinthodonts); and nobody seems to remember the Gymnophiona with their hidden armour. But in some recent Anura an extraordinary development of dermal bone crops up which it would be unfair to discard as neomorph. These bones develop in the cutis of the back or on the upper surface of the head and form plates which, like the scutes of Crocodiles, are sometimes pitted or rugose. Sinking in deeper, they come into contact with the cranium, e.g. with the parietals, and fuse with these dermal bones of a previous invasion, for instance in *Pelobates* and in the Cystignathid *Hemiphractus* and *Lepidobatrachus*. In the latter genus several loose scutes appear on the back; such are especially large and thick in the allied *Ceratophrys*, the right and left fusing with each other into a distinct dorsal shield which is an inch across in old specimens and very thick, though it does not fuse with the inner skeleton. Comparison with the much longer compound shields of *Cacops*, *Aspidosaurus* and *Dissorophus* (Cope's "veritable

Batrachian Armadillo") suggests itself. It is the same material and the same process which has produced the characteristics of the Stegocephalia. It is also common in many modern Reptiles, in some of which it seems possible to prove that it is a case of a repeated process, independently developed according to the same principle. A second invasion produced the osteoderms on the top of the present Lizard's secondary cranium, a still later outburst may account for the mammalian armour. Whether we call it a revival of latent germs, or a lingering tendency, the fact remains. It is just as certain as that the antlers of Deer start as cartilage on the top, quite independently of the parietals and frontals, before these antlers are invaded and supplanted by the rapid growth of dermal bone.

Evidence from the vertebrae strongly points to the direct development of the Anura from the Rhachitomi.

Chapter XIX

THE RELATIONSHIP OF AMPHIBIA AND REPTILIA

The terms Ichthyopsida and Sauropsida were intended to emphasise the close genetic alliance between the Fishes and Amphibia, and that between Reptiles and Birds. Although this introduction of two superclasses, each backed by a formidable array of characters, widened the gap between them, the legend arose that Fishes, Amphibia and Reptiles represented one unbroken straight line of descent; in particular the Reptiles were held to be the direct descendants of the Amphibia, and the Stegocephalia were hailed as an easy proof of this.

According to Howes, *Sphenodon*, p. 70, I myself seem to have been the first who has made the "sagacious remark that the Amphibia and Reptiles do not form a continuous line of development, but are two divergent branches of a common stock of Palaeozoic Tetrapoda". In this remark, quoted from *Phil. Trans.* 1896, p. 23, I spoke of a common stock of "quadripartite Tetrapoda" and continued: "If evolution be true, there must have lived countless creatures with quadripartite vertebrae, which were a 'rudis indigestaque moles', neither Amphibia nor Reptiles in the present intensified sense of the systematist". The best name for these creatures is Credner's *Eotetrapoda*: many of them came to nothing, but died out as hopelessly aberrant forms; others founded flourishing groups with a long run, culminating in the Labyrinthodonts, or surviving as Coeciliae, Urodela and Anura. But even these have passed their zenith, except the still flourishing Anura; all are offshoots of what proved to be the main stem, which produced, besides other lost forms, the Microsauri, and these contained the germs of Reptiles and thence all the rest of the Amniota. If there ever was a sagacious remark concerning evolution, it is one by Sir Arthur Smith-Woodward, that the success of a new group depended upon the right time and sequence of the introduction of new organic inventions. "Too early," if incompatible with other already existing features, was and is as fatal as "too late". This kaleidoscopic game has been played since the beginning of life, and this is in the long run the supreme and only test.

Howes and Swinnerton made much of this. They refer to Credner's Eotetrapoda and Dawson's Microsauria; their own conclusion being that *Sphenodon* has unmistakable Stegocephalian affinities, and is "the surviving representative of that group of animals ancestral to all the living Sauropsida" and to most of the extinct Reptiles, thus leading us to the question "Are the terrestrial vertebrates the descendants of Batrachian-Reptilia or of Reptilian-Batrachia?" But this is not exactly the question, since the solution is that the problematic, searched-for, ancestral group was composed of creatures which, as Eotetrapoda, had not yet diverged either into the future Amphibia, like Stegocephalia, Coeciliae, Urodela and Anura, on the one side, or into Amniota on the other.

The best proof of this divergence is given by the evolution of the vertebral column. During Carboniferous times there already existed:

(1) Phyllospondyli and Lepospondyli, some of which survived as such well into the Permian; but they also gave rise to the temnospondylous conditions, a great step which initiated the holospondylous vertebra; whence we derive

(2) The pseudocentrous, quadripartite Urodela;

(3) The bicentrous Cricoti, so-called Embolomeri, with their postcentrum composed of ID and IV, and a precentrum of BV carrying the BD. These creatures came to an end in the Permian, leaving no descendants. Meanwhile, already in the Coal age reduction to the tripartite vertebra took place: reduction or complete disappearance of IV, leaving ID as the material for the potential centrum; BD carried by the large BV; so leading to

(4) The notocentrous Rhachitomi, flourishing in the Permian and surviving into the Trias;

(5) Contemporary with these Rhachitomi, the gastrocentrous Amniota evolved by reduction of the ID and increase of the IV.

It is absolutely excluded that this gastrocentrous type could have been derived from the amphibian notocentrous, just as it is impossible to attempt the reverse. There are of course other improvements in some of these Palaeozoic amphibians which are normal in Reptiles and are sometimes spoken of as cross inheritances; this is really a meaningless phrase unless it refers to the same germ material, which, when receiving a suitable stimulus, reacts by producing the same kind of modification or organ. For instance, *Ceraterpeton* of the Coal age has the phalangeal toe formula 2, 3, 4, 4, 3 instead of the normal amphibian 2, 2, 3, 4, 2; it represents, therefore, an approach towards the Reptiles. And the Carboniferous *Sclerocephalus*, one of the large Archegosauridae, has the typically reptilian toe formula 2, 3, 4, 5, 4. Such "anticipations" are sporadic, not connected, not keeping step with the otherwise observable trend of their bearer.

There is a difficulty of nomenclature. *Stegocephalia*, an excellent term, should be restricted to the Palaeozoic and Triassic, armoured, *bona fide* Amphibia, excluding the Phyllospondyli and Lepospondyli, and the Microsauri in their amended sense. Stegocephalous as such is a stage in the evolution of the cephalic dermal armour inherited from the Gano-Dipnoid Fishes not only by the Stegocephalian Amphibia, but also fully retained by the reptilian Cotylosauria and cropping up elsewhere, even in recent Lizards. *Eotetrapoda* exactly fits those vertebrates which have emerged from Fishes before they diverged into Amphibia and Amniota. The name Cotylosauria, for reasons given below, is unsuitable for the earliest of all Sauropsida. We prefer Haeckel's term *Tocosauria*.

Our phyletic diagrams are assisted by the simile of a tree, although such help is precarious, even if we avoid the not uncommon error of making it, which itself is at best an induction, the base for further deductions. The best simile of our vertebrate tree is the top of a growing Spruce fir with its leader coming to grief or sickening, while one of the several branches radiating from its base assumes the important office of new leader. The node represents the level of division; below it the axis is Eotetrapod; the cut short leader above contains the Cricoti and the last survivors of the Branchiosauri. The branches on the left represent the Amphibia; those on the right the Tocosauria, and these contain the new leader destined to become the main stem of the whole adult tree, whilst the amphibian growths are relegated into the appearance of side shoots. In reality there were many branches: some, like *Sphenodon*, are still alive; others became extinct; to a third category belong those which came to an end only apparently because we do not yet know the links connecting them directly with their descendants, which seem to have come, whole orders of them, into a flourishing existence suddenly, without warning. Of course it is our ignorance, a lack of material, not a freak of evolution. *Natura non facit saltum*, not even if we call it with de Vries a "mutation", a term used in a totally different sense from that of its author Waagen. Concerning the "branches", this simile is likewise not correct. These branches are summarised generalisations of what are more like the countless entangled dead and living fibres of the vegetation of a peat bog, as I expressed it in 1898. It has since been called a "Stammrasen", the German for stem and sod or lawn!

If all creatures living and extinct were known with their structure and distribution in space and time, then the pedigree of every living thing could be drawn up, but a general classification would be impossible.

It may be objected to as equivocal quibbling when we say that the Eotetrapoda were neither Amphibia nor Reptiles, and with the same breath assert that they are still in an amphibian stage. But stages in our diagrams are successive, vertical; divergence is contemporary, horizontal. For the sake of simplicity we mostly restrict ourselves to dichotomy. Unless the same change affects all, it must happen to some and not to others. Our systems are full of this notion, which is logically correct, but in reality is only partially true.

The new Tetrapod life of the Carboniferous age was intensified during the Permian and caused an outburst of more essentially reptilian organisation, bewildering by the diversity of specialisation and sometimes bizarre. An astonishing number of new forms continue to be discovered, which have been placed in successive groups, though each group in turn has radiated in various directions adapted to aquatic, terrestrial, arboreal, and even aerial life.

In order to attempt to trace the origins of the Reptilia it will be well to state briefly the chief characters which define the class, and to give the relationships of the various groups into which it can be divided.

The chief characters are as follows:

Vertebrae gastrocentrous. Iliosacral connexion postacetabular. Mandible carried by the quadrate. Exit of hypoglossal nerve is occipital. Sternum, unless lost, is costal. Foot articulation is intertarsal, not crurotarsal. Breathing by lungs only; gills entirely absent. No metamorphosis.

We divide the class Reptilia into two subclasses:

A. *Theromorpha* of Cope, Palaeozoic Reptiles which died out with the Trias, except for that Pelycosaurian-Theriodont branch which contains the dawning Mammalia.

B. *Sauromorpha*, so called because they gave rise to the whole host of recent Sauropsida, besides various extinct groups. These Sauromorpha may again be sorted into three further assemblies:

(1) *Autosauri*, mainly Lizards and Snakes and the Mesozoic extinct Pythonomorpha.

(2) *Archosauri*, von Huene's term for the most impressive, the princes of Reptiles, e.g. Crocodiles, Dinosaurs, Pterodactyls, and an abundance of other, mostly marine, creatures.

(3) *Aves*, glorified Reptiles which have now, at least in theory, been traced so far back that their much and long-favoured descent from Dinosaurs has to be given up, and now it must be looked for in the Pseudosuchian level where Autosauri and Archosauri meet, or rather begin to get disentangled. The Theromorpha and Sauromorpha have to be combined into something immediately ancestral to both; Cope's term *Cotylosauria* seems to enjoy increasing acceptance. But this well-intended term has suffered almost from its birth.[1] Cope (1881) proposed Cotylosauria for *Diadectes* as a new division of the Theromorpha; then (1888) they appear as a suborder of the Anomodontia. In 1892, they were raised to the rank of order and were declared with several other genera to be the ancestral group of all Reptiles. In 1896, "On the reptilian order Cotylosauria" (e.g. *Diadectes*, *Pareiasaurus* and *Elginia*), they were defined as having the following characters: quadrate fixed; temporal fossa over-roofed; amphicoelous; ribs one-headed; episternum present; pelvis without obturator foramen. Cotylosauria were founded by Cope upon the Permian *Diadectes*, soon including such ponderous, uncouth forms as *Pareiasaurus* and *Elginia*, all much better placed with the Theromorpha, as Broili did. Abel admits into the group the small Carboniferous *Eosauravus* and *Sauravus*, but leaves the still more important Microsauri entirely with the Amphibia: while Williston, in 1912, made the well-meant attempt to include the Microsauri with the Cotylosauria, at least those which in 1896 I had pointed out as possibly Gecko-like, and therefore reptilian creatures, but he dropped this idea again in 1916. In short, the value of the Cotylosauria as the common root of both Sauromorpha and Theromorpha is avoided in our textbooks, and the connexion with the whole host of Lepidosauria or Squamata and the Birds is left in abeyance.

All this has happened, although Haeckel in 1895 had established in Tocosauria "die gemeinsame Stammgruppe der Sauropsiden". They are clearly intended for what Williston seventeen years later

1 Case (*Revision of the Cotylosaurs of N. America*, Washington (Carnegie), 1911) gives a fair historical résumé, ignoring completely, however, Haeckel's *Syst. Phylogenie* (1895), and Fuerbringer's elaborate, reasoned-out *System. Jena Zeitschr.* 1900, pp. 597–683.

understood by Cotylosauria. Haeckel had gone much further; by dividing his Tocosauria into (1) Proreptilia as the oldest, (2) Progonosauria, and (3) Rhynchocephalia, he attempted to account for the phylogeny of all Sauropsida. The Mammalia are cautiously referred to the proreptilian "level", as near the Amphibia as possible, no doubt in deference to Huxley and Gegenbaur, and the Amphibia and Reptiles had not yet been viewed as the two divergent branches of a common Tetrapod stock. His diagram pedigrees, e.g. pp. 268 and 281, the elaborate diagnoses in order to arrive at a system of the Reptilia, and the text, are full of suggestions, often revelations; in any case, they show an enormous amount of work, thought and goodwill.

As these Stem-Reptiles pass upwards into Pelycosaurs, and thence into other Theromorpha, it is also essential to trace them up to the Sauromorpha, and to consider in particular forms like *Protorosaurus*, *Palaeohatteria* and Rhynchocephalia, which it is not sufficient to dismiss with a few complimentary remarks as ancient groups of still doubtful standing. After all, if we accept the Cotylosaurs as the stem, this must include *all* Reptiles. Stem is only a convenient simile, good for a tree with a tap-root, but in many cases it is a matter of a bundle of roots.

Haeckel characterises the Tocosauria as follows: Acrodont teeth tending to become thecodont. Integument with horny covered cutaneous osteodermal *lepides*. Abundant abdominal ribs. Skull completely encased, i.e. Stegocephalian, the only holes in it being those for the three principal, earliest sense organs—nose, eye and parietal. The three-pieced pelvis attached by the ilium to one or two sacral ribs. Shoulder girdle complete, with interclavicle, clavicles and cleithra; but still without a costal sternum. Quadrate fixed in conformity with the dermal cephalic encasement; lower jaw without symphysis.

Such Tocosauria are not mere hypothetical reconstructions. They are fairly represented by the unfortunately still few and scarce Permo-Carboniferous creatures called Cotylosaurs. But even these scattered forms show already divergence, and they do not keep step. Some indicate advance towards the higher organisation of the Theromorpha, others show a more Lizard-like trend in an inextricably undecided fashion. Others show undoubtedly Stegocephalian characters, not amphibian legacies, but proclaiming a state or condition Amphi-Saurian or Amphi-Reptilian or Batracho-Reptilian. We can sort them into Amphibia and Reptiles, speak of and diagnose them as such, because we do not know the countless intermediate links which must have existed during countless generations. What this archaic lot was like we scarcely know. But one such Amphi-Reptile was the Upper Carboniferous *Cricotus*, others were the Microsauri, small creatures as such ancestors should be; and still older groups were the Lepospondyli, of which Watson knows an increasing number of Upper and even Lower Carboniferous specimens, which it is to be hoped he will find time to study.

Chapter XX

THE MOST PRIMITIVE REPTILES

THE TOCOSAURIA OF HAECKEL

We assume that the Tocosauria, as the term implies, are the progenitors of the whole class of Reptilia. The first stage of these creatures, after final separation of the Tetrapoda into Amphibia and Reptiles, is represented by the Microsauri; the next stage is that of the Theromorpha and Sauromorpha, both of equivalent rank. From the anthropocentric point of view the former are, however, of supreme importance, since we claim them as ancestors of the Mammalia. The Sauromorpha are more impressive by their vast number of species and orders, the endless variation of which has, so to say, explored, attempted, followed up to the utmost limits well nigh every morphological possibility. There is a whole host of extinct groups, which may be easily subdivided at will into 10, 20 or more orders, and which finally culminates in the class of Birds. *Divide et impera*, according to taste. The more orders, the easier these are to diagnose and the more difficult to connect with each other, to round them up and to cut out all the others to mitigate the confusing mob. *Rodeo* (a round-up) is the best equivalent of the now frequently used expression "Formenkreis", which, by the way, is not a "form-circle", but "forms encircled".

It stands to reason that Theromorpha and Sauromorpha are the radiants of Tocosauria. How many short- and long-lived radii there ever existed is still unknown; certainly many more than the comparatively few Carbo-Permian forms hitherto discovered. We do not know whether Theromorpha or Sauromorpha first became defined.

Since Guenther's anatomical description of *Sphenodon*, this, with the other Rhynchocephalia, was hailed as the root of all Reptiles. After I had drawn attention to the lowness of the Geckos and their possible close affinity with the Microsauri, Fuerbringer emphatically declared the Geckos to be the most primitive of Reptiles, and, further, that *Sphenodon* cannot be placed in the ancestral line of the Lizards in the wider sense, including of course the Geckos. The last bid for the position of the Key-Reptile has been made by *Araeoscelis*, a con-

12-2

temporary of *Desmospondylus* or *Seymouria*. Broili's group Araeoscelidia contains *Palaeohatteria*, *Kadaliosaurus*, *Protorosaurus* and *Araeoscelis*. He cautiously treats them as an appendix to the Rhynchocephalia, which he has definitely removed from the Theromorpha. Still more promising are Broom's South African discoveries and discussions of the Eosuchia and Pseudosuchia[1] (see Crocodilia, etc. p. 291), which show that these Upper Permian and Triassic creatures were still in such a flux that they can be rounded up as dawning Lizards, Crocodiles, Dinosaurs, etc. All are unmistakably on the side of the Sauromorpha, not of the Theromorpha. These two superorders may therefore be taken as having parted company at the latest in the early Trias, the very epoch when so many divergent orders seem to have materialised.

The difficulty still unsolved for want of material is the position of the Araeoscelidia, Rhynchocephalia, and *Protorosaurus*. Treated collectively they remind us of a kind of *Olla podrida*, *olla* being the Spanish pot which holds the week's leavings, bones and other scraps, kept simmering to make a by no means unsavoury stock, hopeless to analyse, its composition depending upon environmental circumstances.

Tocosauria, as progenitors of all Reptiles, were creatures with terrestrial pentadactyle limbs. The vertebral column was differentiated into neck and trunk with two sacrals and a long tail. The chorda was continuous, thickest between the vertebrae; they had paired neural arches with simple fore and aft zygapophyses, still without diapophyses. The neural arches were carried by the BV, which were of considerable size, as bulky as the centra. The BV grasped round the chorda, on the dorsal level of which they meet the neural arches. The centrum is composed of the paired IV, which by mutual fusion and proceeding ossification reduce the chorda considerably, and which to an increasing extent are used as part carriers of the BD; a tendency due as much to the growth of the IV which fill the place vacated by the dwindling ID, as to the waning of the BV with which they do not fuse except when the stereospondyle stage is reached. Ultimately, when the BV becomes reduced to a narrow disk (the thicker ventral portion being thereby reduced

1 Abel connects the Pseudosuchia with "primitive Rhynchocephalia" like *Palaeohatteria* or *Protorosaurus*; therefore the roots of the Crocodilian stem point also to the Rhynchocephalia. The outcome of these views indirectly benefits our conception of Theromorpha, namely a self-contained assembly of mainly palaeozoic Reptiles with world-wide distribution, which survive as Chelonia and Mammals.

to an intercentrum), the ribs, hitherto borne by capitula, lose their contact with the BV and come to lie between their BV and the centrum. The next stage shows the ribs articulating upon the centrum only, which then reacts by producing a parapophysial facet, eventually a knob; and the neural arch loses its carrier, the BV, broadens its base and transfers itself on to the centrum, forming with it a neurocentral suture. This suture may be abolished in the stereospondyle stage. Whether the rib develops a shoulder, a tubercular arm for attachment to the diapophysis formed on the centrum, or whether it is attached to the diapophysis proper of the neural arch, it is only a further and last stage in the adventures of the rib.

The number of vertebrae in the sacral region can vary from one to three. The sacrals centre about the twenty-sixth, it may be one or two in front or farther back, and this number of about 25 presacrals seems to be favoured by most Amniota. It would, however, be rash to take this as indicating the primordial condition. This often occurring number of 25 presacral vertebrae is not, however, a mere coincidence. Although it may sound like a platitude, the reasonable explanation seems to be that about half a dozen units for the neck and a dozen and a half for the trunk make a convenient arrangement, with the length of the tail variable to any extent.

Microsauri is Dawson's term for certain little "Reptiles" discovered within decayed tree stumps. When other identical or similar forms were found in Europe, Credner discussed them as Eotetrapoda, and Zittel grouped them as lepospondylous Stegocephalia, a description which other authors have followed. In 1896 I pointed out the remarkable resemblance of some of these creatures to Geckos, which are not only a very independent but also a very old branch of Reptiles, and that some of the so-called lepospondylous Stegocephali (e.g. *Hylonomus*, *Sparodus*, *Smilerpeton* and *Hylerpeton*) may perhaps not be Geckos, nor their direct ancestors, but are already well-advanced Reptiles and not Amphibia; chiefly because of their broad neurocentral sutures, their large movable chevrons in the tail, and the little intercentral wedges in the trunk. In 1898 I distinguished between lepospondylous Stegocephali, e.g. *Ceraterpeton* and *Urocordylus*, and lepospondylous Reptiles, e.g. *Hylonomus*, *Hyloplesion* and *Smilerpeton*, all of the Carboniferous age. On p. 80 of the *Camb. Nat. History*, 1901, I suggested the following grouping: Stegocephali lepospondyli, suborder Branchiosauri, including *Protriton*, *Pelosaurus*, *Melanerpeton*,

Ceraterpeton and *Urocordylus*. On p. 288 I classed Microsauri as an order of Reptilia. Having recognised *Hylonomus* as a Reptile which is at the same time Dawson's type of the Microsauri, we are justified in retaining this term for a reptilian group, while the amphibian members had to receive another group name: Lepocordyli.

Williston, in 1910, without referring to previous authors, discovered that at least two distinct groups had been called Microsauri, and in 1912 recognises two groups: Lepospondyli with *Urocordylus* as type; and Microsauri, "no matter whether these Microsauri are reptiles or not", together with the recently discovered *Sauravus* and *Eosauravus*, to be "dissociated into a group nearly allied to, possibly identical with, the Reptilia in a wider sense, while the others, of which *Urocordylus* may be taken as type, may remain with the Amphibia". This improvement was spoilt, in 1916, by referring the "Sauravidae" to the Cotylosaurs without mention of *Hylonomus* and *Petrobates*.

Lastly, Abel drops the term Microsauri and invents the new name Gastrocentrophori, as an order of Stegocephalia, for *Microbrachis*, *Hylonomus*, *Limnerpeton*, *Amphibamus*, while *Ceraterpeton*, *Ptyonius*, *Ophiderpeton* are called Pseudocentrophorous Stegocephali. This is rather surprising. Although this author absolutely agrees with my contention that gastrocentrous vertebrae are the exclusive character of all Amniota, and although his Gastrocentrophori "have the correct gastrocentra, the chief mass of their centrum being built by the interventralia, and the basiventralia being independent and loosely attached to the vertebral bodies", he continues that "these creatures belong to the Stegocephalia as a separate order, and cannot be regarded as immediate ancestors of Reptiles because they are too specialised".

Regarding the Microsauria as transitional forms between earlier and later stages, they naturally appear rather sporadic in the available material. This may be due to the state of preservation, or to the age of the individual. Even closely allied forms may differ in one character and agree in another. Such characters are:

(1) *The neurolateral suture.* Preformed in the Phyllospondyli by the meeting of the dorsal and ventral "leaves", it is abolished in nearly all Lepospondyli by their more strongly developed "husk". In *Ceraterpeton* it is doubtful, in *Scincosaurus*, which some authors consider synonymous, it is sometimes clear. In *Sauravus* the dorsal elements seem to be fused to the whole length of the body of the vertebra.

It is unmistakable in *Hylonomus*, *Hyloplesion* and *Microbrachis*; and, to judge from Fritsch's drawings, also in *Smilerpeton* and *Sparodus*.

For comparison with other groups, no such suture can be traced in Coecilians; nor in *Diplocaulus*, though in Broili's Fig. 22 the badly drawn bases of the transverse processes may be mistaken for it. On the other hand in *Lysorophus*, which in some respects is more lepospondyle than the Urodeles, the suture extends over the whole length of the body of the vertebra, between this and the right and left basidorsals, and these are not ever joined together.

(2) Anterior and posterior zygapophyses, and other interlocking derivations, are already present in Branchiosauri; very variable in Microsauri.

(3) *The vertebral body.* The chorda is constricted to a very considerable extent. It may be continuous but for the narrow septum of the so-called chordal cartilage, as in the trunk of *Ceraterpeton*, while in front and behind the non-constricted part of the chorda (see sagittal section figure by Fritsch, *Ceraterpeton*) (Fig. 63) lies a more or less "porous" portion, suggesting that the two together represent the split of the intervertebral cartilage as described by myself in *Triton*, and that these split halves were undergoing calcification, perhaps imperfectly; certainly they were no longer pure cartilage.

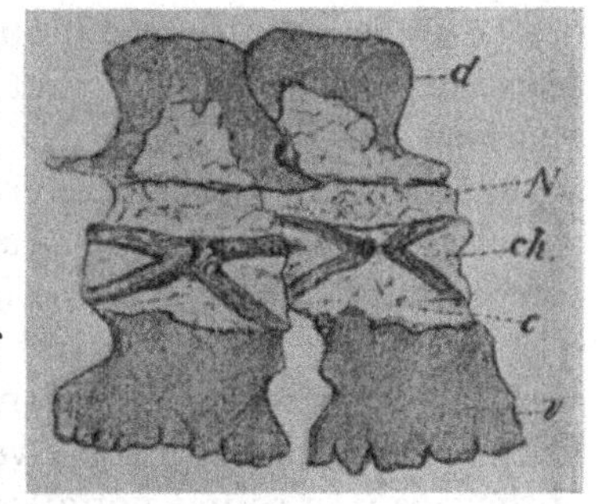

Fig. 63. *Ceraterpeton crassum*. Section of two vertebrae from the anterior half of the tail, × 6. *d*, neural spine; *v*, haemapophysis; *N*, spinal canal; *ch*, notochordal cone; *c*, vertebral body. (Fritsch.)

The body of a typical thoracic vertebra of *Urocordylus* is cylindrical in its ventral half, but in *Ceraterpeton* "we see some kind of laterally compressed lower spinous process instead of the cylindrical shape" (Fritsch). This compression may well be the beginning of the striking deficiency in the ventral half of the trunk vertebrae of Newts, indicating the loss of the basiventral mass, while their derivatives, the ribs, remain.

Sometimes the chorda is completely abolished in the adult, except for the region between the successive bodies, as usually in the tail and trunk of *Urocordylus*. The composition of the vertebral body (i.e. the vertebra minus the neural arch and the ribs) undergoes a most important change, best indicated by the attachment of the ribs. In the original condition the ribs, as lateral outgrowths from the basiventrals, arise midway between the anterior and the posterior half of the "body", being plainly intravertebral, for instance in Newts, and in *Diplocaulus*, *Lysorophus*, Phyllospondyli and various Lepospondyli; in *Ceraterpeton* they lie nearer the anterior than the posterior end; in Gymnophiona, e.g. *Siphonops* and *Ichthyophis*, they are on the average attached to the end of the anterior third; in *Sauravus* they have reached almost the front end, in an almost intervertebral position. It is a step towards the Amniotic intervertebral position which ribs assume when the basiventrals are being reduced to intercentra. Although in none of the Microsauri have the ribs reached this position in the trunk, at least in the tail, the chevrons, as the counterparts of the ribs, are thus intervertebrally placed in *Hylonomus*, *Hyloplesion*, *Petrobates* and *Microbrachis*.

The process which has brought about this change is not a headward shifting of the ribs, but the addition of Gegenbaur's "intervertebral cartilage" (our ID plus IV) to the posterior end of its vertebra, and its subsequent enlargement at the expense of the correspondingly diminishing rib-bearing basiventrals, the hypocentrum or intercentrum. When, as in the Urodela, the "intervertebral cartilage" splits vertically and transversely into an anterior half which adds itself to the hind end of its own vertebra, while the posterior split off half joins the front end of the vertebra following, the relative position of the ribs is of course not affected.

The whole process is obscured, camouflaged in its older phyletic stages by the whole body of the vertebra being ensheathed by that ossifying activity which has produced the typical husk-like lepospondylous vertebra. It is a legacy to these transitional creatures, which they do not get rid of until the arcualia ossify endochondrally each as a separate entity, and thereby initiate the osseous temnospondylous stage. An ensheathing mantle, due to direct ossification of the connective tissue, with or without calcification, still plays a great rôle in the growth of the Urodele's vertebra, and a similar process may possibly cause the completely seamless co-ossification of the vertebrae of Snakes.

The changes described above mean nothing less than that the so-called pleurocentra, namely the interarcualia, are forming those true centra which are destined to take the place of the now old-fashioned pseudocentra. The apparent puzzle has been solved. It dates back to Carbo-Permian times, when truly terrestrial life began to become intense. It is no wonder that Amniotic ontogeny has condensed the evolution of the vertebral column and hurries it on so that it deceives the embryographer into believing that the vertebra is composed of the "arch" (BD plus BV) and the "body" which arises mysteriously in the "perichordal sheath".

(4) *The ribs* of the Lepocordyli have sometimes fairly distinguishable capitular and tubercular portions, attached to normal parapophyses and diapophyses, but if these are united into an incipient

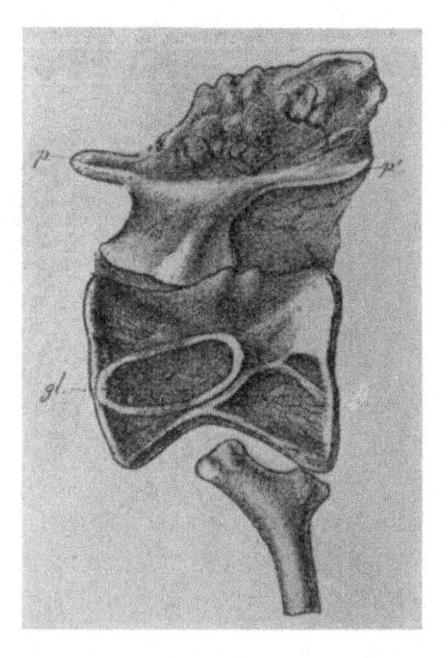

Fig. 64. *Microbrachis mollis*. The twenty-ninth vertebra seen from the left side. *p*, prezygapophysis; *p'*, postzygapophysis; *gl*, joint surface for the rib. (Fritsch.)

transverse process the distinction becomes precarious. The same applies when the base of the rib is broadened into an oval shape which in the absence of a neurolateral suture leaves it uncertain whether the place of attachment is parapophysial or diapophysial, or both. Sometimes, as in the Microsaurian *Microbrachis* (Fig. 64), the trunk ribs articulate with a stout capitular arm upon the side of the anterior half of the body, while a longer and weaker tubercular arm must have been connected by ligaments to the neural arch. For comparison, *Lysorophus* has simple-headed ribs attached across the suture as in Phyllospondyli and Lepospondyli. *Diplocaulus* has double-headed ribs attached separately to the neural arch and to the basiventrals. *Ichthyophis* has double-headed ribs attached by the short and stout dorsal arm to the body and by a longer and slender ventral arm to the huge and forwardly directed parapophysis.

We may now describe in detail the chief characters of the known individuals of this very important group of Microsauri; they still have various characters in common with Palaeozoic Amphibia.

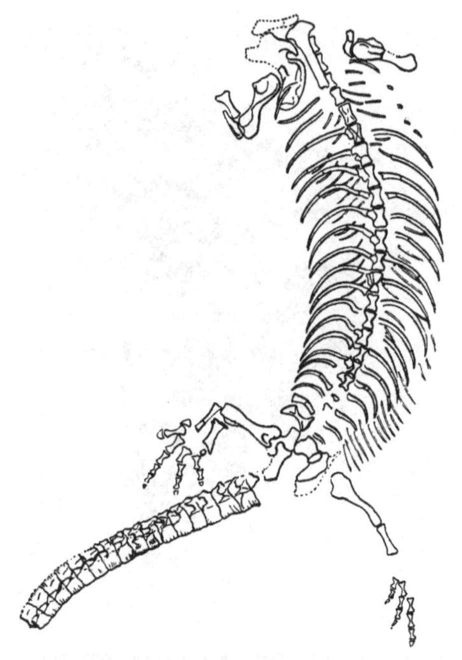

Fig. 65. *Sauravus costei* of Thévenin. From the Stephanian of Blanzy. Reduced.

Sauravus costei of Thévenin. From the Upper (Stephanian) Carboniferous: Blanzy, France (Fig. 65). Notochord persisting, but with a waist. About 23 or 24 presacrals; cervicals not visible, overlaid by the shoulder girdle. Body of vertebrae a bony shell, waist-shaped, extending to the clear intervertebral joints. Low but broad neural arches, which seem to be fused with the whole length of the vertebral bodies. The anterior half of the vertebra is much the stronger, suggesting that the posterior half contains the whole elongated mass of interdorsals and interventrals, reminding one of *Hylonomus*. Williston's remark, that in the "Sauravidae" (*Sauravus* and *Eosauravus*) intercentra are still unknown, has been put more precisely by Thévenin; there are no intercentra in *Sauravus*, and no gaps which could indicate their cartilaginous presence between the clear and clean cut joints. Considering the presence of an unbroken series of intercentra in *Sphenodon*, Geckos and *Seymouria* and their presence at least in the cervical region of other Cotylosaurs, it is not likely that the basiventrals should have vanished completely in these Carboniferous forms. The column must therefore be in the same stage or condition as that of other Microsauri: to some extent still lepospondylous, with the interdorsals and interventrals already preponderant over the basiventrals, with the ribs loosely attached to or near the front end of the vertebra. If we prefer the reduction of the basiventrals, the slender ribs can have been attached by ligaments only. The restored figures of Thévenin admit of either interpretation: there are no "apophyses transverses" for the ribs: all the presacral ribs are very slender and long, without uncinate processes, but consisting of dorsal and ventral halves. There are apparently numerous abdominal, parasternal elements. Two sacral and at least 17 caudal vertebrae carrying large, compressed ventral blade-like processes, which by their striated ridges remind one of *Ceraterpeton* as figured by Fritsch; they indicate the insertion of the well-developed subcaudal muscles and therefore at least a semi-aquatic life. Total length from the missing head to the sacrum is about 10 cm.

Eosauravus of Williston. From the Upper Carboniferous, Ohio. In 1908, Williston published in the *Journal of Geology* an article, "The oldest known Reptile. *Isodectes punctulatus* Cope". In 1909, he described it as *Eosauravus*, to get rid of some confusing synonymy (*Isodectes* and *Tuditanus*) and to emphasise that if the French *Sauravus* is the grandfather, the American *Eosauravus* is at least the great-grandfather of the Saurians. To make sure, it was pointed out that the specimen was found in the Linton horizon of Ohio Coal, of Mid- or Upper Pennsylvanian, being lower than other horizons known to yield Microsaurian remains. A photograph of this specimen in the National Museum at Washington is reproduced in Williston's *Osteology of Reptiles* (Fig. 66).

Here is a shortened description of this specimen: Vertebrae amphicoelous, with persistent notochord, agreeing with those of *Sauravus* and Microsauri generally. Spines rudimentary; no indication of ventral ribs, differing thereby from *Sauravus*. No evidence of hypo-

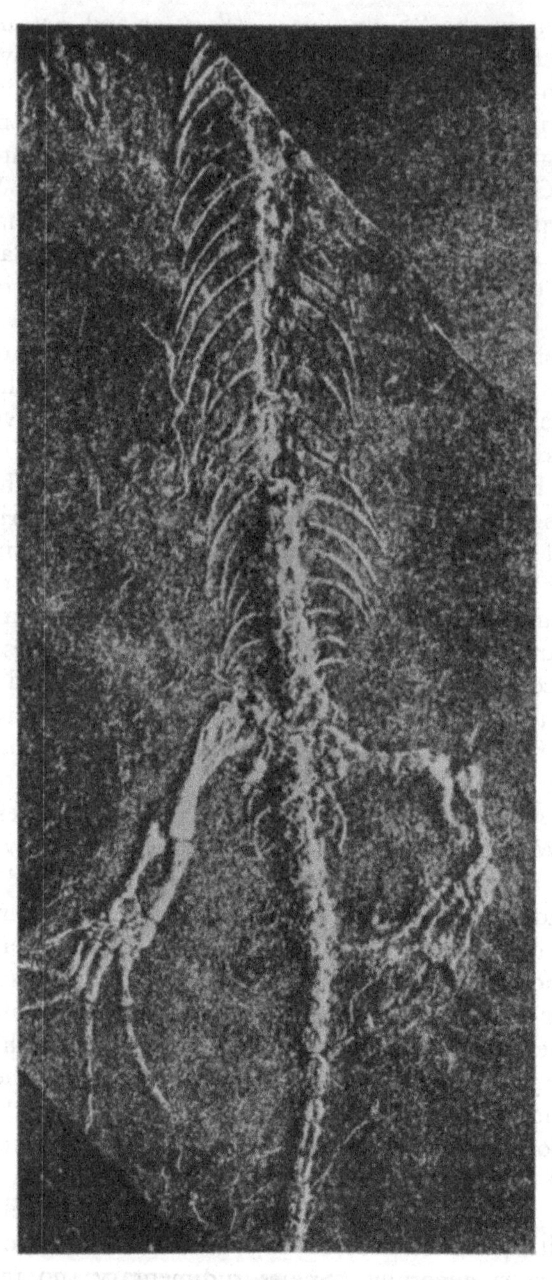

Fig. 66. *Eosauravus*. Part of skeleton, from above. Enlarged. Nat. Mus., U.S.A. (Williston.)

centra. All trunk vertebrae bear rather long, curved ribs, as also do the first three of the caudal series. Two vertebrae intervene without ribs, the two sacrals. The head and neck region of the specimen is unfortunately missing; the well-preserved hind-limbs show clearly the typical reptilian formula 2, 3, 4, 5, 4.

Hylonomus of Dawson. From the Upper Carboniferous of Nova Scotia and Bohemia, extending in Saxony into the Lower Permian (Fig. 67*A*). To understand Dawson's vertebra of *H. latidens*, fig. 20

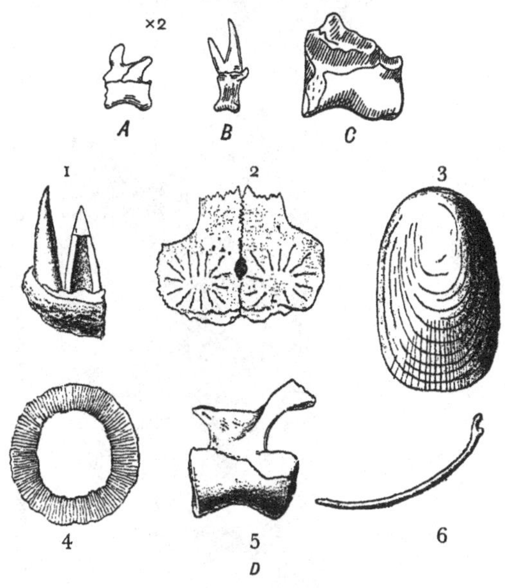

Fig. 67. *A, Hylonomus*; *B, Smilerpeton aciedentatum*; *C, Hylerpeton Dawsoni*. *D, Hylonomus Lyelli* Dawson. 1, teeth; 2, parietal bones; 3, scale; 4, transverse section of tooth; 5, vertebra; 6, rib.

of his Pl. 39, we have to compare it with his *Smilerpeton aciedentatum*, fig. 43, Pl. 40 (Fig. 67*B*), and with his *Hylerpeton Dawsoni*, fig. 81, Pl. 41 (Fig. 67*C*). The body of the vertebra carries two dorsal arches of nearly equal size, the posterior of which, compared with Schauinsland's embryo of *Sphenodon*, reveals itself as the interdorsal element still so fully developed as to enclose the spinal canal; a truly diplospondylous condition. If Fritsch's text-figure, p. 159 (Fig. 67*D*), of a certain American species of *Hylonomus* is correct, its interdorsal seems to be fused with its basidorsal, and reduced in size, instead of being fused with its interventral. In any case the body carrying

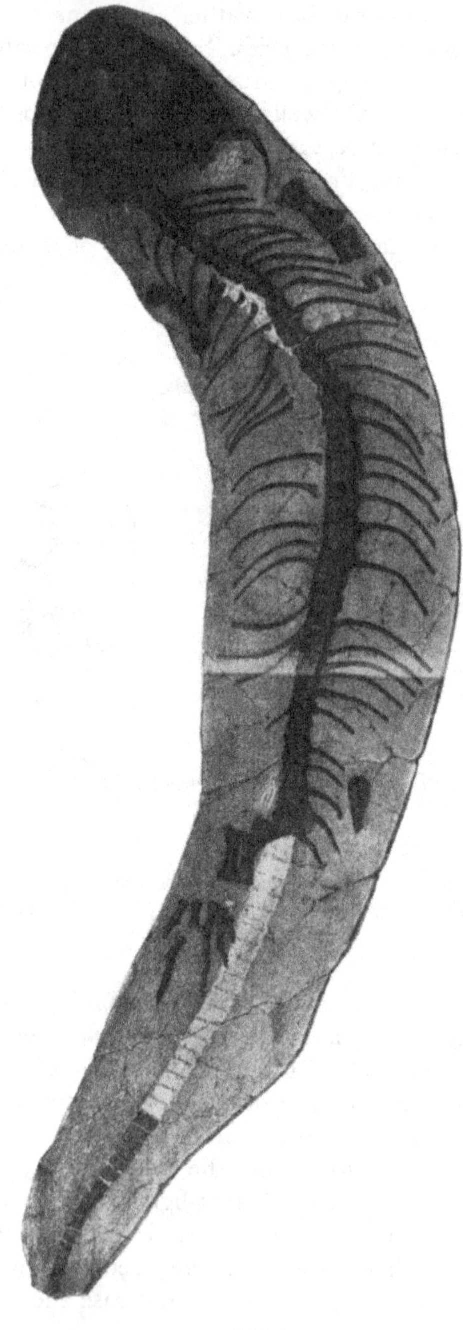

Fig. 68. *Hyloplesion longicostatum.* A complete specimen, enlarged (Fritsch.)

the neural elements may be composed of BV plus IV, a most important consideration, since further enlargement of the interventral and corresponding reciprocal shortening of the basiventral would result in the typical gastrocentrous reptilian vertebra. The chevrons are distinctly placed between neighbouring vertebrae.

Hyloplesion longicostatum. From the Upper Carboniferous of Bohemia; an almost complete skeleton is figured in Fritsch's monograph (Fig. 68). Vertebrae still with complete chorda, or at least deeply amphicoelous. Chevrons clearly independent. The ribs as figured suggest that they arise either from the anterior end or nearly between the vertebrae. It is doubtful whether the atlas carries ribs.

Microbrachis Pelicani of Fritsch. From the Upper Carboniferous of Bohemia. A reconstruction sketch of this aquatic creature, which was about 13 cm. long, reminds one of *Amphiuma*, with its short tail and much reduced anterior and posterior limbs (Fig. 69). The anterior half of the amphicoelous vertebra carries the large neural arch and the double-headed rib; the posterior half carries nothing. Ribs seem to begin close to the skull; they are figured by Fritsch as arising near the anterior end of the vertebra. Recognised by Abel as belonging to his "Gastrocentrophori", by which term the distinguished Austrian morphologist acknowledges them as of reptilian type, while as a systematist he denies them this position.

Fig. 69. Reconstruction of *Microbrachis Pelicani* Fr., from the Upper Carboniferous of Nürschan (Bohemia), natural size. (Abel.)

Petrobates. From the Permian of Saxony. Added by Baur to my non-amphibian but already reptilian lepospondylous creatures (1896). Loose chevrons, placed ventrally between vertebrae like intercentra.

Orthocosta of Fritsch is a doubtful member. On p. 171 Fritsch

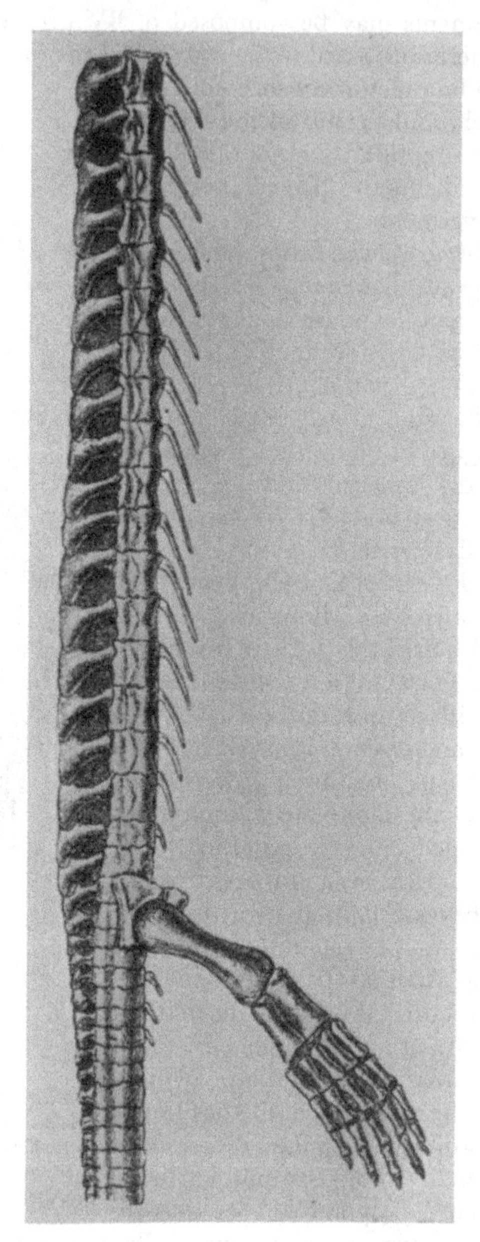

Fig. 70. *Orthocosta microscopica.* Restored and magnified 12 times. (Fritsch.)

figures trunk and tail vertebrae (Fig. 70). In the trunk the dorsal arch is carried by about the anterior quarter or fifth of the body; ribs apparently attached near the anterior end. In the tail the arch is carried midway. Chevrons lie between them.

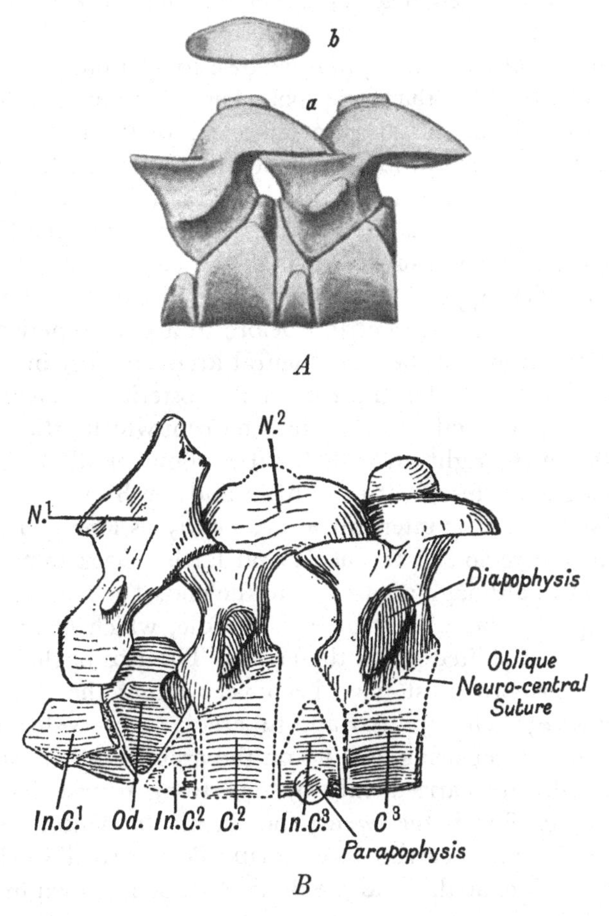

Fig. 71.

A. Desmospondylus anomalus. Posterior dorsal vertebrae; *a*, from the left side; *b*, intercentrum from below. (Case.)

B. Seymouria baylorensis Broili. The first three vertebrae, × 2. Left lateral aspect. *C.*², *C.*³, centra of second and third vertebrae; *In.C.*¹, ², ³, intercentra 1, 2, and 3; *Od.*, odontoid; *N.*¹ and *N.*², neural arches of first and second vertebrae. (Watson.)

Seymouria baylorensis of Broili. Originally described by Williston, 1910, from vertebrae mixed with those of *Cacops*, as *Desmospondylus anomalus* (Fig. 71 *A*), and as such is figured on Pl. 19 of Case's Amphibia and Pisces (1911), without, however, a reference in the text or index. The creature was recognised as a Reptile by Broili and was renamed *Seymouria* after a place in Texas. Several well-preserved specimens of this now celebrated stegocephalous Reptile are known from the Texas Permian.[1]

The latest and best description we owe to Watson.

Total length is less than one foot. The most remarkable features of the vertebral column are the large size of the basiventrals, the apparently large size of the two-headed ribs, and the slanting suture between the very large neural arches and the very high centra. Watson carefully remarks that in the figure in his paper (Fig. 71 *B*) "the atlas is composed of drawings of the elements in my skeleton replaced in what appears to be their natural relations; the centrum of the axis, and the intercentrum before it, are hypothetical".

The atlas consists of the large neural arch, the first intercentrum or BV 1 with a facet for the rib on its posterior surface; and the large quite separate odontoid or first centrum which articulates with BV 1 between the right and left basal portions of BD 1; behind, it is perhaps not yet fused with the BD 2 and the BV 2.

The basiventrals, or intercentra, are nearly as long as the centra, at least in the region of the neck, and farther back in the column they diminish, although of fair size to the end. The second and third have a capitular facet right in their middle, which means that the ribs are not yet shifted on to the centra. The ribs of the atlas, axis and third vertebra must have been very large; the third is even called massive; their capitular and tubercular arms widely separate, the tubercular facet being well up on the arch. Farther back in the trunk the ribs are carried only by the strong neural diapophyses. The sacral vertebra is the twenty-fourth, but in Watson's specimen the twenty-third also carries a sacral rib. Postsacral ribs exist on the fourth and fifth caudal, "and are double-headed, known in no other vertebrates".

The centra are remarkable for their strong slanting, instead of horizontal, suture, and they extend so far up with their posterior dorsal part that they suggest, at least to myself, the possibility of

1 Williston remarks that the so-called Permian of America is part of the Upper Pennsylvanian (top Carbon) and that therefore its reptilian fauna is older than the Rothliegende of the European Permian. Elsewhere (1912) he introduces the term Permo-Carbon for American Permian because it is partly Upper Carbon and partly Lower Permian.

containing the interdorsal units, so that in this primitive Reptile this centrum would still be homologous in composition with the post-centrum of *Cricotus*. Cf. also this unstable region of Microsauri, notably *Hylonomus*, p. 189.

Williston, when trying to bring his *Desmospondylus anomalus* into line with the Stegocephali, was rather unlucky when, *op. cit.* p. 281, he wrote of the large size of these intercentra "suggesting impressively the lower half of the pleurocentra of *Cricotus*", the lower half being of course the hypocentra pleuralia of Gaudry. No wonder he was troubled, p. 267, "why in all true Amphibians, both ancient and modern, the chevrons are an integral inseparable exogenous part of the body of the vertebrae, while in all Reptiles and higher Vertebrates they are the intercentrum or part of it".

Chapter XXI

COTYLOSAURI

THE PRIMITIVE ANAPSID THEROMORPHA

Although this group, if "ancestral to all reptiles", would of course include the Theromorpha and Sauromorpha, it is practically by tacit consent restricted to those Theromorpha which have the cranium still completely roofed over by dermal bones, the only openings being those for nose, eyes and parietal sense organs.

There are other primitive characters, none of which are, however, peculiar to the groups: deeply amphicoelous vertebrae, mostly with continuous although intravertebrally much constricted chorda. The basiventrals, reduced to "intercentra", are in various genera present from atlas to sacrum. There is an incipient division in the cervical and trunk ribs into capitular and tubercular portions, the conjoint facets of which extend in the shape of a long, slender oblique oval. A neck is still practically absent, being restricted to the first and second vertebrae, the shoulder girdle lying close behind the head. The third to seventh ribs tend to become so strong and broad that they overlap and form a support for the girdle; of course not by direct contact of ribs with scapula, but by means of muscles like the levator scapulae and the serratus system of the powerful shoulder girdle. Ribs are in some genera present not only from atlas to sacrum, but exist also in the tail, quite separate from typical chevrons on the same vertebra.

The whole group is restricted to the Carbo-Permian and the Permian proper of U.S.A., South Africa and Russia. The type is Cope's *Diadectes*, which is, as it happens, the least suitable. Other remarkable forms are *Limnoscelis*, *Captorhinus*, *Labidosaurus*, all from Texas or New Mexico; *Elginia mirabilis* from the Elgin sandstone of Scotland; *Pareiasaurus Baini* from South Africa and Northern Russia, a massively built brute, frequently figured.

Diadectes phaseolinus. A powerful beast, more than 3 m. or 10 ft. long, of which more than half belongs to the robust head and trunk; it was vegetarian to judge from the bean-shaped teeth, hence the specific name. About 22 presacral, deeply amphicoelous vertebrae with the chorda not quite interrupted. Atlas with separate right and left basidorsals, loosely carried by the large, united basiventrals (wrongly called proatlantal intercentrum). Centrum odontoid. Second basiventral or intercentrum is small, indicating complete reduction of the others. The ribs of this second vertebra are very small and are carried by its basiventral; the following ribs show a gradual transference on to the centra, until these still holocephalous ribs are carried by the typical neural diapophyses. The third, fourth, and fifth ribs are several inches long, very stout, broad and overlapping; the sixth is suddenly longer, not dilated, and diapophysial. The thoracic and other presacrals are remarkable for additional interlocking joints. They remind one of the zygosphenes and zygantra of Snakes and Iguanas, but here they are in exactly the reverse position, the sphene arising from between the postzygapophyses and the antra sunk into the anterior half of the vertebra (Fig. 72). They were first noticed by Cope in Dinosaurs; later also in *Placodus*. They are absent in *Limnoscelis*. The term Hyposphene used by various palaeontologists being preoccupied, and therefore inapplicable, they are now named Metasphenes and Metantra.

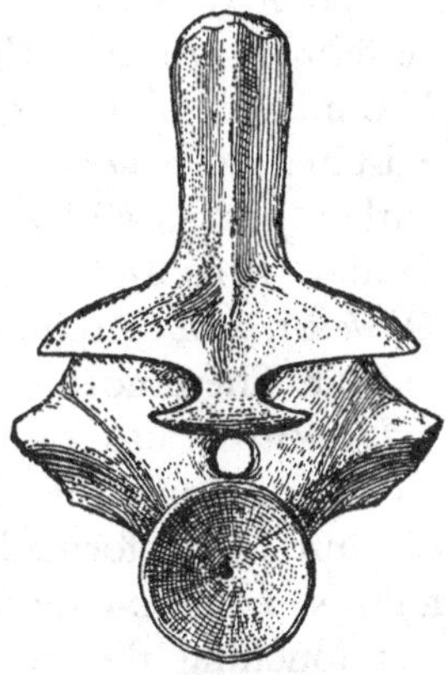

Fig. 72. *Diadectes* (Cotylosauria). Dorsal vertebra, from behind, showing diapophyses, postzygapophyses, and metasphene. (Williston.)

It is doubtful whether the first caudal, following upon the only sacral, carried a pair of ribs. Then four or five of the anterior caudals carry slender, short transverse ribs which cease farther back, being indicated only by short, low stumps. Coincident with the reduction of these caudal ribs is the sudden appearance of typical chevrons, the first of which is wedged in between the fourth and fifth caudal vertebrae.

Limnoscelis paludis (Fig. 73). Complete skeletons of these sharp-toothed creatures are figured and well described by Williston. Total length about 2 m. Tail long.

The vertebrae are deeply amphicoelous; at least 26, perhaps as many as 28 presacrals, one sacral and many caudals. The existence of a proatlas is very doubtful. The atlas has a large basidorsal and a large basiventral, which bears a rib upon a backwardly directed process. There are ribs on all the vertebrae, from the atlas to the eleventh caudal. The cervical ribs are remarkable, like those of *Diadectes*. The second to the fifth increase from 44 mm. to about 88 mm. in length; all broadened, especially distally. Williston distinguished this complex as *notarial ribs* and *notarial vertebrae*. The seventh rib seems to be the last of this compound, the eighth being apparently much longer, 6 in., more like typical thoracics, and these, although not sternal, decrease towards the sacrum to short, feeble lumbar ribs. It may be remarked here, as Williston has pointed out, that typical ribless lumbar vertebrae are not the prerogative of the Mammalia, since several such vertebrae exist in *Labidosaurus* and *Captorhinus*. The same author has carefully described and figured caudal ribs and chevrons of *Limnoscelis*, *Diadectes*, and of their contemporary Pelycosaurian Theromorph, *Varanosaurus*.

Limnoscelis has only one strong, typical sacral vertebra, the connexion of ilia with column being formed by stout ribs. Then follows a sacro-caudal, much weaker and just touching the pelvis. It naturally suggests a transitional stage in sacral formation. Then follow in the same horizontal plane, transverse, slightly curved ribs, attached by sutures to the diapophyses of the second to eleventh caudal, where they cease, but leave little stumps on the respective arches. Quite separate from these undoubted ribs arise the typical chevrons; the first between the third and fourth caudal, the others continued on the long tail. The "anterior chevrons are quite sessile, with broad sutures on the ventral side of the centra"; as in many Reptiles, these chevrons are attached to the posterior end of the preceding vertebra, to which they do not belong genetically. Farther back,

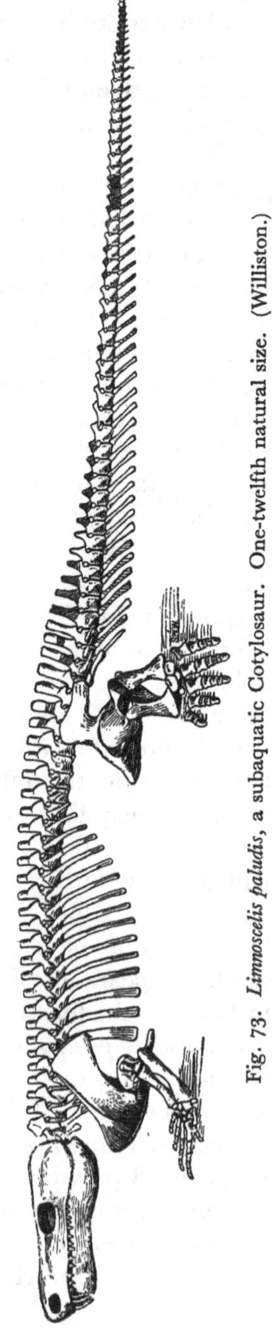

Fig. 73. *Limnoscelis paludis*, a subaquatic Cotylosaur. One-twelfth natural size. (Williston.)

Fig. 74. *Pareiasaurus Baini* (Seeley).

the chevrons are simply wedged in between the centra. *Diadectes* seems to have only four caudal ribs, and the first chevron lies between the third and fourth vertebra; and, since there is only one sacral, no sacro-caudal, this genus may in these respects be considered more specialised than the generally more primitive *Limnoscelis*.

Pareiasaurus Baini (Fig. 74). Mid-Permian in South Africa, Upper Permian in Russia. A massive, heavily built creature with a proper neck of eight to nine vertebrae; 18 presacral, two or three sacral and about two dozen stunted tail vertebrae. Some of the posterior cervical ribs are about 18 in. long, straight and turned horizontally backwards, and so much broadened as to overlap each other. Pelvis tilted into preacetabular position.

Cope (*American Naturalist*, 1880, p. 304), writing about the skull of *Empedocles*, one of the synonyms of *Diadectes*, refers to certain cup-like facets, *cotyli*, in the occipital joint, which struck him as rather Mammalian, "and the character of this articulation is so distinct from anything yet known among vertebrated animals that I feel justified in proposing a new division for the Theromorpha to include the Diadectidae, to be called *Cotylosauria*". The doubtful feature was later explained away and dropped out of the definition; but the term remains apparently as a vague allusion to the "stem" out of which sprouted the Reptiles like the cotyledons from a bean.

Chapter XXII

CHELONIA

THE MOST PRIMITIVE EXISTING THEROMORPHA

The chief characteristics of this group can be described as follows. The broad trunk is encased by a shell composed of a ventral plastron and a dorsal carapace formed by numerous osteodermal plates, overlaid by horny shields which are rarely vestigial or may even be absent. Toothless; four limbed; pelvis complete with ischiad and pubic symphyses. No cervical ribs; the others with capitulum only.

More than two hundred recent species exist; the order is cosmopolitan. The earliest known specimens, which occur in the Mid-Trias, were already typical terrestrial Tortoises. During the Cretaceous epoch some began to adapt themselves to an aquatic life, transforming their hind-limbs, and especially the enlarged fore-limbs, into flippers; these earlier marine forms died out, but a second attempt in the Tertiary epoch became established as the group of Turtles.

The number of vertebrae in recent Chelonians is eight cervical, nine thoracic presacral, two sacral, and a variable number of caudal; total precaudal, 19. These numbers are very constant (Fig. 75). The centra are elongated, longer than they are broad. Traces of the chorda remain longest in the middle of the centra. The intervertebral joints are in a very unsettled condition, amphi-, opistho- or procoelous, even biconvex vertebrae occurring in the neck. Intercentra are in most cases restricted to the atlas and axis, though sometimes found also on the third cervical vertebra; they reappear in the tail as paired or unpaired nodules, or as short chevrons which tend to fuse on to the caudal end of the preceding vertebra. Fibro-cartilaginous intercentral disks occur regularly in the neck and tail; they disappear in the immovable trunk.

The atlas and second vertebra are of great interest because of their diversity. The atlas of many remains throughout life ideally temnospondylous: the large BV 1, IV 1 and IV 2 of the Trionychidae remain free and all lie in the same horizontal level (Fig. 76). BV 2 is absent in *Chitra*; present but fused on to the end of IV 1 in *Trionyx hurum*. In the pleurodirous *Chelys*, the *Matamata*, BV 1 is very small and its BD fuses completely with IV 1. In the allied genus *Platemys*, BV 1 is lost and the neural arch is firmly sutured on to its centre, that is IV 1 (Fig. 35). Thus this atlas is reduced to two pairs of elements and forms one physiological unit, and the basioccipital portion of the condyle articulates directly with the centrum, which looks more like a normal centrum than an odontoid. The lateral occipital condylar portions always articulate with the first neural arch. The first

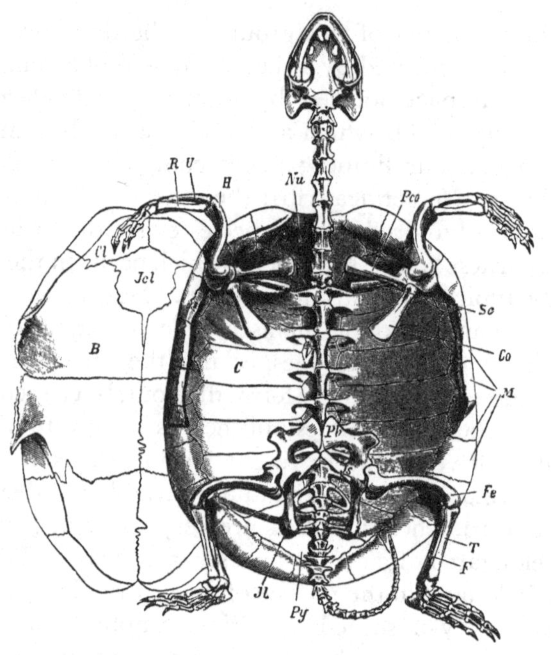

Fig. 75. Skeleton of *Cistudo* (*Emys*) *europaea*. *V*, vertebral (neural) plates; *C*, costal plates; *M*, marginal plates; *Nu*, nuchal plate; *Py*, pygal plates; *B*, plastron (ventral shield); *Cl*, clavicle; *Jcl*, inter-clavicle; *Sc*, scapula; *Co*, coracoid; *Pco*, acromial process (pro-coracoid); *Pb*, pubis; *Js*, ischium; *Jl*, ilium; *H*, humerus; *R*, radius; *U*, ulna; *Fe*, femur; *T*, tibia; *F*, fibula. (Claus-Sedgwick.)

trochoid or pivot joint of *Platemys* is absolutely intervertebral. The Trionychidae, with their loose opisthocoelous centre, have a double trochoid joint, which is intervertebral between IV 1 and IV 2, but intravertebral between BV 1 and the odontoid. The latter joint is strictly homologous with that of the Mammalia, even to the absence of the BV 2. Two widely different groups, like the soft-shelled Turtles and Mammalia, have hit upon the same combination and permutation of the same number of component units with which all

Amniota must have started. If the game is played long enough, any result within morphological reason is sure to happen repeatedly.

The last, or eighth, cervical vertebra forms elaborate joints owing to the retractable neck. Its centre fits with a knob into a cup of the ninth vertebra, and the postzygapophyses form broad, curved articular facets for the anterior zygapophyses of the ninth, which vertebra is in its turn fixed immovably on to the shell by the strong ribs of the tenth vertebra (so-called first ribs) and sometimes by its own short ribs. These zygapophysial joints are most elaborate in the Trionychidae, where they form the only articulations with the ninth vertebra, for the centra do not join but become separated by partial resorption. Inspected alone they are puzzling specimens. In the Turtles all joints of the non-retractile and short neck are much reduced.

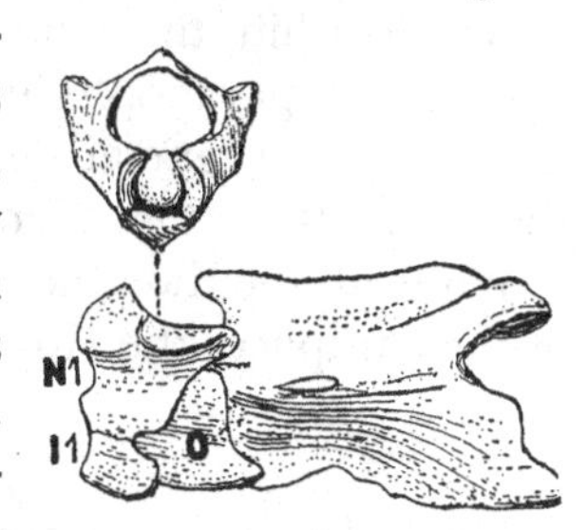

Fig. 76. *Chrysemys* (Chelonia). (Williston.) *N* 1, basidorsal 1; *I* 1, basiventral 1; *O*, interventral 1.

The neural arch of the ninth vertebra rests upon its own centrum, but those of the other trunk vertebrae have extended their base over two adjoining centra, retaining like the ribs their original intercentral position. In most cases the neurocentral sutures remain throughout life. Ribs are entirely suppressed on the eight cervicals, but "transverse" processes exist in the Pleurodira, i.e. those Tortoises which hide the neck by tucking it sideways into the gap between the upper and lower shell, the space also of the withdrawn fore-limb. In the Cryptodira, which withdraw the neck "straight" into the shell, such processes are vestigial or absent, according to the development of the musculus lateralis cervicis and the musculus cervicalis ascendens. Owing to the suppression of the intercentral disks in the thoracic region, the ribs naturally come to lie in a pronounced intervertebral position, all the more since they are reduced to their capitula, which are broadened out and wedged in ventrally, on the right and left sides, between the otherwise closely touching centra; the tubercula are completely abolished as no longer useful, or as impossible owing to the formation of the rigid carapace.

The first pair of thoracic ribs, those borne by the ninth vertebra, are peculiar. They arise from the anterior portion of the centrum; they are much reduced, sometimes to mere threads of bone, and they lean against the anterior rim of the second pair of ribs, mostly without reaching the carapace. They are vestiges of the transitional stage when the carapace began to consolidate itself, with its neural and costal dermal plates getting hold of the spinous processes and of the ribs. There is no lumbar region, and the posterior thoracic ribs gradually shift their base backwards upon their respective centra, upon short parapophyses which creep dorsally until in the tail they pass the neurocentral suture; and the short ribs are then carried by diapophyses on to which they fuse.

While all the thoracic ribs are intimately involved in the formation of the carapace, the ribs of the two weak and short sacral vertebrae

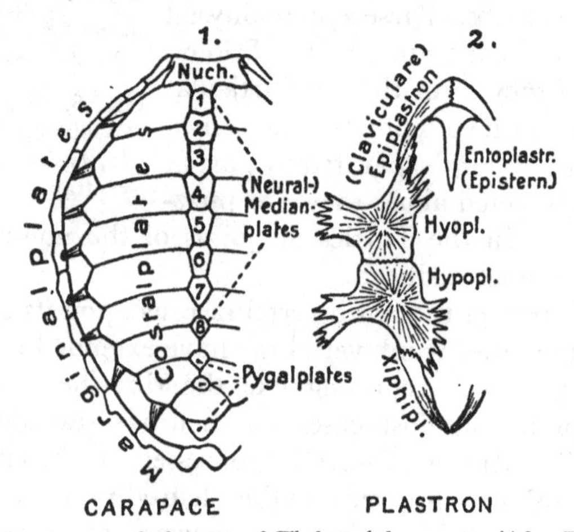

Fig. 77. Carapace and plastron of *Thalassochelys caretta*. (After Buetschli.)

sometimes remain quite independent throughout life, just touching the upper ends of the iliac bones; but since these find a much more effective support in the shell, the distal ends of both sacral vertebrae come to fuse on to the inner surface of the eighth, so-called last pair of costal plates; so called since individual variation indicates that this pair itself is the compound of originally an eighth and ninth pair.

The developing carapace profoundly affects the vertebrae and ribs (Fig. 77). It arises from three distinct sources:

(1) A median, longitudinal series of neural plates which are the horizontally broadened spinous processes, implying the suppression of the zygapophyses from the tenth vertebra backwards, all of which fuse with each other and the shell.

(2) The so-called costal plates, in reality the ribs which in their

fullest development are horizontally broadened or flattened, so that the successive ribs are in sutural contact with each other throughout their length (e.g. *Testudo*); or the broadening and contact is restricted to the proximal third (*Toxochelys*) or even less (*Archelon*); while in *Protosphargis* and *Sphargis* the ribs are of almost normal shape and widely separated from each other. The distal ends of the ribs may or may not reach the marginal plates.

(3) The marginalia, which are true dermal bony plates formed laterally. In some fossil and recent genera the ribs are too short to reach these marginalia and the resulting gap is then filled in by a series of additional dermal bony plates known as the supramarginalia. It has been and is still frequently supposed that similar larger dermal plates overlaid and came into contact with the ribs, surrounded, invaded, and supplanted their cartilaginous endoskeleton, and thus accounted for their broadening into the costal plates. It could scarcely be assumed that this process of conversion affected the whole rib, including its plainly discernible neck and capitulum. However, the more or less blurred ontogeny often gives the impression as if a bony plate had fused with and partly made a cast of the rib.

A secondary result of the formation of the stiff unyielding box-like carapace is the suppression of all axial, intercostal, and similar muscles, not because these were squeezed out of existence by the sinking in plates, but because the whole affected region was made immovable. Yet considerable traces of these muscles still exist in young specimens. In many recent Tortoises, especially in the large Land Tortoises, the vertebrae and the capitular portions of the ribs are reduced to mere bony outlines; the reduction to thin, paper-like bony lamellae proceeds with age.

The number of costal plates is very constant, eight pairs. Some fossils have nine or even ten, the reduction having since taken place in the sacral and the postsacral region. A similar shortening of the carapace may account for the fact that the ninth vertebra, with its vestigial pair of cervical ribs, has no corresponding costal plate of its own. Its neural plate is called the nuchal and is often held to be not serially homologous with the other neurals, apparently because it is assumed that the neurals as parts of the vertebrae broaden at the top and are transformed into the plates, just as the costals are founded upon the ribs, while the nuchal and the various pygals are taken to be only pure dermal ossifications; cf. Zittel, new edition.

The nuchal plate, about which so much uncertainty prevails, lies mostly in front of the ninth vertebra; it overlies the last cervical ribs, with which it is connected by ligament only, but the posterior corner of the plate often fuses with the spine of the ninth vertebra. In the Chelydridae or Snapping Turtles, a distinctly ancient family, and still more in the equally ancient Trionychidae, the nuchal sends out a pair of long rib-like processes which either extend laterally beyond some of the marginals, or their ends overlap those of the second thoracic vertebra, or they in turn are overlapped by the first costal plates, e.g. in *Cyclanorbis*. Such rib-like processes are present also in the Cryptodirous *Cinosternum*. The explanation may therefore be that the nuchal plate in reality represents the neural plate of the eighth vertebra fused with the otherwise wanting costal plates of the ninth, which in this case has, by dorsal growth, squeezed out the neural plate of the ninth vertebra. That neurals can be squeezed out of existence by the mutual approach of right and left costal plates is amply shown by many Trionychidae; in *Cyclanorbis* the whole neural series is actually reduced to two. The inferred compound nature of the nuchal is strongly supported by the description of two nuchals in *Clepsydropsis carinata*, a Miocene relation of *Chelydra*.

It is not hopeless to reconstruct upon the above data the hypothetical primordial Chelonian, contemporary with Pelycosauri and Triassic Parasuchia.

Beyond the eight cervicals, at least 12 metameres were distinguished by greater development of the dermal plates, the series of which originally reached, Crocodile fashion, from head to tail. In the trunk region the median series fused on to the spinous processes, forming with them the neural plates. The first pair of costals was then suppressed or perhaps never had a chance of further development, owing to the concentration or shortening of the forming carapace. This concentration fore and aft suppressed even the three last costals, leaving their neural plates undisturbed. The ribs of the eighteenth to twentieth vertebrae were retained in their original condition for the attachment of the pelvis. The latter in time moving forwards caused the posterior neural plates to turn into "suprapygals" or "supracaudal plates". This forward move of the pelvis accounts further for the occasional occurrence and weakness of a third sacral rib. In fact, the whole partly co-ossified and very variable "synsacral" region is an atavistic reminiscence.

Sphargis or *Dermochelys coriacea*, the Leathery Turtle, deserves special discussion because its peculiar shell intimately concerns our conception of the evolution of the Chelonia as an order and the systematic position of *Sphargis* itself.

This Turtle, of which fossils are not known, is cosmopolitan in the tropical seas, centred in the Indian and Western Atlantic oceans, whence it strays occasionally to the coasts of Europe. It is the largest of recent Turtles. Instead of being covered with horny shields it is naked, and instead of being encased in a carapace composed of neural and costal bony plates which intimately affect the vertebral column and the ribs, the whole trunk is wrapped up in a shell-shaped coat of thick corium, the middle stratum of which contains hundreds of dermal ossifications, which form a mosaic of platelets (Fig. 78). This shell is not in contact with the internal skeleton; its inner and outer strata are leathery cutis, the outside of course being covered by the epidermis, but this is devoid of horny scales or shields. The mosaic platelets of the rather flexible case are not of uniform size, but arranged in 12 longitudinal series, or ridges, separated by smaller platelets: a median dorsal row reminding one of neurals; three right and left lateral rows; on the underside two pairs of lateral rows and one ventral median row. This arrangement is best studied in very young specimens. Near the front end of the plastral region and armpits is an irregular number of smaller additional platelets to fill up the gap. The same applies to the chief lateral, dorsal and ventral rows. The navel has actually prevented the otherwise single median row from joining, so that there is in the young a partly paired median row. In the tail end of the dorsal shield the total number of rows decreases from seven to five to two, and only one on the very short tail. It is quite possible that both the dorsal and ventral median

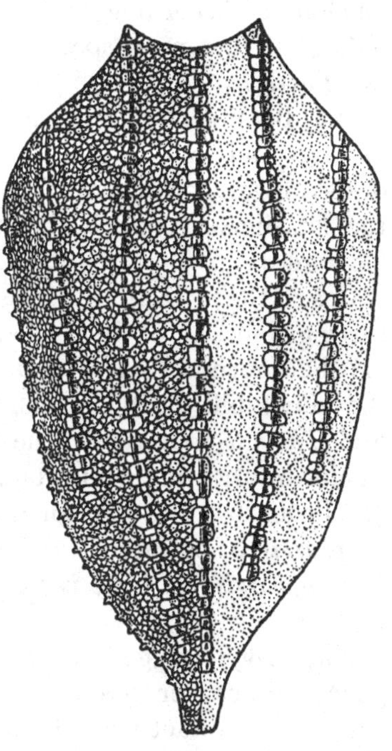

Fig. 78. Superficial surface of the carapace of *Dermochelys*: the small plates and the outermost row of larger plates are only shown on the left side. (After Versluys.)

rows were originally paired, just as the spinous processes are the result of fusion of the prolonged right and left basidorsals.

Sphargis has become the subject of a still growing, somewhat animated literature.

Cope, Dollo, and Boulenger hold that *Sphargis* is the sole living representative of a primitive athecal group, in opposition to all other Chelonia. According to Baur, the late Chelonological enthusiast, *Sphargis* is the most specialised member of the recent Chelonidae, i.e. Turtles.

van Bemmelen (1896) modified this view in so far that both *Sphargis* and the Chelonidae have arisen from two different groups of terrestrial Tortoises, with parallel or convergent evolution, so that the Leathery Turtle is the most specialised Chelonian but not a Chelonid.

Case (1897) inclines to Baur's view, and further considers that three groups, all of them Chelonidae, started in different directions: (1) The normal *Chelone viridis* or common Turtle; *Chelone imbricata*, the Hawksbill Turtle, which yields the tortoise-shell of commerce; and *Thalattochelys caretta*, the Loggerhead Turtle, carnivorous, with shields imbricating while young, smooth and juxtaposed in the adult. (2) From *Lytoloma*-like forms through the extinct *Psephophorus* and *Eosphargis* to *Sphargis*. (3) The extinct *Protostega*, *Protosphargis* and *Pseudosphargis*. If Case is right, the athecal condition is not fundamental.

Hay (1898) suggested that the osteoderms of *Sphargis* represent a generation earlier than the bony plates of the *Thecophora*, which have come into contact with the neural arches and ribs. The osteoderms of *Sphargis* are homologous with the dermal osseous scutes of the Crocodiles, and since this Turtle has no other shell than its mosaic layer, it must either have lost, or perhaps it never possessed, horny shields, nor can it have developed a second or later set of plates.

Versluys (1914) makes the mistake concerning the meaning of "theca". The carapace is formed by dermal elements which arise out of (1) the deeper layer of the cutis, now called thecal (e.g. the neural and costal plates, including nuchal and pygals); (2) the more superficial layer called epithecal, e.g. the marginals.

Jaeckel (1914) accepts Goette's view (1899) that the neurals and costalia belong to the inner skeleton, and that the costal plates are nothing but the broadened-out ribs themselves, while the neuralia are the broadened neurapophyses. Here we may remark that

Goette's interpretations of his own splendid and numerous observations were not readily accepted by Gegenbaur's school.

Abel (1919) accepted Dollo's (1901) suggestion that, after the ancestral terrestrial Tortoises had acquired the typical carapace, some of their descendants adapted themselves to aquatic, ultimately oceanic, life; a change which caused weakening, eventually suppression, of the deeper osteodermal layer (called thecal by Abel, following Versluys). Then some of these marine forms returned to littoral life with its surf, and substituted a new, more superficial osteodermal layer, the "epithecal" mosaic for the lost original carapace. As a strict believer in his own law the famous Belgian palaeontologist was obliged to resort to this assumption. Abel winds up the adventurous story with the following sentence, here translated: "But *Dermochelys* (*Sphargis*) has again become an inhabitant of the High Seas, therefore requires the secondary bony shield just as little as the primary, and therefore has lost also the second". Now, if the creature loses both kinds of shell, how can its complete mosaic shell be accounted for?

The whole question has become strangely complicated. Every one of the different views contains some reasonable points of importance. Even the assumption of two osteodermal layers seems clearly proven by the subsequent discovery of the Upper Cretaceous marine *Toxochelys bauri* with three unmistakable osseous buckles fixed on to the top of three undoubted neuralia; the bases of the buckles overlap the interneural sutures of vertebrae 12/13, 14/15 and 17/18. In the likewise Upper Cretaceous marine *Archelon ischyros*, about 12 such additional buckles make a continuous series. They cannot be incipient plates, and there are no costal plates, only ribs. The additional elements must therefore be vestigial.

We are driven to the following solution. All Crocodilian and Chelonian osteoderms belong to the same category; scutes consisting of an osseous base or core, invested by horny shields. Those of the Crocodiles are quite independent of the internal skeleton excepting the head. It may be noted that the various kinds of Caiman differ from the closely allied Alligators by a weak ventral armour composed of thin, overlapping bony scutes, each of which is formed of two parts united by a suture. This armour is continuous with the scutes on the sides and back of the trunk. Further, in the Jurassic Teleosaurs the abdomen is protected by a large oval armour, a mosaic of extra strong bony units. The armour of the Chelonia marks an advance: the endoskeleton, the vertebrae and ribs have been drawn into the composition of a carapace, of which it forms the series of median or neural, and lateral or costal, plates. All the rest of the carapace is composed of typical osteodermal plates, to which belong the marginals, the supramarginals which may be eventually developed, and the additional buckles upon the neural plates, still preserved in *Toxochelys* and *Archenteron*.

In some of those Chelonians which have become marine, flattened out, the shell or armour is weakened, this weakening involving both the endoskeletal and the osteodermal elements. The latter form the mosaic layer which, weakened in strength, becomes reduced to the middle layer of the corium. Its thinness was compensated by the completeness of the mosaic layer. Its withdrawal from the surface, owing to its reduction in thickness, further reacted unfavourably upon the formation of the horny shields, which have ceased to exist in *Sphargis*, as is also the case in the mud-loving, freshwater *Trionyx*, where the loss of tegumentary protection has reached its limit.

A detailed account of the progressive reduction and elimination of the costal and neural plates, and of the shields which originally agreed in their numbers with the 10 to 12 or more still existing marginals, has been published in Willey's *Zoological Results*, pt. III, 1899, under the title of "Orthogenetic variation in the shells of Chelonia".

If too much was made of Crocodilian resemblances in reconstructing the ancestral Chelonian, this was done because these are the only orders which allow direct comparison, and terrestrial Triassic Tortoises were still unknown in 1899.

In most of our recent schemes of pedigrees of the Reptiles, the Tortoises and Ichthyosaurs are treated as if they had no ancestors. For some time the Dicynodont Theromorpha were the favourites. Recently they have got into the "Formenkreis" of the Tocosauri or Lepidosauri with Lizards and Snakes, which seems an almost impossible arrangement; authors have perhaps been lured on by the Rhynchocephali. The Chelonia fit much better into the Archosauri and need not go lower than the Pseudosuchian level. Reluctance to place them with the Diapsids is not difficult to overcome. The widely open temporal region is the result of various reductions. Judicious horizontal splitting of the double bridge of *Emys*, for instance (the double bridge being formed by the postfrontal and the squamosum and by the maxilla plus jugal plus quadratojugal plus squamosum), could result in an upper and lower temporal fossa. Analogous changes apply to the post-temporal bridge, which is composed either of squamosum plus supraoccipital or of squamosum plus parietal. Such a bridge exists in *Sphenodon*, spanning a wide fossa; it is obvious in *Belodonts*; in Crocodiles the thickening bridge squeezes out the passage, in adults almost beyond recognition. On the other hand, in *Hydromedusa*, a South American *Pleurodire*, the bridge is formed by a long and slender arm of the squamosum meeting the supraoccipital. In other Tortoises the bridge is gone, so that the temporal and post-temporal fossae are confluent. And if, as in *Cistudo*, the whole jugal arch vanishes, the supra-, infra- and post-temporal fossae are joined into one great unit. The extreme opposite of such a skull is that of *Chelone*, with its false roof formed by the broadened squamosal plus post-frontal meeting the parietal, and thus producing a secondary Stegocephalous skull, at least in outward appearance; the moving cause being the masticatory muscles, which require a wider area for attachment.

Lastly, the Crocodilia and Chelonia are distinguished by the unpaired, ventral, anterior penis and a longitudinal slit-like outer cloacal opening; "orthotrema" of older zoologists. *Sphenodon* has no special copulatory organ, an ancient condition to judge from the structure of its cloaca. In sharp opposition are the Lizards and Snakes, with a transverse cloacal opening, hence "plagiotrema", and with a right and left penis, reversibly withdrawn into the tail; therefore lateral and post-trematic. The male copulatory organ of some Birds is a semi-penis, asymmetrical, the right being abolished; in the majority it is very vestigial or absent.

Chapter XXIII

PELYCOSAURI

THE PRIMITIVE SYNAPSID THEROMORPHA

This group consists of Reptiles with a single temporal fossa, namely the lower one, which is bounded above by the postorbital and squamosal, and below by the jugal and squamosal bones. When, as in *Varanosaurus*, the jugal bridge is lost, the arrangement somewhat resembles that of the Sauropterygia and Lepidosauria. The remaining cranial bones have a smooth surface, no longer sculptured like those of the anapsid Cotylosauri. The quadrato-jugal is present, connecting the jugal with the typically reptilian quadrate.

The vertebrae are amphicoelous; the basiventrals are reduced to intercentra, mostly small, which are present from the atlas to the end of the trunk, absent in the sacral vertebrae, which are usually two in number, but reappearing behind these as chevrons. The proatlas articulates with the atlas and with the lateral occipitals. Ribs, if clearly two-headed, have the capitulum still intervertebral and the tuberculum carried by the diapophysis.

These Pelycosaurs are a tolerably natural assembly when taken in the present restricted sense. They seem to be essentially American, Lower Permian; only the highly specialised genus *Naosaurus* is known also from Central Europe. Apparently they left no descendants. A flourishing group of short duration: their average length from nose to sacrum was 2 to 5 ft. with a very variable tail.

Four typical examples will now be described in detail.

Varanosaurus (Fig. 79*A* and *B*). The same shape as *Monitors*; total length about 3 ft., with a long tapering tail. About 27 presacral, two sacral and about 50 caudal vertebrae; amphicoelous. Intercentra, BV, present in the neck and the rest of the trunk, but apparently absent between the second sacral and the first caudal and between the first and second caudal; they reappear as typical chevrons between the second and third caudal and then are continued far into the tail.

There are at least six fair-sized cervical ribs, increasing in length from the atlas backwards; that of the seventh vertebra is obviously thoracic. In the lumbar region the ribs become vestigial; those of

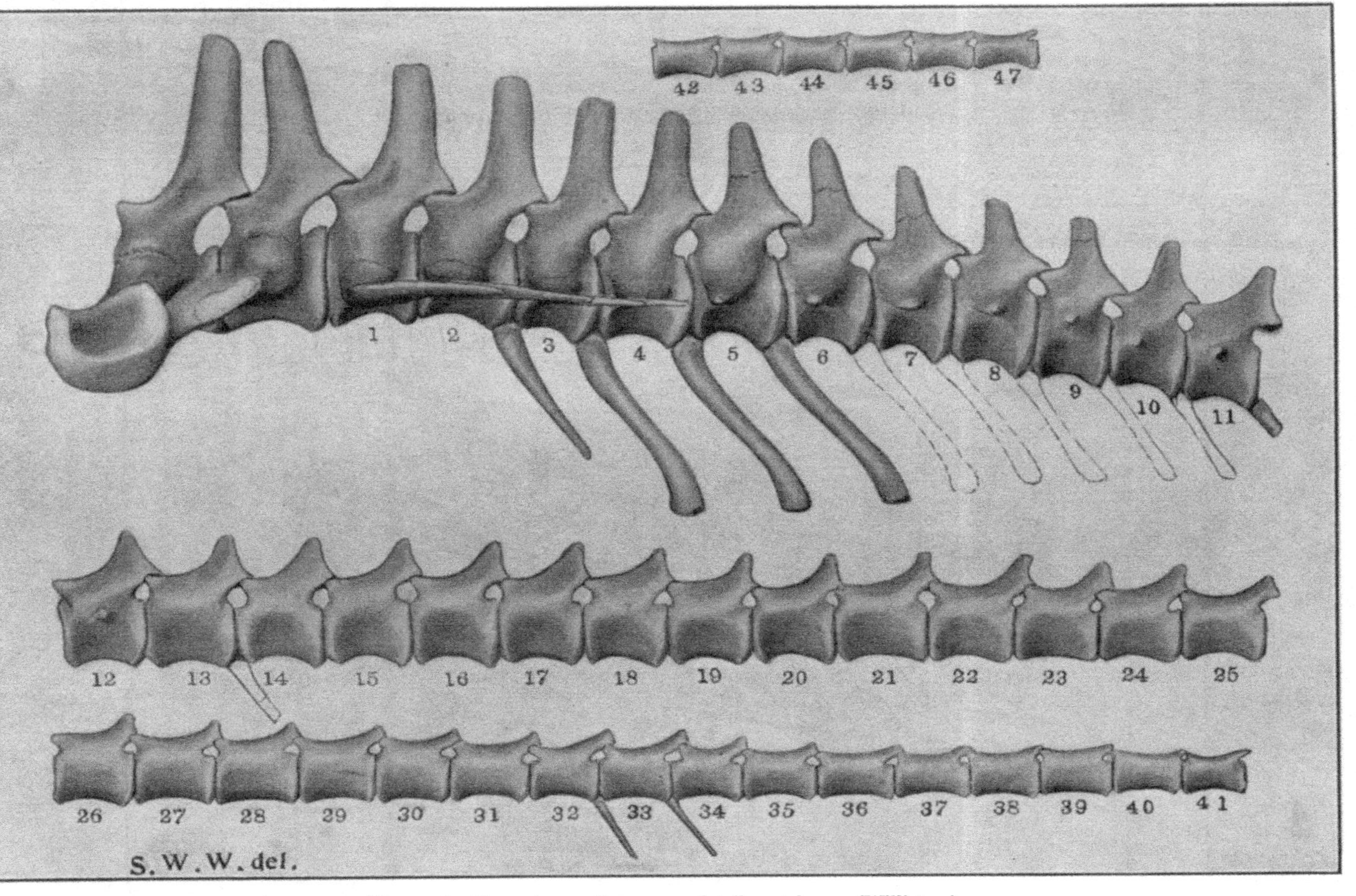

Fig. 79 *A*. *Varanosaurus*. Sacrum and tail vertebrae. (Williston.)

Fig. 79*B*. *Varanosaurus*. Restoration and skeleton. (Williston.)

the sacrum are stout. The first to fourth caudal vertebrae likewise carry ribs, slender, horizontally extending backwards, and, to judge from Williston's careful drawing (Fig. 79 *A*), arising from the centrum, well below the neurocentral suture. From the fifth to the twelfth caudal vertebrae he figures what may be vestigial transverse processes. The only articulations in the column are typical zygapophysial ones.

Ophiacodon (Fig. 80). Nearly twice the length of *Varanosaurus*, with the same vertebral formula. Ribs still holocephalous, those of the sixth or seventh cervicals increasing in length to the thorax, decreasing to vanishing point in the lumbar region, reappearing on the anterior caudal vertebrae and followed by typical chevrons on the rest of the long tail.

The atlas region (Fig. 81) requires careful revision and interpretation. There is a proatlas, unusually complete, since it contains a pair of neural elements loosely articulating with the basidorsal of the atlas and possibly with the occiput. These neurals are widely separated from their ventral complement, which is a small apparently median piece attached to the fair sized BV 1, and the latter is fused with the BD 1 of the atlas. These neural arches of the atlas embrace the large epistropheus, which carries a strong, blade-like spinous process. The first rib, that of the atlas, still arises from between the ventral posterior corner of BV 1 and the centre of the atlas, which is sutured to, not fused on to, the centre of the epistropheus. The second basiventral behaves like a typical intercentrum, the third is quite hidden, the others are freely exposed. The second rib is in a state intermediate between holocephalous and dichocephalous; the capitular portion is still connected with its BV 2, although only perhaps by ligaments, while there is a stout shoulder which articulates with the diapophysis of the epistropheus. The third rib is carried entirely by the diapophysis, as are the following ribs.

Dimetrodon (Fig. 82). One of the most bizarre creatures because of the neural spines which, although slender, are rapidly elongated from the second vertebra onwards until those of the middle of the trunk reach a length of 3 ft. and then steadily decline towards the tail. The creature itself measured about 8 ft., of which more than half belongs to the skull and the comparatively normal tail. Looked at as a whole, these spines irresistibly suggest a sail, which, however, could not be lowered, the neural arches being firmly sutured or even fused with the vertebrae and the apparent absence of good zygapophysial joints suggests stiffness. The whole structure was unfit for

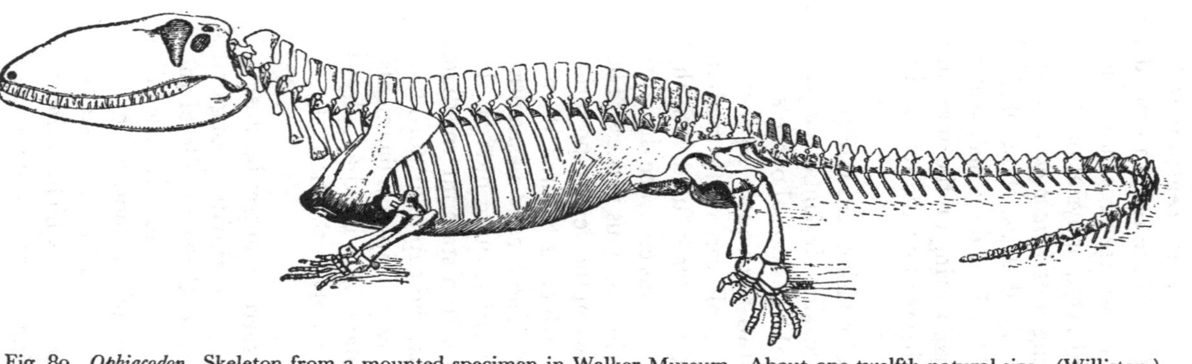

Fig. 80. *Ophiacodon*. Skeleton from a mounted specimen in Walker Museum. About one-twelfth natural size. (Williston.)

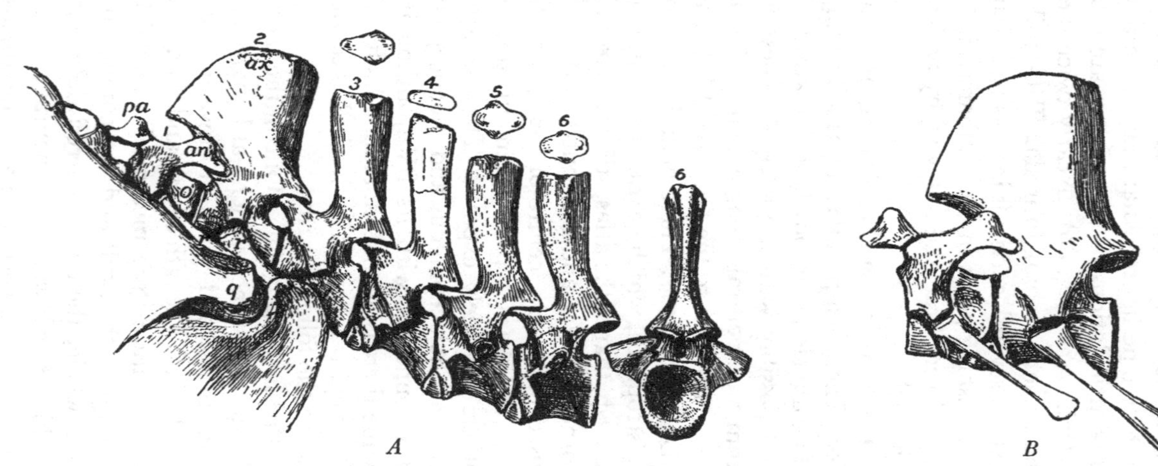

Fig. 81. *Ophiacodon*. *A*. Notochordal cervical vertebrae, with intercentra. *pa*, proatlas; *an*, arch of atlas; *o*, odontoid; *ax*, axis; *r*, rib; *q*, quadrate. *B*. Proatlas, atlas, axis, and ribs. (Williston.) Cp. Fig. 31.

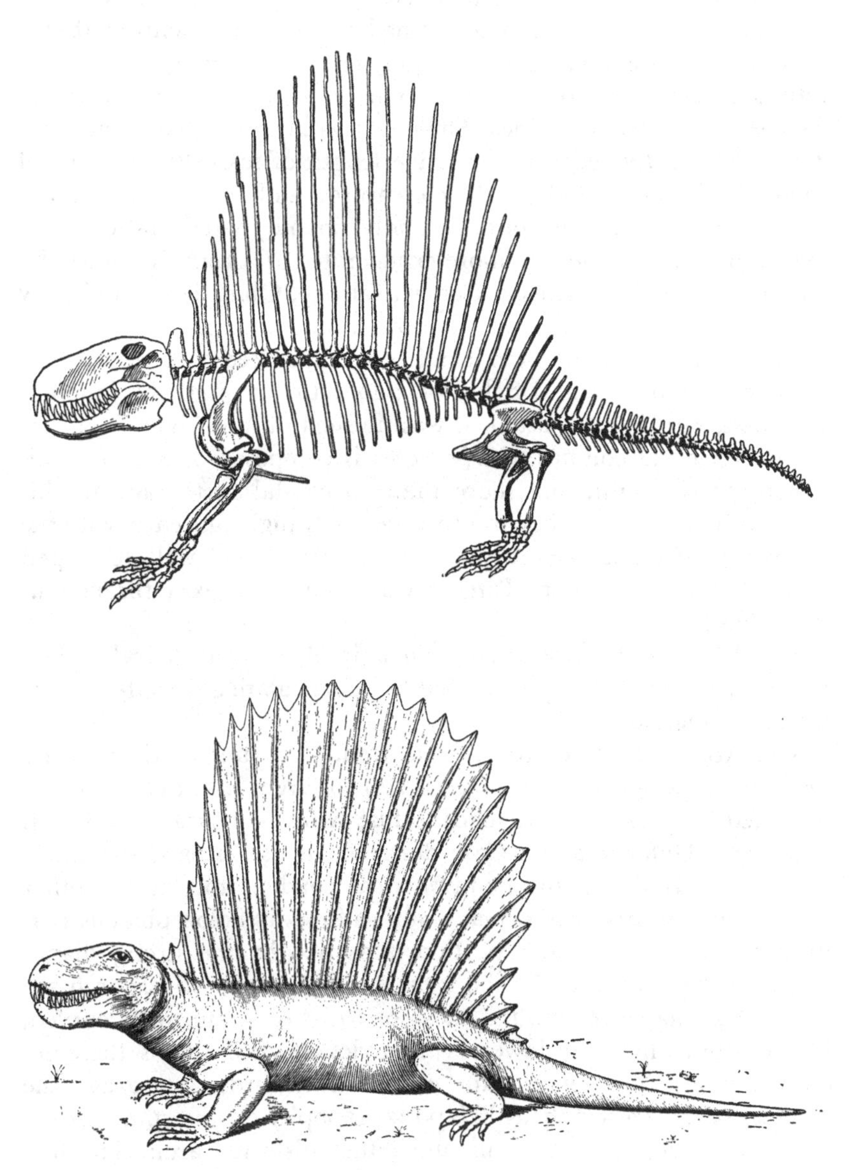

Fig. 82. *Dimetrodon.* Skeleton and life restoration. About 8 ft. long. (Williston.)

offence or defence, an excellent case of hypertely, the result of cumulative inheritance having gained so much momentum that a once possibly useful structure, as spiny crests are, brought about the ultimate destruction of its owner. This view, though reasonable, is hampered by two difficulties. The neural spines of the contemporary genus *Sphenacodon* reached a relative length of scarcely one-sixth of those of *Dimetrodon*, and yet these creatures, and others less endowed than *Dimetrodon*, also died out. Secondly, not only had the latter a very long innings, but the still more exaggerated *Naosaurus* is known by various species from Russia, Central Europe, Ohio, Texas and New Mexico; thus there was a distribution in time and space of what was more than an ephemeral and localised freak, indeed a successful type in which it took Natural Selection a long time to find the inherent weakness, but then, as usual, it was finished off quickly.

The vertebral column of *Dimetrodon* is composed of 27 presacral, two or three sacral, and more than 30 caudal vertebrae; amphicoelous. The neural arches, mostly excessively high, fuse early with the centra; prezygapophyses and postzygapophyses and well-developed diapophyses are present. Intercentra or chevrons exist throughout the column.

The Proatlas is represented by a pair of vestigial neural arches, which seem to have a right and left facet articulating with the arch of the atlas (Fig. 30).

The Atlas consists of the paired arch, large unpaired or rather fused BV 1, which median piece is still so large that it can scarcely be called intercentrum. It tends to fuse with the neural arch which it carries. Unfortunately, Case described it as "proatlantal intercentrum". It lodges the hemispherical condyle and at the other end the first or atlas rib, which is still so broadly holocephalous that its tubercular or shoulder portion is carried by the strong diapophysis of the neural arch. The right and left halves of this arch are of peculiar shape, resembling those of *Ophiacodon*, but more curved, like a boomerang; the lower arm carries the rib, and is therefore diapophysial; the upper arms extend backwards and grasp the pedicles of the epistropheus like postzygapophyses; the spine proper seems to be suppressed in this altogether distorted arch. The first centrum carries nothing; since it is so large in the dorso-ventral direction and does not reach the occiput, it suggests that it is still a centrum in an ideally complete condition, containing both interdorsal and interventral elements, cf. the centra of *Seymouria*.

The Epistropheus is, like that of *Ophiacodon*, remarkable for its strong, blade-like spinous process. The zygapophyses are well developed; the diapophyses are strong and directed downwards, carrying the tubercular portion of the second rib, the capitulum of which has left a facet on the postero-ventral portion of the fair-sized BV 2 or intercentrum.

Case figures and describes an "anterior cervical vertebra", the intercentrum of which is so large that it forms a considerable portion of the intervertebral joint facet, the vertebra being markedly procoelous. Of course amphicoelous vertebrae are concave at either end. Nevertheless, it is valuable information, showing how these joints can be moulded on a vertebra in accommodating the centre to the size of its own basiventral. In this case the intercentrum acts as a swivel to its own centrum and seemingly we have, in this part of the neck, the rare instance of an intravertebral joint.

All the ribs are remarkably long, not dilated, curved down and steadily increasing in length, so that there is no distinction between cervical and thoracic, especially since there is no trace of costal portions, nor of a costal sternum, possibly because these parts were still cartilaginous. The strong scapula is slung on to the region of the sixth to ninth vertebrae, so that a physiological neck may be said to comprise the first five vertebrae. All the ribs show most instructively how they become alienated from and independent of the basiventrals, their progenitors. The capitula of the cervicals still articulate by facets with the basiventrals, the tubercle separately with the diapophysis. Farther back the capitula move upwards so as to lie between the adjacent centra; they still touch the head of the intercentra, but there is no longer a facet. This means clearly that the connexion has become ligamentous, drawn out and weakened. Towards the lumbar region the capitulum "has leapt suddenly from the notch between the vertebra and the intercentrum to the articular facet which appears on the edge of the centrum". In fact, the capitulum has been shifted from the original intercentral position on to a typical parapophysis of the centrum. The intercentra are still of fair size in the trunk, although reduced to the ventral third or less of the complete basiventral; beyond the second or third sacrals they reappear, revived and modified as chevrons.

Naosaurus or *Edaphosaurus*. Various species, Carbo-Permian of Europe and North America. The neural spines are much elongated as in *Dimetrodon*, but carry from one to about ten transverse processes, the lowest sometimes several inches long, the others irregular and dwindling (Fig. 83). The intercentra gradually disappear beyond the thorax. Case describes and figures a vertebra of *Naosaurus*, apparently

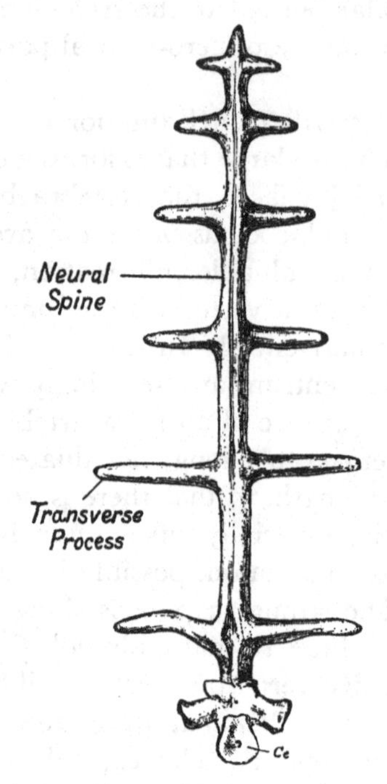

Fig. 83. *Naosaurus claviger*. Anterior view of dorsal vertebra. *Ce*, centrum. (After Cope.)

the epistropheus, with a pair of "centrosphenes" below the anterior zygapophyses; these are supposed to have fitted on to facets of the first neural arch. He also remarks that on some vertebrae the medullary canal is not closed dorsally by the neural arches. Such a gap is left between the pedicles in other Reptiles and Mammals only by the atlas ring.

Chapter XXIV

THE THERAPSIDA OR THERIODONTIA

MORE ADVANCED SYNAPSID THEROMORPHA WHICH INCLUDE THE MAMMAL-LIKE REPTILES

Theromorph Reptiles with one fossa, the upper temporal, which is large and bordered below by the squamosum. Every character which might be used to differentiate this group from the Pelycosauria seems liable to break down in the case of some or other members of this perhaps ill-assorted assembly, since they do not keep step, or they break off with some peculiarity which may or may not be an ancestral reminiscence. However, by the average total of all the available characters, they reveal themselves as having reached a higher plane than the Pelycosauri. Such advances are:

(1) There is no longer any sculpturing of the dermal bones roofing the skull, an indication that they have sunk in, being covered by the cutis instead of being still in the closest juxtaposition to the cornifying epiderm.

(2) Besides the large temporal fossa there is sometimes a smaller lower fossa below the squamosum of very variable configuration, perhaps a reminiscence of the lower and only fossa of the Pelycosauri.

(3) The intercentra are usually restricted to the cervical and, as chevrons, to the caudal region; they are gradually suppressed in the trunk.

(4) The vertebrae are still amphicoelous, but the chorda seems to be completely suppressed within the large centra, which alone carry the neural arches. There are often more than two, even four or more sacrals.

(5) The acetabulum is a closed cup.

(6) The phalangeal formula is 2, 3, 3, 3, 3.

(7) Elevation of the neck and trunk apparently well above the ground when walking.

(8) Division of the teeth into incisors, canines and "molars".

(9) The quadrato-jugal, though present in *Delphinognathus*, is usually suppressed, leaving a gap between jugal and quadrate. The latter is held firmly by the downgrowing flange of the squamosum and shows in the various genera a steady reduction in size, becoming finally squeezed out or wedged between the squamosum, pterygoid and the underjaw.[1]

Fig. 84. *Moschops capensis* (Dinocephalia). Skeleton, in American Museum. One twenty-second natural size. (After Gregory.)

Certain members of this group will now be considered in detail.

Moschops capensis. A Mid-Permian, South African, Dinocephalian Theriodont. A large beast; total length from nose to end of ilia about 5 ft.; skull 1 ft. long (Fig. 84).

Vertebral column: six or seven cervicals, total of presacrals 29,

1 In *Phil. Trans.* and in *Anat. Anz.* 1901, I held this firmly fixed and reduced condition of the quadrate to be incompatible with direct ancestral descent of the Mammals from these Theromorpha, because of the erroneous equation Reptilian quadrate = Os tympanicum. Since that time Broom has suggested complete suppression of the quadrate for the Mammals; Palmer has definitely proved that the tympanic bone is modified from one of the dermal bones covering the Reptilian jaw; and lastly Wortman has discovered in various Insectivors what can scarcely be anything else than the last vestige of the otherwise missing quadrate. The last gap between the Reptiles and the Mammals is thus filled satisfactorily. This agreement of the quadrate, the phalangeal formula and the differentiation of the teeth cannot well be mere homoplastic coincidence.

three sacrals. Vertebrae plane, short but broad, with neurocentral sutures. Intercentra restricted to the atlas and epistropheus. The atlas ring is large and formed by the BV which carries the BD and this articulates with the prezygapophysis of vertebra II, and seems to have carried a rib. The odontoid or centre 1 is united with centre 2. It articulates with the neural arches, which are separate from each other; also with the occipital condyle by a central eminence, and by two facets with the large BV 1. The second BV is still so large that it touches the BV 1 and thereby encroaches on the ventral extension of the odontoid, cf. *Dimetrodon*, *Sphenodon* and Crocodiles.

Fig. 85. *Moschognathus whaitsi* (Broom). First five vertebrae.

Considering the size of the first two basiventral masses, the absence of other intercentra in the trunk is all the more remarkable. In the figure by Broom of the allied *Moschognathus whaitsi* (Fig. 85) the fourth cervical vertebra is distinctly weaker than the third and the fifth, reminding one of a similar reduction in *Eryops*.

Free, movable ribs are present from the atlas to the tail. The capitula fit upon facets on the anterior margin of the centra, just as they should when intercentra have been suppressed; the tubercular portion abuts apparently against the underside of the prominent diapophysis. The first of the three sacral ribs is greatly expanded, articulating broadly by its capitulum with the centrum; the tuberculum also articulates extensively with the diapophysis. This rib therefore lies across the suture.

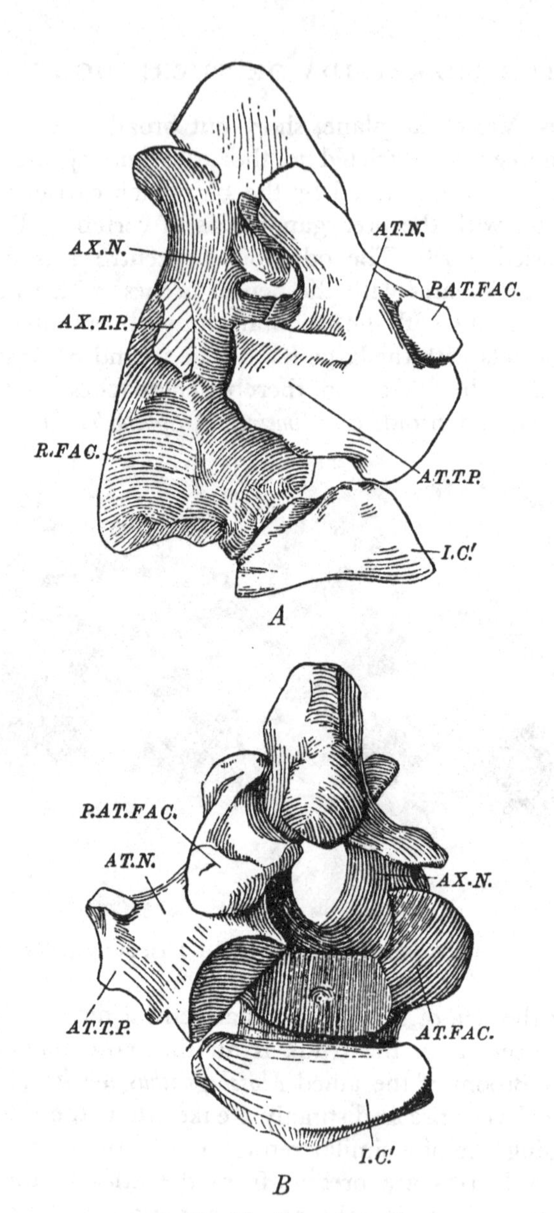

Fig. 86. *Kannemeyria.*

A. Right side of axis-atlas of mounted skeleton. One half actual size. Partially reconstructed from left side. *At.Fac.*, facet for neural arch of atlas; *At.N.*, neural arch of atlas; *At.T.P.*, transverse process of atlas; *Ax.N.*, neural arch of axis; *Ax.T.P.*, transverse process of axis (broken); *I.C'.*, first intercentrum; *P.At.Fac.*, facet for proatlas; *R.Fac.*, facet for first rib.

B. Axis-atlas of mounted skeleton from in front. One half actual size. Left half of atlantal neural arch missing. Lettering as in *A.* (H. Pearson.)

Gregory's remark that the pelvis shows a marked advance in the forward growth of the ilium is not quite clear, since, on p. 193, he holds that the first and strongest of the three sacral ribs corresponds to the single Amphibian sacral rib.

Kannemeyria. A beautifully preserved atlas and epistropheus of this likewise large Theriodont has been figured and described by Helga Pearson (Fig. 86).

The BD 1 and the BV 1 are large, slightly touching each other. The neural arch has a pair of prezygapophyses for articulation with the lateral occipitals, a feature reminding one of *Eryops*. Miss Pearson, however, suggests that these zygapophyses were intended for the proatlas. The atlas and epistropheus seem to have formed one physiological complex. The much restricted chorda runs right through the middle of centrum 1, which seems to be fused with centrum 2, and this is certainly amalgamated with BV 2; cf. Crocodiles and *Sphenodon*. The BV 2 carried a pair of ribs, but atlas ribs are not indicated. The condyle fits on to large facets on the neural pedicles and probably on to the first basiventrals, which form one very broad median piece.

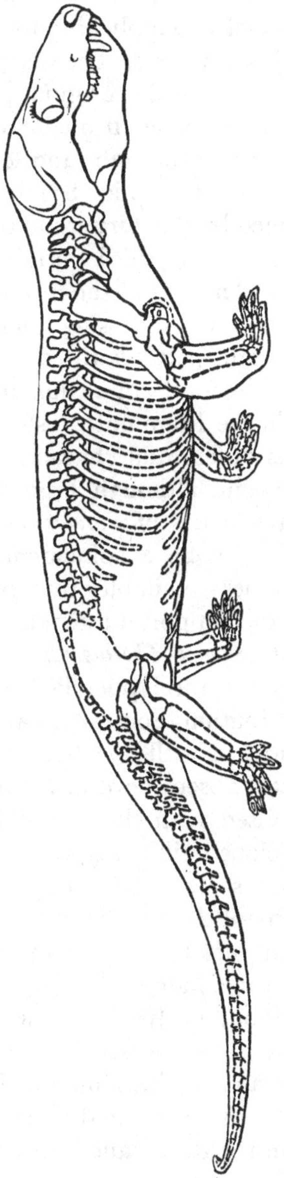

Cynognathus. This may be taken as a sample of Therocephalian Theriodonts; South African, Lower Trias. As the name of this group of rather numerous genera implies, their skull is the most mammal-like of all Reptiles. According to Broom, the leading expert, the closest approach is the skull of *Sesamodon*. These creatures are not only Theria by dentition and skull, but veritable Theromorpha; certainly not Theromora, no matter

Fig. 87. *Cynognathus.* Skeleton, one-fourteenth natural size. (Gregory.)

whether this term be intended to mean clumsy, sluggish or stupid; especially inapplicable to some of the smaller members which were agile carnivores. The occipital condyle has become double owing to reduction of the original share in it of the basioccipital, which has been shortened from behind forwards so that the atlas articulates only upon the right and left lateral occipitals.

Cynognathus (Fig. 87) has six cervical vertebrae. The sacrum is formed by the thirtieth to thirty-second, with or without participation of the thirty-third vertebra. There are no intercentra beyond the BV1. The ribs of the lumbar region overlap each other by horizontal expansions, enclosing thereby odd-looking intercostal spaces. The powerful pelvis is tilted, pivoting upon the sacral vertebrae, so that the iliosacral connexion lies in a distinctly preacetabular position, as in the Mammals, instead of postacetabular, as in other Reptiles. This differential character seems to have been appreciated only by Piveteau, although I pointed it out many years ago. The equally heavy pelvis of *Pareiasaurus* agrees in this respect with *Cynognathus* and may be a coincidence, but when more Theriodont material becomes available this preacetabular feature may well become another important genetic link with the early Mammals.

Dicynodont Theriodonts. A specialised group of several genera; it began in the Upper Permian and survived into the Trias with wide distribution: South Africa, India, Russia and Scotland. The occipital condyle is still one tripartite unit. Vertebrae amphicoelous. Intercentra absent beyond the atlas which alone carries no ribs and may be fused with the axis. The other presacral vertebrae carry well-developed ribs. Apparently there are seven cervical vertebrae, the broad scapula lying opposite the eighth to the tenth vertebra. The ilium of *Lystrosaurus* is so much expanded that it covers at least eight vertebrae, from the twenty-sixth to thirty-third. To judge by Watson's figure (Fig. 88), the whole pelvis seems scarcely tilted, the acetabulum lying at the level of the twenty-ninth and thirtieth vertebrae. The sketch strongly suggests expansion of the ilium by fore and aft instalments. The tail is composed of about one dozen short, rather stunted vertebrae, with small, plainly reduced chevrons from the third caudal onwards.

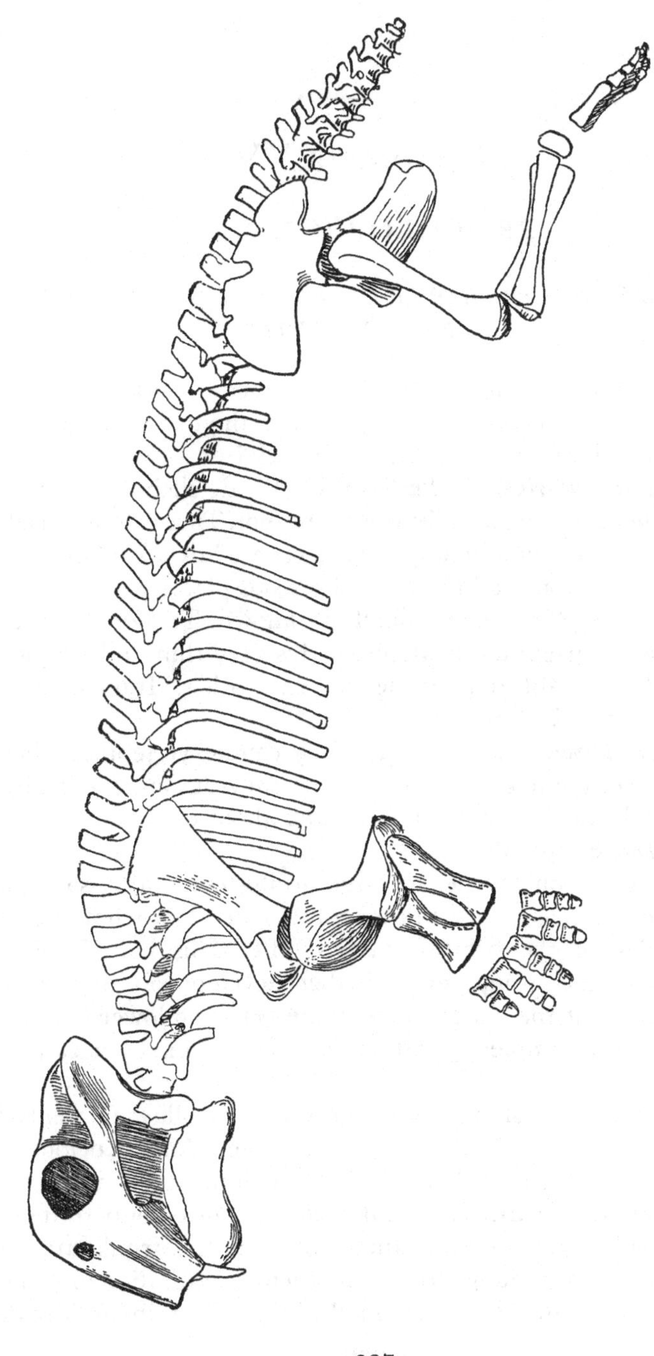

Fig. 88. *Lystrosaurus*. Skeleton, as restored by Watson, slightly modified. One-fourth natural size. (Williston.)

Chapter XXV

PROCOLOPHONIDAE

THEROMORPH REPTILES OF UNCERTAIN RELATIONSHIP

There are three, possibly four, closely allied genera of the Lower Trias of Europe and South Africa, easily distinguished from all other Reptiles by a large united orbito-temporal fossa.

Sclerosaurus (*Labyrinthodon Ruetimeyeri* of Wiedersheim), Switzerland. Length of head and trunk a little over one foot. Three sacral vertebrae, tail stunted. No abdominal ribs. Back covered by a kind of carapace composed of six longitudinal rows of osseous plates.

Koiloskiosaurus, Thuringia. Skull not known. Two or three sacral, many caudal vertebrae. Abdominal ribs very thin and simple, not compound, the right and left meeting each other. Intercentra seem to be indicated. Length less than 1 ft.

Telerpeton elginense. Two sacral, many caudal vertebrae. The first ten caudal vertebrae carry ribs; the first chevrons are carried by the third caudal. Length of head and trunk about 6 in.

Procolophon, South Africa.

According to von Huene the large orbito-temporal fossa is partly due to the absence of a supratemporal bone. The whole posterior bony border of the orbit is also absent. This hole is either a neomorph of the group, or a further development of the upper temporal fossa of the Theriodonts, there being no lower hole within the strong and complete jugal bridge. Transitional stages are not known.

Other characters common to the group are as follows: Twenty-four presacral and two or three sacral vertebrae. Amphicoelous, with continuous notochord, although this is much narrowed in the sometimes well-marked waist. The trunk vertebrae possess only the normal, rather broad zygapophyses. Intercentra in the trunk are either completely lost, or to judge from von Huene's restoration of *Koiloskiosaurus* (Fig. 89), reduced to very small vestiges. The ribs are decidedly holocephalous, with the characteristic oval, oblique articulation upon the centrum and low diapophysis; which shows that the ribs have already lost contact with their progenitors, the basiventrals. On the tail are ribs and chevrons, certainly a low, rather Theromorph feature, although equally present in *Araeoscelis*, *Sphenodon* and Crocodilia, therefore not helpful in our search for the Permian ancestry of the Procolophonia. Another feature, possibly more promising, is the phalangeal formula, in which we may discern the trend towards the Theriodont but not to the Sauromorph condition:

Telerpeton and *Koiloskiosaurus*: 2, 3, 4, 5, 3.

Sclerosaurus (and *Pareiasaurus*): 2, 3, 3, 4, 3.

Theriodonts and Mammals: 2, 3, 3, 3, 3.

Further, it is at least remarkable that no Chelonia have more than three phalanges, but are sometimes reduced to two on all fingers and toes; the very order about which prevails so much uncertainty whether they are of Theromorph or Sauromorph, not to mention Cotylosaurian, ancestry.

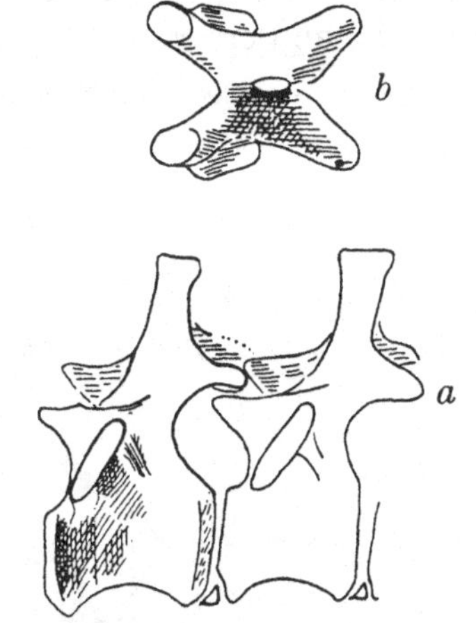

Fig. 89. *Koiloskiosaurus coburgiensis* H. Reconstruction of thoracic vertebrae: (*a*) seen from the left; (*b*) from above. Slightly enlarged. (Von Huene.)

The thin, simple abdominal ribs, the absence or presence of intercentra, and the amphicoelous notochordal vertebrae are no criteria that the Triassic group of Procolophonia must be older than the Theriodonts, which, essentially Triassic, have given rise to veritable Sauro-mammals, besides such specialists as Dicynodonts and other ephemeral side branches. The average small size and the shortness of the neck of our little group are certainly in favour of its ancient standing, but many "living fossils" show that very old characters can be retained long after others have been evolved which raise their bearers to a much higher standing. It is by the new features that we recognise a new group. To say this may be a platitude, but it is not uncommonly ignored that the neomorphs date their bearers geologically, not their most primitive characters like a persistent chorda, which is a legacy from the beginning of vertebrate life. The morphological and taxonomic value of almost every organic feature may attain its fullest all-important status, then decline and drop out, superseded by others. Gills give way to lungs, and even these may be dispensed with. The constant competition between cartilage and bone is another example. If these changes kept step in their sequence in the countless big and small groups of animals, taxonomy would be easy, but every group has established its own custom, the result of trial and error.

Chapter XXVI

MAMMALIA

The Mammalia are gastrocentrous vertebrates, they have an exclusively dentary-squamosal masticatory joint, and their sternum, unless much reduced, is serially segmented.

The known Mammals date from the Upper Triassic; the earliest seems to be *Tritylodon* of the Karroo of South Africa, which is represented by an almost complete skull, very stout and nearly 2 in. long (Fig. 90). The others are very imperfectly known, mainly only by mandibles and scattered teeth, e.g. *Microlestes* of the Rhaetic of Europe and *Dromatherium* of Carolina; they apparently were small creatures of about the size of a Mole or a Rat. They begin to get more common in the Jurassic and become abundant in the Eocene, when most of the main groups were being evolved.

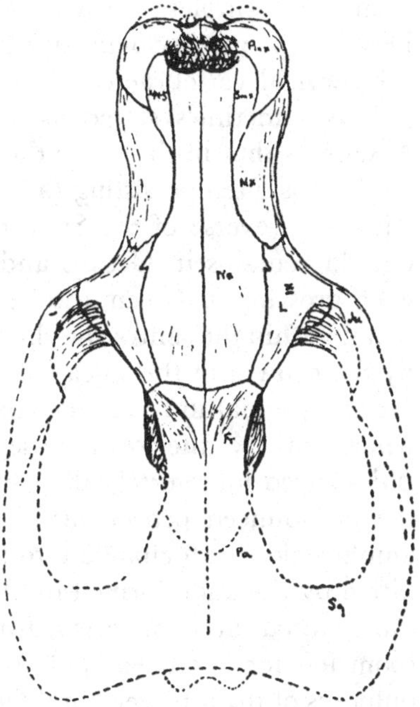

Fig. 90. Skull of *Tritylodon longaevus* (Owen). The front portion is very slightly restored from the only known specimen. The back portion is entirely hypothetical. (Broom.)

The mammalian vertebral column has reached the morphologically highest stage, the tendency initiated by the Reptiles of forming highly efficient bony vertebrae by fewer ossifying units at the expense of others having reached a perfection greater than in Birds. The vertebrae are absolutely gastrocentrous, in a solid, stereospondylous condition, and are composed mainly of the neural arch and centrum only; interdorsals are completely absent; osseous basiventrals are also suppressed with the exception of the "hypochordal element" (BV I) of the atlas and the frequently occurring vestigial chevrons in the tail. With these exceptions, therefore, intercentra are normally absent.

The atlas alone usually preserves the ancient temnospondylous condition and its centrum is added as the odontoid process to the second vertebra. Even the first pair of basiventrals can disappear as an osseous entity in various Marsupials, e.g. the Wombat, *Phascolomys*, the Koala, *Phascolarctus*, Phalangers and Kangaroos, so that their atlas ring consists of the basidorsals only, joined by connective tissue.[1] The gradual suppression or loss of the basiventrals concerns, however, only their ventral ossifying portion, the axial portion forming a complete fibro-cartilaginous disk containing in its middle the last remnant of the chorda as the "nucleus pulposus". The menisci formed by, and representing, the axial portion of the originally complete basiventral elements act like pads or washers between the usually plane or very slightly concave ends of the centra: the cervical vertebrae of Ungulata are somewhat opisthocoelous.

It is sometimes stated as a difference between Sauropsida and Mammals that in the latter the chorda disappears first in the centra, whilst it is often persisting (as the nucleus pulposus) intervertebrally, "just the reverse of the Sauropsida". This is not correct, since the chorda remains in Geckos and *Sphenodon* practically unaltered and ever growing between the centra, there being no proper joints, whilst within the centra the chorda is first interrupted, then destroyed, by the growth of the so-called chordal septum. Further, Crocodiles are exceptional in retaining small rests in the middle of their centra, which may be due to the absence of such a septum. If the chorda is destroyed intervertebrally, this is done by the big knob formed by the pronounced procoelous or opisthocoelous vertebrae. Even the amphicoelous can change into the procoelous or opisthocoelous condition by the knob formed by the centrum, which stamps a hole into the meniscus and converts it from a disk into a ring; and there is no room left for a nucleus pulposus. It is possible that this "nucleus pulposus of the intervertebral ligaments" is confused with the "intervertebral Knorpel" of the Newts. The intervertebral, in this case intercentral region, is further complicated by the development of

1 It is impossible to endorse Gegenbaur's conclusion (Vol. I, p. 257) that the atlas of the Marsupials represents a step lower than that of the Reptiles in which the ring had already been closed by the middle piece which is still wanting in various Marsupials. The appearance of a middle piece in some Marsupials (e.g. *Thylacinus*) indicates how in the Placentalia ossification of such a middle piece may have been brought about from the lateral pieces of the neural arch.

separately ossifying rather thin, disk-shaped epiphyses on the anterior and posterior ends of the mammalian centra with which they ultimately fuse.

Confusion concerning the homology of the intervertebral disks or menisci is still occasionally found. Gegenbaur stated in his *Lehrbuch*, p. 247: "The separation of the continuous Anlage into the vertebral bodies is effected in Lizards and Snakes by the division of the intervertebral cartilage into a posterior knob and an anterior cup, by which feature these creatures are connected with the Amphibia. In Crocodiles and Birds these cartilages between the centra are turned into a special apparatus: either it remains as such (Crocodiles), or it forms intercartilage, menisci, which are separated from the centra by joint cavities (Birds)". He, and not a few later authors, did not recognise in the menisci the axial portion of the basiventral mass, but confused them with the intervertebral cartilage of the Newts which are the ID plus IV mass. Through the eventual reduction of the menisci the successive centra meet end to end but for the synovial membranes, and even these vanish wherever concrescence of vertebrae takes place. Schauinsland (p. 525) states that the middle portion of the thickenings, which appear in the intervertebral slits of the early embryo, form the "menisci or fibrocartilagines intervertebrales", and he suggests for these the new term of "primary intervertebral bodies". On p. 526 he observes that the lower arches (intercentra or hypochordal clasps, our basiventrals) are not formed at the level of the primary vertebral bodies as in the Anamnia, but have been pushed cranially, to lie on the underside of the menisci. This is of course true, because the meniscus is the axial portion of the whole basiventral, which has been squeezed between the bodies of the growing vertebrae, leaving room for the ossifying cartilage only on the ventral side. They have not been pushed cranially; they have always been at the head end of their vertebra, but as these bodies or centra have grown and become greatly elongated, having made common cause with the neural arches, it looks as if the basiventrals had been pushed forward. Elsewhere (p. 497) he is fully aware, with Gegenbaur, that the intervertebral Knorpel becomes much reduced, especially in Gymnophiona and Newts. This is the frequently occurring confusion, due to lack of appreciation of the fact that "intervertebral" is a relative expression which does not imply homology. It depends upon which units of the quadripartite vertebra form its body or Koerper, the rest then must assume, or be squeezed into, the intervertebral position. In the Amniota the body or centrum is formed by the interventrals which carry the neural arch; the body or axial bulk of the Rhachitomi is formed by the basiventrals which carry the neural arch. This does not mean a change from one series of units to another, but a divergence from the original, primary condition in which the units were all equal in size, each unit being the half of a skleromere.

Lastly, some authors speculate about the fact that all the above changes are observable when the whole column is still in the membranous stage, which shows already from the earliest blasteme the future definite segmental arrangement; so that it is therefore not the future cartilage which is responsible for these changes. This is an instructive instance of the psychology of the embryographer from the old Malpighian standpoint. It is all there, earmarked, planned, preformed, predestined. Shortened recapitulation, the result of stored-up morphological reminiscence of what the ancestors have gone through, is nowhere. We prefer the much maligned Goette, who referred all these changes to phylogeny.

The ontogeny of the mammalian vertebral column has been well described by Froriep; most of the salient features found in *Bos* have been corroborated by Macalister in human embryos. Subsequent investigators have considerably enlarged our knowledge, notably Schauinsland; and Gegenbaur in his *Lehrbuch* gives a concise general account.

The chorda is at first complete, but from the beginning is much reduced in comparison with Amphibia and Reptiles. Froriep found in the Cow embryo of 8·7 mm. body length that the separation of the sklerotomes from the myotomes was still incomplete; the spinal nerves in this precartilaginous stage lying behind his "primitive Bogen", the space of the future bodies of the vertebrae still occupied by embryonic connective tissue. In the embryo of 8·8 mm. the thickness of the chorda amounts to 0·03–0·05 mm., its constrictions coinciding with the cranial ends of the primitive arches. The chordal sheath is only 2μ in thickness; it becomes still thinner in 12 mm. embryos, and it becomes partly invisible, especially in the region of the centra, in 18·5 mm. embryos, which correspond with those of Man of two months. The chorda itself is mostly shrivelled up.

An important mammalian feature, although to a lesser degree already observable in Reptiles and Birds, is that the precartilaginous matter in the skleroblastic tubular layer makes its appearance as a continuous "perichordal" tube, instead of beginning as separate serial cartilaginous units so clearly visible in Newts. In the Mammalia, therefore, the first indication of the future vertebral segments is a richer growth of intercellular substance (the usual way of describing the origin of cartilage), the material for the primitive arches "which in turn in the lower vertebrates formed the body".[1] Gegenbaur explains this reversed condition by the consideration that "bereits jene perichordale Wirbelanlage von dem Material der Bogen entstand" (that this perichordal mode of laying down the vertebra—or appearance of the vertebra within the perichordal tube—had itself already arisen out of the material of the arch). Whatever this sentence may mean, he continues as follows: "that what was in the Amphibia a secondary mode of making vertebral centra, has in the Mammalia been transferred into an earlier stage; and this temporal shifting is a cenogenetic process connected with the lesser volume of the chorda". This I suppose means that, the mammalian chorda being reduced, the space gained invites the perichordal cartilaginous matter to close round the chorda and then to form the body or centrum earlier than in the lower vertebrates, since the latter devoted their perichordal material to building up the bulk of the vertebra out of the basidorsals and basiventrals. In any case, it is a surprising assumption that the vanishing of the chorda should be the *causa movens*, instead of explaining the reduction of the chorda as being caused by the growth of the neighbouring sklerotogenous layer, which, in the Mammalia and other Amniota, produces the interventralia or gastrocentra; the latter are met by the basidorsals above, because these have shifted back upon the so-called postcentra or true centra, as is demonstrated by *Cricotus*, p. 160. Interdorsals being unknown in the Amniota, this caudal shift of the basidorsals brought them all the easier into contact with the gastrocentra.

These gastrocentra were discovered by Froriep. In Cow embryos of 12 mm. the "primitive arch" is differentiated into a neural arch (BD), an intervertebral disk and a "hypochordal mass" (BV), the first and last being already cartilaginous. Tailwards from the hypochordal mass, and below the chorda, has appeared a mass of hyaline cartilage, the beginning of the future centrum, bilaterally symmetrical, with indications of being composed of a right and a left

[1] This sentence of Gegenbaur's is rather mysterious. By body or "Koerper" he can mean only the axial bulk of the nascent vertebra. This would amount to an acknowledgment that the Rhachitomous centre is part of the whole arch, namely the BV, which is correct, but this is not the centre of the Amniota!

piece. This most important fact has been corroborated by other investigators, who describe this mass correctly as of paired origin, but they avoid mentioning its essential position in the same horizontal level as the hypochordal; they therefore leave in abeyance the question whether it represents a pair of interventral or of interdorsal elements, i.e. whether gastrocentra or notocentra: cf. also the clear account of *Sphenodon* by Howes.

The growth of the ossifying neural arch (each half with its own nucleus) preponderates over that of the centrum with its own pair of nuclei. Thus it comes to pass that the neurocentral suture approaches the dorsal level of the chorda. These sutures are abolished long before maturity, as in Birds and Snakes. Gegenbaur remarks, p. 256: "The process of ossification extends from the arch upon a not inconsiderable portion of the vertebral body, so that this in its osseous condition may be considered as being formed by a portion of the arch". This conclusion is unjustified: cf. the *Ichthyosauri*, with centra so high that they alone form the whole axial portion of the column. "Body" and "Centrum" should not be used promiscuously in such critical cases. In connexion with the neural arch of the Mammals, Schauinsland, p. 552, remarks: "Different from the Reptiles is Man, since in his dorsal spinous process appears a special centre of ossification. Such a centre is not found in *Sphenodon*". It is not impossible that these osseous centres are the last latent vestiges of supradorsalia, so common in Fishes, apparently absent in Tetrapoda with the exception of the Stegocephalians in the cervicals, of which distinct sutures between the spine and its pedicle have been described and figured.

The Atlas, cf. also the special chapter, is of especial interest, since it retains the original temnospondylous condition. The atlas ring is formed as usual by the first pair of basidorsals (neural arch), and by the first pair of basiventrals, i.e. the first joined hypochordal cartilages or so-called bony ligamentum transversum. The first centrum joins that of the second vertebra; owing to the complete and early suppression of the second hypochordal elements, the caudal portion of the first centrum broadens out considerably, spreading over the whole surface of the second centrum. Consequently, as Froriep was the first to point out, the ring of the atlas articulates directly with the first centrum, instead of with the second hypochordals and their axial disk (meniscus 1), as in the Birds and some Reptiles.

To emphasise this important feature, apparently characteristic of the Mammals, we can express it as follows: The first trochoid joint, namely that between the atlas and the epistropheus, is an *inter*vertebral joint in the Sauropsida, but an *intra*vertebral joint in the Mammals.

The following joints of the mammalian column are all intervertebral, since they pass between the caudal surface of the morphologically entire anterior vertebra and the cranial surface of the entire vertebra behind; the meniscus or pad belonging genetically to the cranial end of its vertebra. It would be pedantic to say that the whole joint is double, namely intervertebral in front and intravertebral behind, because of the synovial cavities or slits in front and behind the meniscus.

The basiventrals of the Epistropheus do not always disappear, but develop sometimes into a pair of ossicles between the rudimentary rib and the second centrum, behind the ventri-lateral corners of the first centrum or odontoid. Macalister described and figured such a pair as "hypochordal epiphyses" in a child of 28 months. For further abnormal human vestiges, see Chap. XIII. Such basiventral vestiges cease beyond the second vertebra, but they occur sporadically mostly as median ossicles of irregular shape, not always easily distinguished from detached bits of epiphyses. In the tail, however, these elements occur normally in many Mammals, e.g. Cetacea, Edentates and Marsupials; they are known promiscuously as chevrons or haemal arches, more properly haemapophyses, i.e. the ventral complements of the original basiventrals. They are often loosely attached to the intervertebral region, sometimes to a meniscus, with a tendency to fuse on to the posterior end of the vertebra next in front, to which they do not belong genetically. These right and left chevron halves may remain separated, or they fuse distally, occasionally diverging again, according to the muscles they carry. They are perhaps best developed in *Manis*, where the right and left halves are each movably attached to the centra of neighbouring vertebrae by an anterior and a posterior arm which enclose a wide gap, bridging over the intervertebral joint. Intermediate conditions occur in the various species of *Manis* and in *Orycteropus*.

The vertebral column in all Mammals can be divided into cervical, thoracic, lumbar, sacral and caudal regions. The cervical and thoraco-lumbar will be described in this chapter, the sacro-caudal region in the next.

CERVICAL REGION

The normal number of cervical vertebrae is seven. The few exceptions are all the more interesting. The majority of Edentates—Armadillos, Anteaters, Pangolins and *Orycteropus*—have the normal number, but the various kinds of Sloths have six, seven, eight or nine. The great number of nine in *Bradypus* clearly means elongation of the neck, a secondary feature brought about by conversion of two originally thoracic vertebrae into cervicals, the ninth and sometimes the eighth still bearing a pair of short movable ribs; the eighth cervical, but not the ninth, is even perforated by the vertebrarterial canal. The Two-toed Sloth, *Choloepus didactylus*, has seven; *C. hoffmanni*, a closely allied species, has only six; in both species the vertebrarterial canal ends with the sixth or seventh cervical respectively. The only other abnormal number occurs in the genus *Manatus* with six, while the other Sirenia, like *Halicore*, the recently extinct *Rhytina*, and the Miocene *Halitherium*, are normal. In *Manatus senegalensis*, the second and third are sometimes co-ossified. A specimen of *M. americanus* (Cambridge Museum) has the vertebrarterial canal incomplete in the second to fourth, absent in the fifth, but again complete in the sixth, which carries a distinct movable rib. Synostosis of cervical vertebrae is carried to the extreme in the Right Whale, *Balaena mysticetus* (Fig. 91), where the second to sixth arches are completely fused, to a lesser extent also with the first and seventh, but the nerve holes are not interfered with; all the vertebrae are much shortened and form one physiological unit.

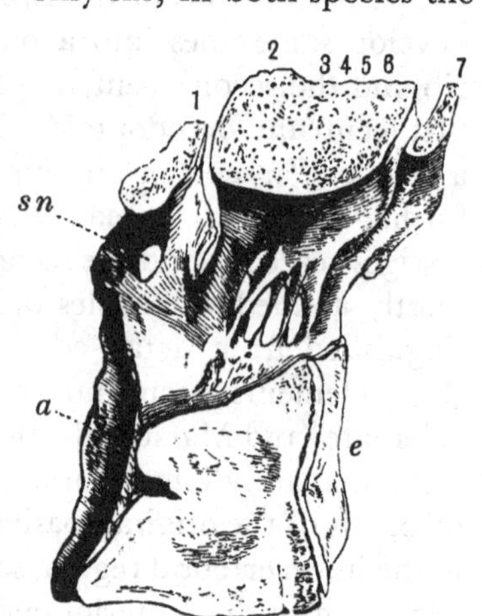

Fig. 91. *Balaena mysticetus.* Section through middle line of united cervical vertebrae. *a*, articular surface of occipital condyle; *e*, epiphysis on posterior end of body of seventh cervical vertebra; *sn*, foramen in arch of atlas for first spinal nerve; 1, arch of atlas; 2, 3, 4, 5, 6, conjoined arches of the axis and four following vertebrae; 7, arch of seventh vertebra. (Flower.)

Concerning the cervical nerve exits in the Mammalia: the first, really the proatlantic or suboccipital, is usually enclosed by an osseous bridge from the anterior end of the neural arch of the atlas. The second issues between the atlas and epistropheus, and so forth, i.e. *the spinal nerves issue from behind the neural arch of the vertebra to which they belong.* In the trunk the exit lies either in a notch, or this is bridged by bone and converted into a foramen near the posterior end. This arrangement varies much in the Monotremes. Whilst the cervical nerves pass behind their vertebrae, those of the trunk pass either right through the middle of the arch, although nearer the posterior end, or they are incompletely enclosed.

Cervical ribs have left no trace in the mammalian atlas, but they are still clearly developed in the other neck vertebrae of the Monotremes as broad but short bony elements either sutured or fused with the vertebrae. They are not divided into a capitular and a tubercular portion, being attached chiefly to the centra below the neurocentral suture and in a distinctly intervertebral position. In most other Mammals the cervical ribs are in a more or less vestigial condition, recognised as independent units by their centres of ossification, but their cartilage fuses early with that of the basal portion of the neural arch (the ribs having phyletically given up long ago their origin from the basiventral or hypochordal mass and having transferred their attachment dorsally, above the neurocentral suture, on to the neural arch). In fact the cervical ribs now, especially in the higher Mammalia, notably Man, are represented by that bar which ventrally borders the foramen transversarium; therefore in human anatomy rightly called pars costalis. This costal part is much smaller, difficult to recognise, in the seventh vertebra, thereby emphasising the abrupt separation of neck from thorax; this is a significant difference between the Mammalia and the Sauropsida, testifying to the remote date of the divergence of the two classes.[1]

[1] Here is an unfortunate case of nomenclature, cf. A. Robinson in Cunningham's *Anatomy*, 1917. The foramen transversarium is bordered dorsally by the "root of the neck of the neural arch", dorso-laterally by the posterior tubercle, ventri-laterally and ventrally by the anterior tubercle which is homologous with a rib in the thoracic portion of the column. "The posterior and anterior tubercles are connected by the transverse process." In reality the two tubercles are secondary features for muscular attachment. In any case "transverse" is here used for a distinctly dorso-ventral connexion, whilst the transverse rib-carrying projection, as typically developed in the lumbar region, conveys to the comparative anatomist or morphologist something extending in the transverse horizontal plane; all the more if these lumbar transverse processes are composed almost entirely of the ribs.

THORACO-LUMBAR REGION

The ribs of the thoracic and lumbar region are usually divided into:

(1) True thoracic ribs which join the sternum directly.

(2) False ribs which join the sternum indirectly, their distal portion being loosely attached to the rib in front.

(3) Floating ribs which have lost their sternal half and are free.

(4) Lumbar ribs which mostly do not look like ribs at all, being firmly sutured or even fused on to the centrum or to the diapophysis, so that the combination appears like a flat horizontally expanded transverse process.

These categories pass gradually into each other and are subject to much variation in formation and numbers. In the majority of Mammals the true, complete ribs are composed of a dorsal or vertebral, and a ventral or sternal half. To speak of these as dorsal and ventral "ribs" leads to slovenly confusion. Only a few Mammals have retained the Sauropsidan intermediate or third rib element; it is best developed in the Monotremes, less clearly in some Edentates like the Armadillos, and it is almost vestigial in Toothed Whales. Uncinate processes are, however, unknown, although in a few genera most of the ribs are so broad or flattened that they overlap, without synovial contact. The greatest number of ribs is 24 in the Two-toed Sloth; next comes *Hyrax* with 22; the lowest number is nine in some Toothed Whales, and in some Whalebone Whales the first or manubrial rib is the only one which reaches the sternum. The normal, complete ribs articulate with their vertebrae by their capitulum and the tuberculum. The tuberculum, no doubt of secondary development, forms a synovial articulation with the underside of the diapophysis. The capitulum, usually the much longer portion and in reality the rib proper, is variably attached. Either it articulates with synovial joint facets between the centra of neighbouring vertebrae, or it forsakes this original position and transfers its head tailwards and dorsally upon the centrum of its vertebra, which is induced to form a kind of parapophysis for the capitular support. This may be formed entirely by the centrum below the neurocentral suture, or the attachment may lie right across this suture, or ultimately it may move dorsally so that the articulation is formed entirely by the neural arch. The resulting more or less transversely elongated diapophysis carries both capitulum and tuberculum in this way, so that the capitular facet has now come to lie in the same horizontal plane but headwards

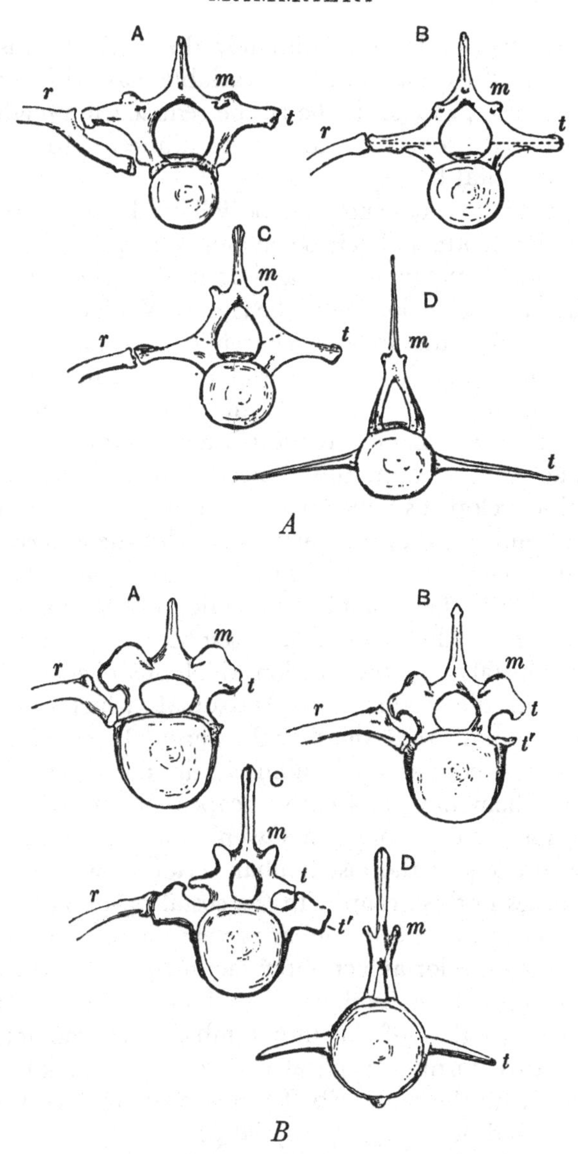

Fig. 92. *A*. Anterior surface of vertebrae of Dolphin (*Globicephalus melas*), 1/7. A, fifth thoracic; B, seventh thoracic; C, eighth thoracic; D, first lumbar; *r*, rib; *m*, metapophysis; *t*, transverse process. The dotted lines indicate the position of the neurocentral suture.

B. Anterior surface of vertebrae of Sperm Whale (*Physeter macrocephalus*), 1/24. A, eighth thoracic; B, ninth thoracic; C, tenth thoracic; D, fifth lumbar; *r*, rib; *m*, metapophysis; *t*, upper transverse process; *t'*, lower transverse process. (Flower.)

from the tubercular facet. Ultimately the distinction is abolished, and the reduced rib may fuse in a variable way with the distal end of the transverse process. If the rib be reduced to nought, only this transverse process remains, and may continue to grow into a broad or thick horizontal blade.

These progressive changes can be followed up in the column of one suitable individual. Their behaviour is instructive also in Whales and Dolphins, cf. the figures 20 and 21 in Flower's *Osteology* (Fig. 92 *A*, *B*). The tubercular portion of the ribs of Whalebone Whales is but loosely attached to the diapophysis, while the capitular portion and neck are so short that they do not reach the centrum, being connected to it by a ligamentous band. On the other hand, in Monotremes the tuberculum is absent and the rib articulates with the vertebra only by means of its capitulum. The abnormal occurrence of a more or less well-developed seventh pair of cervical ribs is rather rare, but if they are complete they may articulate with the anterior end of the manubrium, in which case it is entitled to the dignity of possessing an extra thoracic rib. The complete reduction of such an additional rib may reduce it to a dorsal portion, resembling the cervico-dorsals of Birds, and a distal or ventral portion which phylogenetically furnishes the material for the prosternum of Marsupials, which is the equivalent of the suprasternalia of Rodents or the sterno-clavicular disks of Man.

The following condition prevailing in the Perissodactyls and being peculiar to them has apparently escaped notice in the textbooks, excepting for a short notice in Owen's *Anatomy of Vertebrates*. The transverse process of the last lumbar vertebra, which represents its rib, articulates in this group with the costal portion of the sacrum.

Tapirus. The fifth or last lumbar vertebra articulates by a synovial joint with the anterior aspect of the sacral rib in a vertical plane.

Old cart-horse. Five lumbars. The fourth lumbar "transverse" process overlaps that of the fifth lumbar with contact. This fifth lumbar in turn forms a vertical end to end synovial articulation, 2 in. broad, with the sacral rib. There are several irregular exostoses and even a kind of osseous epiphysial pad.

Zebra. Six lumbars. The last pair of the horizontal transverse lumbar processes articulate end to end with the sacral ribs behind, and in front with the fourth pair of lumbar transverse processes.

Equus Przewalski. Six lumbars. The sixth articulates synovially with the sacral rib and with the fifth lumbar process. The first right lumbar typically transverse process carries a distinct but thin rib,

3 in. long; on the left side the process alone is present, without any indication of having carried a rib.

Rhinoceros. Four lumbars. Last lumbar processes articulate with the sacral rib and with the third lumbar.

These intercostal joints are all distinctly opisthocoelous, most pronounced in the Rhinoceros, in conformity with the central joints which at least in *Rhinoceros* and *Equus* are opisthocoelous throughout the thoraco-lumbar region. Such intercostal joints apparently do not occur in any other order of Mammalia, but they recall the at least analogous condition of the additional jointing of the last lumbar of American Edentates, cf. p. 61. They are completely absent in the Ruminants and Pigs, but a Cambridge specimen of *Hippopotamus* has an incipient sacro-lumbar articulation, and this is distinctly procoelous, the other vertebrae being plane.

Chapter XXVII

THE SACRO-CAUDAL REGION OF THE MAMMALIA

The pelvic girdle, carrying the hind-limbs slung around the trunk by muscles and ligaments, has, in the Tetrapoda, gained direct skeletal support upon the vertebral column by means of ribs, an arrangement which has profoundly influenced the whole region between loins and tail and has culminated in the formation of the sacrum.

The number of sacral vertebrae began with one in the Amphibia, increased to two in most Reptiles, and to still more in some extinct Reptiles and in the Birds. The number is most variable in the Mammals, where it may range from one to thirteen, owing to secondary addition and eventual reduction. Naturally anatomists have tried to ascertain the homology of the first or primary sacral vertebra in the various classes; whether the single one of Amphibians is homologous with the first or second of Reptiles, etc. There the trouble began and has not yet been settled, and it cannot be settled since it is a purely a priori problem. Just as there never was one Adam and Eve, there was no first pair of Tetrapods but many potential ones, whose pelvis floated, so to speak, to and fro, until it settled according to the exigencies of life, with many subsequent alterations. Speaking broadly, the whole limb girdle as such is homologous in all Tetrapods. Certainly the ilium is now monomeric, although it arose polymeric in the Fishes, cf. the lateral fin theory, which after many fights has at last superseded the ingenious, bolder, but typically idealistic theory of Gegenbaur.

In the Tetrapods the pelvic muscles are still polymeric, derived from a considerable number of myotomes; such muscles are a secondary acquisition of the pelvis and ultimately determine the region of its attachment to the axial skeleton with which the ilium originally had nothing to do. Whether the ilio-sacral connexion happened at the level of the twentieth or the fiftieth vertebra is a question of detail, and we do not know the cause and means by which the number of protovertebrae has been, and still is, decided. One group of creatures may have increased that number to one hundred before their limb girdle was phyletically advanced enough for axial connexion; another group of the same class may never have possessed more than half that number. The mechanism of increasing the number, whether by fission or by sprouting at the ends of the series, is not known. The reduction is simple, beginning at the tail end and progressing forwards.

The change from one to more sacrals can in the simplest case be brought about by broadening the ilia, so that, as described in Birds, an additional vertebra in front or behind the primary one is drawn into the complex for support. If the ilia become very broad or long, as in some Dinosaurs and in Birds, a considerable number of former lumbar vertebrae can be converted into typical sacrals, provided the lumbar ribs had not been previously lost; in the latter case physiological, but morphologically incomplete, sacrals would result; the same applies to the posterior additional sacrals of the Sauropsids, in which the posterior portion of the ilia is greatly extended.

Gegenbaur states that in the Mammals the foremost sacral is the typical sacral vertebra, since it alone has a large proper pars costalis, although on the next following vertebra costal portions (namely ribs) are indicated by separate centres of ossification. He uses this fact in support of another faulty notion, that the mammalian sacrum is to be derived rather from amphibian (Anura?) than from reptilian conditions; an idea which the old master was induced to accept from authors who were disappointed in their attempt to trace the origin of feathers and hairs from reptilian scales. And yet, although he was fully aware of the difference between the Sauropsid and the mammalian pelvis, he concluded that the eventual increase of the mammalian sacrum is due to later addition behind, caudal to the first or primary sacral vertebra.

The account of the changes in the sacral region as given in Gegenbaur's textbook is rather confusing. It is one of the unfortunately frequent instances where the running account in ordinary print represents the original, rounded off, clearly written chapter, while the smaller print indicates later additions of more recent authors' discoveries and views, which are not only at variance with each other, but contradict the text. Very difficult indeed has been the incorporation of the gist of articles from the three volumes of his *Festschrift*, which was published when his grand textbook was still in the making.

He states on p. 262 that if, of the Carnivora with normally 20 thoraco-lumbar vertebrae, *Hyaena* has 16 thoracic and four lumbar, and *Felis* has 13 thoracic and seven lumbar, the loss of ribs cannot be due to a forward move of the sacral connexion, but must be due to greater required freedom of the loin region. Reduction of the number of lumbars is, however, due to a forward move of the sacral connexion. The third statement that a vertebra with the same serial number may, in different species, have changed from a thoracic into a lumbar, this into a sacral, and this "finally" into a caudal, is also correct when taken by itself, but the statements taken together make confused reading, especially since the author accepts Rosenberg's discovery that in Man at least one vertebra, the twenty-fifth, shows ontogenetic recapitulation of its conversion from a lumbar into a sacral. This observation of Rosenberg's is certainly true; the last presacral becomes thereby the first sacral, but therefore Gegenbaur is not justified in stating (p. 259) that this, our present first sacral, is *the* primary sacral and that all the rest are secondary posterior additions. The initial fault is his declaration that the mammalian "pseudosacrals" are secondary additions, the increase in numbers proceeding tailwards. This seems still to be the prevailing view, provided the problem is not ignored.

In a Foal of a few months, or in an equally young Camel, the pelvis is loosely attached by one sacral rib only; all the following vertebrae are in a peculiar, almost embryonic condition, quite free from each other, each consisting of a quite separate centrum, more or less recognisable rib portions and the pair of dorsal arches widely separated from each other, without trace of the spinous process which later is represented merely by junction, and without a special centre of ossification. Much later synostosis forms the pseudosacrum with the gaping incisurae ischiadicae. Laterally the series of transverse blades become connected by their distal ends with each other, forming a continuous rather sharp-edged blade. The edge thickens with age and in an old cart-horse in the Cambridge Museum this compound edge of the last four or five vertebrae has been converted into a ledge 3 in. long with its free lateral surface forming a vertical plane surface half an inch high, with the most perplexing appearance as if it were but recently dislocated from the ilium. How is this to be explained, if not as a senseless faithful recapitulation of three successive phyletic stages? First, a pelvic-bearing function of the posterior segments, which office induced or taught them synostosis. Next, the addition of more anterior segments with reciprocal release of posterior units, which, however, continued to repeat the synostosis but with reduced lateral edges. Lastly, these edges or ridges, although for untold generations never any more in contact with the ilium, repeat with the progressing osteosis of old age a sham articulating facet. This is an instance of purposeless, futile reconstruction.

In 1879, we, Gegenbaur's staff and special pupils, approved of the master's sneering remark that according to Claus the pelvis of the eel-shaped Perennibranchiates was so "rudimentaer" and placed so far back "because it had got out of control". Intercalation of vertebrae was not allowed, and tailward migration of the pelvis was at least disapproved for obvious reasons. Loss of control is, however, in this case the intentionally frivolous rendering of "reduction by disuse". The formation of a sacral connexion is directly correlated with the use of the pelvis as a support, which weakens and ceases in so many swimming Tetrapods, be they amphibian Sirenidae or mammalian Sirenia. Firm sacral connexion prevents, or at least makes it difficult, for the hind-limb to be shifted forwards or backwards, and in so far it may well be said to be under firm control. A loose reduced ilio-sacral connexion renders the girdle more yielding to other impulses, as for instance the forward momentum of the trunk, which at the same time is checked by the girdles slung or fixed on to the trunk skeleton. It is a complicated process, which has to be considered in detail, each case upon its own merits.

Fuerbringer brilliantly correlated the long neck of Cranes, Storks and Swans with their slow and steady beat of the wings during long-sustained flight. Each beat is a forward impulse to the heavy body, which, suspended upon the wing planes, translates this friction through the muscles and bones of the wings and shoulder girdle to the trunk, i.e. upon the sternum and ribs, so that ultimately the forward shoot of the body converts thoracic into cervico-thoracic ribs, resulting in the increased number of cervicals. This idea, logically sound, is based upon reasonable mechanical premises, and its author has much laboured the theoretical detail concerning the implied alteration of the muscles and nerves and their transfer on to neighbouring segments. The same phenomenon, apparent forward move of the column, in reality the sliding back of the girdle, applies to the eel- or snake-shaped creatures in which the posterior limb complex has been shifted near the cloacal region, i.e. as far as it can go; unless it has been reduced to nought during the process. Such creatures do not walk but wriggle, and their limbs, if not used, are in the way, therefore literally brushed back. There are various compromises, obvious to those who are not above examining for instance those Skinks which quickly undulate through the loose sand whilst the weak, reduced limbs are extended backwards and pressed against the trunk, where they are received by a right and left longitudinal niche so as not to be in the way; or let us look at the claws of a Python.

All the above cases may be called negative or passive shifting of the limbs towards the tail, comparing, if we want to study the process, allied species or genera which possess the same number of presacral vertebrae. As pointed out elsewhere, it is still an unsolved problem how the increase of the original, archaic total number of vertebrae has been brought about; we cannot possibly presuppose an absurdly great number upon which to draw on demand. *But there is also positive or active shifting of the limbs headwards, amounting incidentally to a shortening of the trunk.* In these cases, also, the posterior limb complex is the moving agent, but by positive push instead of resistance.

This phenomenon can be studied in the Tetrapoda with normally well-developed hind-limbs, which not only support the body but push its whole weight forwards. The critical level of stress coincides with that of the increased contact of the girdle-bearing segment and next vertebra in front. The ilium, having established the first sacral vertebra, may reasonably be expected to increase the iliosacral contact by use, so that the rib portion as well as the ilium become stronger. The ilium is induced to broaden by the attached trunk muscles, and, assisted by the incipient pre- or post-acetabular slant of the pelvis, becomes broad enough to draw into its service the muscles of the segment in front and behind, thus establishing three sacrals, but the push introduces a bias towards the segment in front, so that a previous lumbar is converted into an additional sacral or "dorso-sacral". This same bias acts upon the third or posterior vertebra in the opposite way; it tends to be relieved of its pelvic-bearing function as this is transferred forwards, so that ultimately the third becomes a postsacral, returning thereby to the status of a caudal.

This addition in front, with compensation behind, amounts to a creeping forwards of the sacral region, or *migration of the pelvis*. According to the exigencies of life, two sacrals may be enough and three sacrals may be too many, as for instance in the majority of Reptiles. Theoretically this migration phenomenon may continue almost *ad infinitum*. It is best demonstrated in those individual cases of Reptiles with three sacrals; the middle one is the strongest, with the ribs transverse; the anterior is in the process of being annexed, the rib slanting obliquely backwards, while the ribs of the posterior slant obliquely forwards, indicating the different pull. In fact, one is not yet completely converted into a sacral, the other not yet discarded. The sacral region is in a flux, arrested in *statu nascenti*, so that what is abnormal in the present generation demonstrates a reversion to normal ancestral conditions. Recapitulation by ontogeny of phylectic acquisition. Instructive are also those cases in which the right and left halves of the pelvis have not kept step. They date back to the embryonic stage, in which the halves were not yet united ventrally.

The proper numbering and naming of the adult sacral vertebrae is rather puzzling, even in closely allied species with their individual variations.

Supposing *S* designates the original primary sacral and is in all cases the thirtieth. Anterior additions to be labelled *A*, *B*, etc.; posterior additions α, β, etc.

Then the three following sacral combinations are known to occur and may be described as follows:

1.

	B	
29	*A*	1
30	*S*	2

2.

30	*S*	1
31	α	2

3.

29	*A*	1
30	*S*	2
31	α	3

etc.

1. Sacrum = *S* (30) and *A* (29), naturally referred to as the first and second sacrals, *S* being in this case the second.

2. Sacrum = *S* (30) and α (31), referred to as first and second sacrals, *S* being in this case the first.

3. Sacrum = *S* (30), *A* (29) and α (31), referred to as first, second and third sacrals, *S* being again the second.

Help may be afforded by the nerve plexus, the nervus bigeminus to the plexus ischiadicus and plexus pudendus being supposed to issue between *S* and *A*. But these plexuses change likewise, although not keeping step with the changes in the pelvis.

The proper definition of the sacrum is "all those vertebrae which are more or less synostosed because they carry or have carried the ilium". This definition includes the so-called primary vertebra and the later anterior and posterior additions, provided they act or have acted as bearers of the pelvis. Theoretically, there is no limit to the number of vertebrae drawn into the sacrum. For instance, in Birds the number of additional anterior sacrals, i.e. converted lumbars, may be five to nine, and that of the secondary postsacrals, i.e. former caudals, may be increased to seven. On the other hand, it must be left to common sense whether we apply the name of sacrum to the single primary vertebra of a Newt, or to the solitary remaining one of *Semnopithecus*, which certainly is not the homologue of the primary one of the Monkey tribe.

Migration can naturally only take place if the pelvic connexion is not hopelessly fixed as in Birds, where the forward and backward extension of the broadened or elongated ilium establishes permanent anchorings fore and aft. A similar hindrance is met with in Tortoises, with their carapace and plastron, to both of which the ends of the ilia and ischia can become attached, forming in the Pleurodira vertical struts; the dorsal ends of the ilia prefer the carapace for better support and may eventually release the sacral ribs.

The composition of the sacrum by forward migration has been carried to the extreme in the Mammalia because, their forwardly tilted pelvis having no posterior extension, all the implicated vertebrae are of necessity presacrals, whilst there can be no posterior additions, such as occur in Birds and various fossil Reptiles. It was a mistake of Gegenbaur's to declare the synostosed complex behind the one or two primary sacrals as secondary, posterior additions, called by him *pseudosacrals*, and therefore homologised with those of Birds. The vertebrae with their ribs are indeed homologous, being in both classes ilio-sacral, but the respective parts of the ilia are different; this is a warning to distinguish between complete homology and isotely. The explanation of the mammalian pseudosacrals is that they have undergone synostosis, because they all had acted as bearers of the pelvis, which service rendered them immovable. Even now in those Mammals in which they are most numerous, there is often a sharp, sudden difference between them and the caudals. All this means nothing less than that *the last of these pseudosacrals was originally the first, oldest, primary sacral of all*, and their numbers show the extent of the forward migration of the pelvis. Otherwise it would be difficult to account for their synostosis.[1]

Analysis of the number of thoracic vertebrae yields no useful results. They range from 9 to 24, with the arithmetical mean of 16 to 17, which coincides with the Monotremes; elsewhere this mean number is very rare, occurring only in the Three-toed Sloth and an Anteater. The number of thoracic plus lumbar is scarcely more promising; it ranges from 13 to 30, with 21 to 22 as the mean. Twenty-two itself seems scarcely to occur; above it, with 23, stand Elephants and Perissodactyls; *Choloepus* varies from 25 to 28, Hyracoidae from 28 to 30. With the exception of these few, and omitting the Armadillos, all the other mammals range from 21 (a few Insectivores) to 16 (most Chiroptera). No better result can be obtained from the total number of presacrals, which, ranging from 20 to 37, with a mean of 28 to 29, leaves again the Perissodactyls, Elephants and Hydracoidae, *Halicore*, Sloths and *Orycteropus* above the mean. The number of seven cervicals is doubtless the normal typical mammalian condition; the excalation of one vertebra within the series of *Manatus* is as clearly a secondary specialisation as is the unsettled number in the neck of the Sloths. Therefore, if we want to understand the trend of the building up of the trunk proper, we had better avoid searching for the elusive arithmetical mean and accept the *status plurimorum*. Such a practical mean coincides with the 26 to 27 vertebrae and is realised by such important groups as Monotremes, Marsupials, Rodents, *Manis*, Artiodactyls, Carnivora, the majority of Insectivores, *Tarsius* and *Chiromys*. While we reasonably expect the higher numbers as ancient conditions, it is fairly certain that the lower numbers, with shortening of the trunk, have been brought about by a forward shift of the pelvis, culminating in the Monkeys and the Bats. The Lemurs still range from 30 to 25, the Catarrhines centre about 26, Apes from 25 to 23, Man normally with 24 presacrals. The Edentates, admittedly an old assembly, exhibit the greatest range of all: *Choloepus* with 34, *Tatusia* with 22 to 20, having the shortest presacral trunk of all Mammals. O. Thomas has well distinguished the Edentates as Paratheria, compared with

1 A totally different kind of addition to the sacral compound is caused by the elongation of the distal ends of the ischia coming beyond the ilia into contact with some genuine caudal vertebrae, which leads to ischio-caudal synostosis, and this in turn spreads forwards by ligamentous ossification and reduces the extent of the ischio-sacral gap to a small foramen. A distant analogy occurs in *Rhea*, cf. p. 315, in the co-ossification of the distal portions of the ischia with those of the ilia, an unique feature.

which the Marsupials are a compact group, invariably with the normal number of 26 presacrals.

THE SACRUM OF MAN AND THE LARGE APES

The sacrum of Man and the large Apes is an epitome of the morphogeny of the mammalian sacrum; moreover, it stands as a type, for it combines low with advanced characters, and nearly the whole true tail has become vestigial. These characters are correlated with the upright gait.

It is surprising that the standard textbooks have not thoroughly applied all the ascertained facts to its morphology. To a great extent this reluctance is due to the lack of appreciation of migration of the limb girdles; this is a principle, which, overdone in support of the now discarded derivation of the limbs and girdles from visceral arches, has lost interest.

Analysis shows that the human sacrum is normally formed by five sacral synostosed vertebrae, the twenty-fifth to twenty-ninth. These are usually followed by four, sometimes three or five coccygeal vertebrae, of which the thirty-third and thirty-fourth are very vestigial. That most of the original tail has been lost is clearly indicated by the long embryonic filum caudale, mostly chordal, which soon vanishes. The vertebral portion of the remaining tail looks as if it were withdrawn into the body; in reality it is arrested in its further development and is overgrown by surrounding tissue, so that the tip appears withdrawn into a little dimple or fovea towards which the hairs converge spirally.

"Transverse processes" are still distinguishable on the thirty-first vertebra and are larger on the thirtieth or first coccygeal, which usually preserves its articular surface with the sacrum. The twenty-ninth vertebra (fifth sacral) is the primary sacral with a lateral groove for the fifth sacral nerve (the thirtieth). This pair of grooves is the last complete representation of the laterally open intercostal space between the ribs of the first caudal and the last sacral vertebra. The space between the twenty-ninth and twenty-eighth, and the spaces as far as that between the twenty-sixth and twenty-fifth, are converted into the "sacral foramina" with passage for the fourth to the first sacral nerves. The lateral masses are homodynamous with the partes costales (alae and transverse processes of the first sacral) which have coalesced, but indicate their nature by their separate centres of ossification. Each complete sacral vertebra has five centres of ossification, viz. two neural, one for the centrum and two for the original ribs, which like the neural arches are carried entirely by the centrum, whilst the intervertebral disks are involved in the general co-ossification during adolescence.

The twenty-fifth vertebra is genetically the latest addition to the sacrum. Since it possesses the best-developed costal elements, the ilium must have drawn this vertebra into its service at a phyletic time when the lumbars had not yet lost their ribs. If the twenty-fourth vertebra is drawn into the sacrum, the ilium now comes into direct contact with centre and arch. On the other hand, the presence of six lumbars is a case of *epistasis*, i.e. the orthogenetic forward trend of sacral formation has been arrested; a rather misnamed "reversion" to an ancestral stage.

The first sacral is, in Man, the principal carrier of the pelvis and as such is developed more rapidly than the second, which now comes into contact with the ilium later in development. The others have progressively withdrawn and synostose still later with each other. This behaviour is an instance of cenogenesis, in this case a reversal of the phyletic order. What now bears the greatest stress and strain, and is therefore the most important part of the whole structure, is laid down first and hurried on; the rest can wait, and if their "mneme", the organism's reminiscence of its ancestral acquisitions, has become dim, the respective vertebrae may conceivably revert to their original independence and shape. Such a reconstruction or recovery would be next to impossible for a sacrum which has been slowly building up into a well-synostosed series; but quite understandable if the forward shift of the ilium has been phyletically rapid, so as to leave no grave alteration behind, especially if great agility and flexibility are of greater importance than heavy support. Thus alone can we understand how the long-tailed Indian Monkey, *Semnopithecus patas*, has reduced its sacrum to the lowest possible limit, namely a single sacral vertebra.

Intimately correlated with the shift of the pelvis and alteration of the sacrum is the extent and relative position of the plexus sacralis and the plexus pudendus, discussion of which here would lead too far. It has been ascertained that neither the nervus furcalis nor the nervus bigeminus are reliable zero data. Such formations likewise belong to the category of "imitatory homologa" (Fuerbringer's term), under the spell of regional adaptation of which they are part, specialised to meet certain requirements.

The same factors have caused the lumbar region to arise with, on the one hand, its abrupt transition to the sacrum and, on the other, its gradual metamorphosis from the thoracic region, whether such region extends between the twentieth and thirtieth segments, or is shifted ten segments backwards or forwards. Loin or flanks are not thorax, neither of them being the realisation of an abstract idea, but they have become what they are through division of labour or function, with due regard to organs which must not be interfered with.

The pelvis, as such, may reasonably be considered as homologous throughout the Vertebrates. It is immaterial whether it floated up or down along the column for aeons until it connected somewhere with the axial skeleton, which independently had been shortened or lengthened. The whole limb apparatus could then initiate the phenomenon of secondary migration; it is only this later process which we can follow up by analytical induction and even by ontogenetic observation.

RÉSUMÉ OF THE EVOLUTION OF THE SACRUM

The ilio-sacral connexion has passed through five phyletic stages.

Stage 1. No connexion. There is none in the Fishes. It is secondarily lost in certain Tetrapoda with reduced limbs and pelvis.

Stage 2. The vertically upright ilium gains attachment to one vertebra by means of one pair of ribs. *Normal Amphibia.*

(*a*) This first, primary ilio-sacral vertebra lies in the same transverse-vertical plane as the acetabulum. Hence the sacral position is "*acetabular*": *Urodela*, *Lepospondyli*, *Stegocephali*; ontogenetic repetition in most Amniota.

(*b*) The ilium is much elongated, and the ventral half of the whole pelvis appears to be tilted tailwards and upwards, so that the ilio-sacral connexion lies in a plane in front of that of the acetabulum; hence "*preacetabular*": *Anura*, *Anomodontia* (Dicynodon).

(*c*) The ilio-sacral connexion is "*postacetabular*"; the tilt of the pelvis is forward, the pivot in both cases being the iliac attachment: *Reptiles* and *Birds*.

Stage 3. More vertebrae are added to the sacrum, the broadening ilium getting additional support from vertebrae with their ribs either in front, or behind, or both. The result will be two or three ilio-sacrals, which thereby change their numerical standing. If the original first or primary is designated *S*; if an added anterior, hitherto lumbar is called *A*; and if an added posterior, hitherto the first caudal, be called α, then the formula of the whole sacrum may be $S + A$, but now described as composed of two sacral vertebrae, no. 1 and no. 2, since it is customary to count from head to tail; or the formula may be $S + \alpha$, also described as sacral no. 1 and no. 2, although the serial number has changed; or, lastly, it may be $A + S + \alpha$, in which case *S* becomes no. 2. This process of adding to the original single sacral vertebra begins already in Amphibia, e.g. those Newts with individually two sacrals. It is also common in Lizards, Crocodiles and Chelonians with abnormally three more or less complete sacrals; also in Mammalia.

Stage 4. Increased number of sacrals by continuation of this process leads obviously to sacral groups.

(*a*) Either forward extension by forward growth of the ilium, the formula being $C + B + A + S$; e.g. *Dicynodont seeleyi* with four sacrals, of which *A*, *B* and *C* are clearly preacetabular.

(*b*) Or, forward and backward increase of the ilium with additional assimilation of former lumbar and former caudal vertebrae; *Sauropodous Dinosaurs*, and independently resorted to and carried to the extreme by all Birds, in conformity with their upright gait. Most instructive are the formulae of *Sauropoda*: *Morosaurus* and *Diplodocus* with five, *Brontosaurus* with six sacrals, the last of which in all cases is a postacetabular caudo-sacral. Birds with from five to nine anterior, and from four to seven posterior or caudo-sacrals. The Bird's sacrum is greatly complicated by a more or less complete reduction of ilio-sacral ribs in the region of the very variable and much developed and condensed sciatic plexus.

Stage 5. *Mammals*; increase of the forward and backward enlargement of the ilio-sacral complex is checked or compensated by withdrawal of the vertebral-iliac connexion behind, whilst new additional lumbar vertebrae are annexed and converted into sacrals. If the respective lumbar ribs are still present, the ilio-sacral connexion is by the pars costalis; if the ribs have already been reduced, the connexion is directly ilio-vertebral, i.e. by the centra or by the transverse processes.

The withdrawal of the ilio-sacral contact begins at the posterior, caudal end and proceeds headwards, resulting in the formation of a right and left gap between them and the ilium, bridged by ligamentous connective tissue, which in old specimens may become ossified. These "pseudosacrals" of Gegenbaur are homologous with those of Birds only in so far as the material is concerned, but they are totally different in their relation to the ilium. In the Birds they are caused by the contact of originally çaudal units with the posterior, originally postsacral half of the ilia to which they remain attached, and their increase in numbers proceeds from the primary sacrals tailwards. In the Mammalia they are transformed by the anterior extension of the ilium and therefore they increase from the primary sacrals forwards. They are in fact previous lumbars, like the presacrals or anterior additional pseudosacrals of Birds; and this process of headward assimilation and conversion of lumbars into dorso-sacrals, and these into sacrals, has been active in the mammals to a very great extent and is still going on. This process has gained momentum, but when the economic or physiological limit is reached compensation asserts itself by progressive reduction at the posterior end. All this amounts to a forward migration of the whole pelvis, the initiative resting with the use of the posterior limbs.

Chapter XXVIII

PRIMITIVE SAUROMORPHA OF UNCERTAIN AFFINITIES

The earliest ancestors of the modern reptilian group and the Birds, which form collectively the Sauromorpha, first make their appearance during Permian times. The examples which are to be described in this chapter, namely, *Araeoscelis*, *Kadaliosaurus*, *Protorosaurus* and the Rhynchocephalia, certainly represent early stages of Sauromorph development, though their exact relationships are difficult to define.

Araeoscelis gracilis. Lizard-shaped, Permian of Texas (Fig. 93). Vertebrae amphicoelous, seven elongated cervical, about 20 others presacral, two sacral and many caudal. Some of the anterior caudals have long curved ribs besides chevrons. No abdominal ribs. Intercentra present. Long, lizard-like running legs; phalangeal formula 2, 3, 4, 5, 3. An unsolved difficulty is the skull; there is one pair of fair-sized upper temporal fossae, well separated from the orbits, bordered by parietal plus supratemporal and squamosal plus postorbital. The squamosal is very large and extends forward as a thin bony plate along the lower border of the postorbital to the jugal, which forms the whole lower border of the orbit. A typical large quadrate is apparently fixed; quadrato-jugal is absent. If the supratemporal, which extends from the parietal to the squamosal and quadrate, forms a bridge or arcade, this would bear a striking resemblance to a third arch, so strongly developed in *Sphenodon*, and present, though reduced almost to vanishing point, in the Crocodiles.

Kadaliosaurus, Permian of Saxony, claims affinity mainly by its limbs; abdominal ribs absent; the all-important skull is not known.

Protorosaurus, of the Permian of Germany and the magnesian limestone of England, has been redescribed by Seeley (Fig. 94) and by Williston. Many specimens are now known, but the only skull known, and that imperfect, is that of the first example to be described. Williston refuses affinity with Rhynchocephalia in favour of Lepidosauria. Abel leaves it as a separate order, but on p. 539 he says that the Pseudosuchia lead back to the primitive Rhynchocephalia like *Palaeohatteria* and *Protorosaurus*. On the whole it appears to be closely allied to the Rhynchocephalia.

Fig. 93. *Araeoscelis gracilis* (Protorosauria). Skeleton and restoration. About one-fourth natural size. (Williston.)

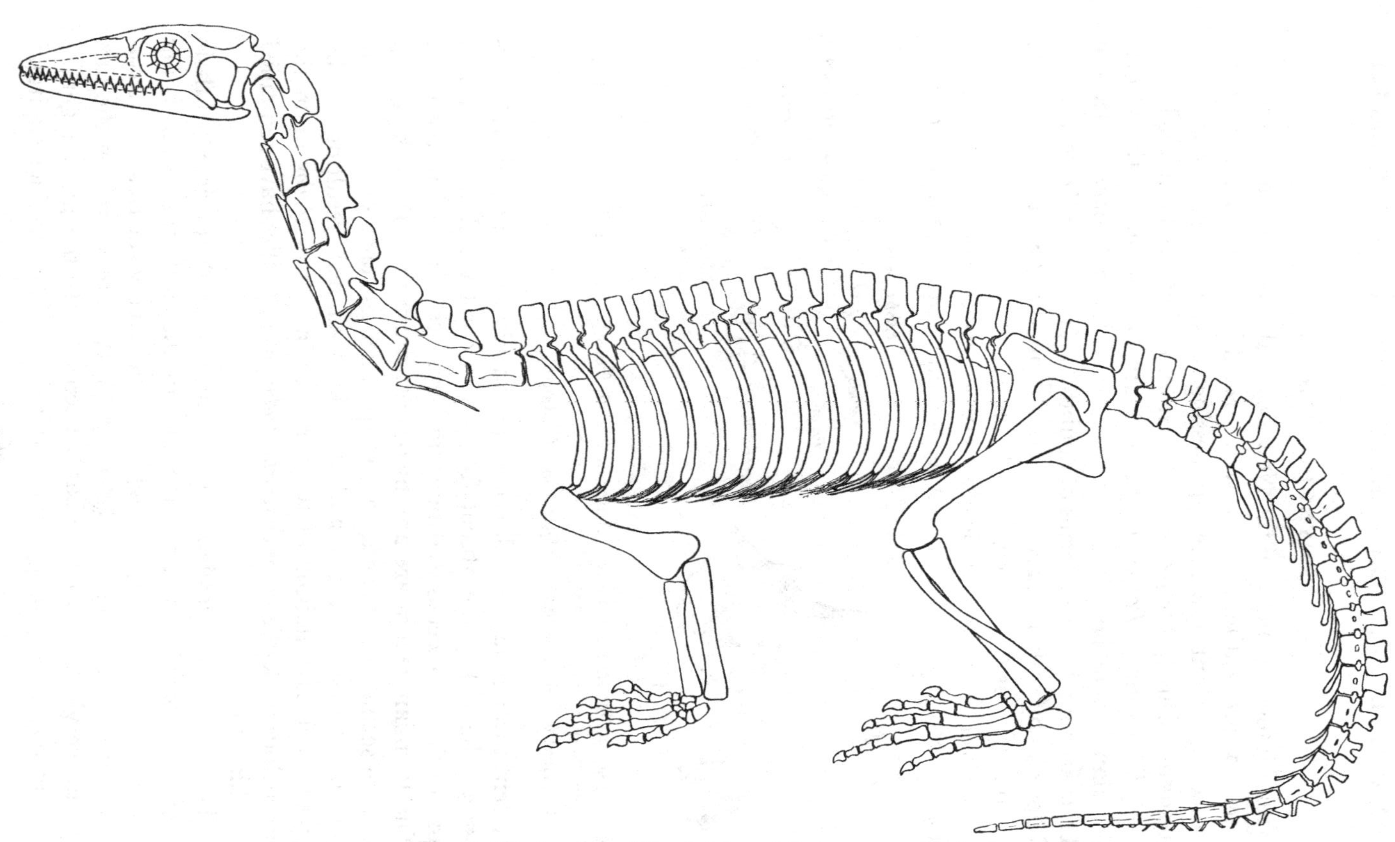

Fig. 94. Restoration of the skeleton of *Protorosaurus*. (Seeley.)

RHYNCHOCEPHALIA

With the sole exception of the New Zealand *Sphenodon*, the Rhynchocephalia are extinct. They were lizard-like, unarmoured, four-footed, acrodont creatures, possessing a still cartilaginous costal sternum. Chiefly Mesozoic, they first appear in the Permian Lower Red, for example *Palaeohatteria*, and are next found in the Upper Permian magnesian limestone, for example *Protorosaurus*. In the Trias are found marine forms of doubtful affinities like *Thalattosaurus*; followed by other marine creatures in the Cretaceous–Eocene period, such as *Champsosaurus* and *Simoedosaurus*.

The vertebrae are typically gastrocentrous, composed of basidorsals, centrum and basiventrals or "intercentra", the latter being

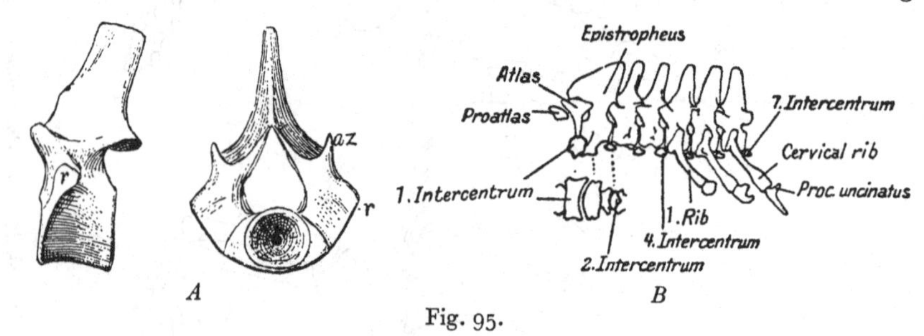

Fig. 95.

A. Sphenodon (Rhynchocephalia). Anterior dorsal, from the side and front. *az*, prezygapophysis; *r*, rib. (Williston.)

B. Cervical vertebral column of *Sphenodon*. (Osborn.)

present throughout the column except in *Rhynchosaurus* and *Hyperodapedon*, where they are absent from the second vertebra backwards, and in the Cretaceous marine forms they are restricted to the neck. The Permian genera are amphicoelous; during the Trias the group as such seems to have entered an experimental stage, one end of the centrum becoming plane and then knob-shaped, resulting in the opisthocoelous condition of the cervicals in *Hyperodapedon* or in the procoelous neck of *Sauranodon*. *Sphenodon* and its allies have remained amphicoelous.

The ribs of the Permian genera seem to be still restricted to their capitular portion, the neural arches of the neck and trunk being devoid of diapophyses, although the eight cervical ribs of *Protorosaurus* are rather long. In *Hyperodapedon* ribs are absent on the first to fifth cervicals, having probably been lost. In the Jurassic genera and in *Sphenodon* (Fig. 95) the cervical ribs have distinct capitular facets on the basiventrals, or between the centra, the corresponding tubercula being attached to short protuberances well on the middle of their centra. In the trunk, both the capitulum and tuberculum are in the usual way transferred on to the neural diapophyses, therefore "one-headed". For detail, cf. *Sphenodon*, Crocodiles and Tortoises. It is important to note that the Rhynchocephalia had still a primitive costal sternum, this being entirely cartilaginous, the osseous keel of *Sphenodon* excepted, unless this contains the dermal episternal elements. The articular processes are restricted to the usual prezygapophyses and postzygapophyses.

The regional number of vertebrae in certain examples of this group is as follows:

Palaeohatteria:	cervical 6,	trunk 20,	sacral 2;	first sacral	= 27th.
Protorosaurus:	„ 8,	„ 16,	„ 3–4?;	„	= 25th.
Sphenodon:	„ 8,	„ 17,	„ 2;[1]	„	= 26th.
Pleurosaurus:	„ 5,	„ 40,	„ 2;	„	= 46th.

The tail in this group is long, and in *Sphenodon* is capable of regeneration. The second ribs of *Sauranodon* and *Homoeosaurus* are forked distally.

The ontogeny of *Sphenodon* has been studied by Howes and Schauinsland. The following account, together with a more detailed description of the adult condition, may therefore throw some light upon the meaning of some differential features mentioned above.

The chorda of the embryo is interrupted by septa of so-called chordal cartilage (Fig. 96), caused by the formation of the centra which, increasing in thickness, compress the chorda until, with ensuing ossification, solid, hourglass-shaped centra are formed. The amphicoelous shape is partly due to the basiventrals forming not only a ventral semi-ring, but a complete ring, the so-called intervertebral disk which does not encroach upon the chorda. The disk, together with the enclosed remnant of the chorda, acts throughout life as an articular pad between the vertebrae, cf. Geckos. In sagittal sections of the first and second vertebrae the dorsal half of the ring is markedly thinner than its ventral bulky portion; farther back in the column this difference disappears.

Howes has found the component cartilages to appear in the following order. First, the basiventrals, which by fusion form the future

1 Howes has found three in an embryo; the third or posterior pair being much weaker and rather asymmetrical; an indication of a forward shift of the pelvis.

"intercentra"; second, the right and left basidorsals or neural arches, which are at first separate and later fuse together; third, the interventrals. Here he made the important discovery that the odontoid or first centrum arises from paired cartilaginous clusters, which he unhesitatingly recognised as the first pair of interventrals. Schauinsland, especially, has shown that the cranial end of the atlas centrum has a more or less separate continuation (Fig. 96). In sagittal section it clearly shows even its chordal septal cartilage connecting the dorsal and ventral portions. He reasonably attributes this anterior piece to

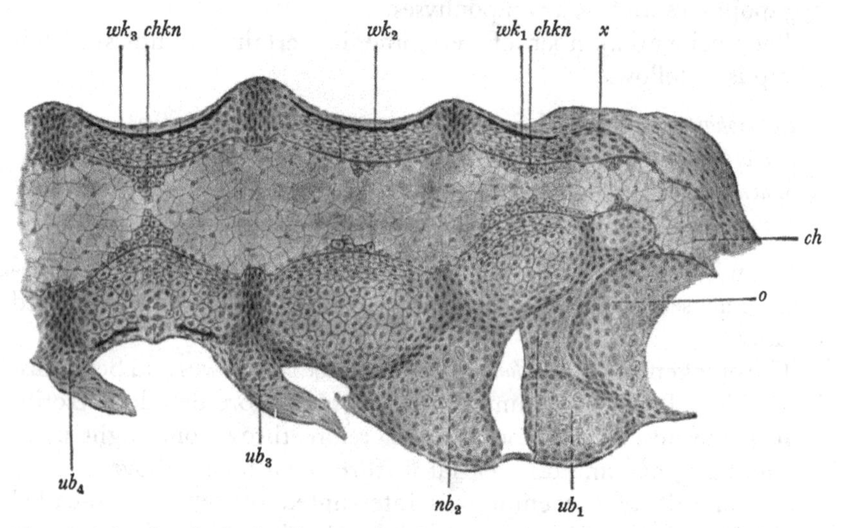

Fig. 96. *Sphenodon.* Sagittal section through the first three cervical vertebrae of a *Sphenodon* embryo. *ch*, notochord; *chkn*, chordal cartilage; wk_1, centrum of atlas which later fuses with wk_2 to form the odontoid process; *x*, anterior, more or less separated portion of the atlas centrum, presumably the proatlas; ub_1, ub_2, etc., "untere Bögen" or hypochordal bows, i.e. basiventrals; *o*, occipital. (Schauinsland.)

the body of a proatlas. The neural arches of this proatlas are, shortly before hatching, represented by a right and a left small roundish cartilage, each of which is connected in front with the exoccipital, behind with the first pair of basidorsals.

At first the BD 1 and BV 1 do not meet each other on the side of the odontoid. Later, by their junction, the atlas ring is formed; it then literally slips backwards over the finger-shaped odontoid like a ring. Thus it comes to pass that BV 1 and BV 2 come into contact with each other and that the first centrum is pushed dorsally, away from the ventral side of the column. The first centrum fuses with the second united pair of BV 2; also with centrum II, but only peripherally, thanks to the outer non-cartilaginous portion of the skeletogenous layer, which invests the two centra like a mantle. In sagittal section the core or bulk of these centra remains unaffected by their ossification in the immature and even in the adult, the discontinuity being probably due to lingering remnants of the "intravertebral" disk or axial portion of BV 2. The first centrum having partly slipped into its atlas ring, it has no lateral apophyses for ribs; from the second onwards such are present in the adult as small tubercles.

The chevrons have been recognised as the direct derivatives of the paired intercentra. Howes and Swinnerton have devoted much time to the origin and behaviour of the "intercentra" of *Sphenodon*. They present a great puzzle; at first they are paired, then median or fused cartilaginous elements, and in development they extend from the atlas far into the tail. Of this set only the first four or five of the neck, and the chevrons, are retained by the adult. From the middle of the neck to the first chevron the first set vanishes, to be replaced in the ripe embryo by a secondary set, which is retained in the adult as a series of little short-horned osseous menisci, usually median and transverse, i.e. unpaired, wedged in between the trunk vertebrae. The vanishing of the first set is described as implying "no loss of substance but merely histological change, leading to a loss of individuality" which is later recovered by ossification. For detail we have to refer the reader to Howes' text, pp. 19–20, 25, and a table on p. 27. The phyletic meaning of this extraordinary behaviour of the intercentra is not thereby settled. It was known already that in *Lacerta* some of these wedges remained paired, right and left.

A fairly adult specimen in the Cambridge Museum shows the following: vertebrae 1–8 are cervical, 9–25 trunk and 26 and 27 sacral. Intercentra are present from the first, which, as the rib-bearing ventral half of the atlas ring, is of course the first basiventral mass, to the thirtieth, the last wedge lying between 29 and 30. The first chevron is attached to the caudal end of the thirtieth. An important feature is the occurrence of triple intercentral pieces: between the fourth and fifth, and again between the sixth and seventh, lies a small pair of nodules in front, a larger behind. In those between the twenty-third and twenty-fourth the proportions are reversed, the anterior pair being the wedge, the median piece being a nodule.

For the solution of this problem we have to appeal to the principle of dispondylism, according to which both the anterior and the posterior skelerotome half of a vertebra originally produced, or perhaps more likely retained the power of producing, the neural and haemal arches, so that in the present case the anterior pair belongs genetically to the previous, the median piece to its own vertebra. Since the unpaired or primary set is clearly more broadly and firmly attached to the cranial end of its own vertebra, it follows that this set is the true, usually occurring, intercentral or basiventral element; whilst the secondary set, the irregularly cropping up paired elements, belong to the posterior end of their vertebra and are serially homologous with the chevrons. This would explain the frequent tendency of chevrons to fuse on to the posterior end of a vertebra, not only in *Sphenodon*, but in many other Amniota.

The chevrons proper belong genetically to the centrum of the Amniota, i.e. to the interventral units; the usual intercentra of the trunk and neck are then the ventral remnants of the rib-producing basiventral units, and as such are continuous with the non-ossifying fibro-cartilaginous so-called intervertebral disk of the Amniota. In other words, ribs and chevrons are not serial homologues, as has been believed by the majority of morphologists. Ribs and chevrons can coexist on the same vertebra without need for the suggestion that they are produced by a kind of bifurcation, followed by the separation of the more lateral and more ventral arms. This accounts for the occurrence of three or really four pieces in the trunk and neck of *Sphenodon*. Last, but not least, it settles the vexed question of dorsal and ventral ribs.[1]

The first two or three ribs are represented by, or reduced to bands of, connective tissue without bone or cartilage, such being suppressed even in the embryo. The first pair of these ghost-like costal vestiges is attached to the first basiventral mass. The second arises from the BV 2, and its tubercular portion is indicated by strands which are attached to a small osseous tubercle on the middle of the second centrum. The third behaves similarly. The vertebral arteries and the lateral sympathetic chain pass through these double basal attachments, capitulum and tuberculum, of the reduced ribs. From the

[1] Considering that the chevrons are genetically part of the IV, i.e. the centrum of Amniota, and therefore of the intervertebral cartilage of the Amphibia, while the ribs and the true intercentra are part of the BV which in the Amniota are reduced to an intercentral and physiologically intervertebral position, utter confusion is wellnigh unavoidable. Coloured diagrams are the surest preventive measure.

fourth backwards the ribs are osseous. The short capitula retain their partly intercentral attachment, while the longer tubercula are carried by low processes of the centra, parapophyses. In the trunk the capitulum and tuberculum become shorter, merging into one facet at first dumb-bell-shaped, farther back oval, articulating directly upon the centra, which send out no processes. At the same time the facets gradually shift backwards and dorsally upon the midcentrum, as they are in the neck, until in the posterior thoracic region they lie right across the neurocentral suture. The stouter sacral ribs articulate with their centra and neural arches upon very short processes, which therefore represent united diapophyses and parapophyses or transverse processes. The distal ends of the two sacral ribs are united by non-ossifying cartilage, which thus forms a kind of compound epiphysis for the iliac bones. The caudal ribs are rapidly declining into short, pointed stumps, co-ossified with the vertebrae; they disappear beyond the third or fourth caudal, which happens to be the first in which the split into an anterior and a posterior half begins to show itself.

Lastly, some of the ribs carry uncinate processes, usually lost in maceration. The best account is given by Howes, who traced them through the whole thoracic region, from about the last cervical to the twelfth poststernal rib: "...they arise, after the ribs have become chondrified, by independent concentration of the cells which go to form the intermuscular septa. They chondrify independently when the ribs of the embryo are well ossified. Jeffrey Parker has already called attention to the similar independence in *Apteryx*". He leaves the question of their morphological standing in abeyance.

A tenable point of view is to look upon all these uncinates of Birds, Crocodiles, *Sphenodon*, Urodela, and above all the Gymnophiona together with the "Rippentraeger", as having arisen out of the tendinous portions of muscles which extended, and partly still extend, between the ribs and the diapophysis or transverse process of the vertebrae. Such muscles must have been movers of the ribs, e.g. levatores costarum, which are really parts of the intercostales externi, which in their turn lead transitionally to the system of intertransversarii. The external intercostal muscles of Birds are actually divided by the uncinate processes into a dorsal and a ventral half. Further, since the free ends of their uncinates are fixed to the overlapped, next following, rib by band-like connective tissue, the uncinates represent here a skeletal link between neighbouring ribs. Again, as scalene muscles, themselves representing levatores costarum, in the cervical region extend from the cervical lateral processes to the free ends of the more posterior ribs, we can see the way to the formation of ligaments extending from one transverse process to the body of the next following rib. It is no wild assumption that such ligaments, the remnants of muscles, should calcify, ossify by direct process, or even become cartilaginous. Examples of such conversion are common enough.

Chapter XXIX

GECKONIDAE

THE MOST PRIMITIVE SURVIVING AUTOSAURI

The Geckonidae are four-footed, absolutely terrestrial Reptiles with deeply amphicoelous, gastrocentrous vertebrae; the Eublepharinae form an exception, since they have advanced to the procoelous stage. Basiventrals are present as intercentra throughout the column. Post-thoracic ribs are slender and prolonged as cartilaginous strings which, as in Chamaeleons, meet in the medio-ventral line. The skin of trunk and tail usually has numerous, often merely granular but typical dermal calcifications or ossifications. The cranium is broadly depressed.

The Geckos, about 270 species, with their cosmopolitan distribution in all tropical and subtropical countries, and even in oceanic islands, are a very independent and without doubt a very old, if not the oldest surviving, branch of the Sauromorpha. Fossils are still unknown, but their unmistakable resemblance to the Palaeozoic Microsauri are at least remarkable. Fuerbringer pleads, with good reasons, for affinity with the highly specialised Chamaeleons in opposition to all the other Lizards with which they will of course always be classified, just as this happened for many years to *Sphenodon*. They are, in fact, an important key group.

The vertebral column contains from 50 to 60 vertebrae, of which the twenty-seventh and twenty-eighth or twenty-eighth and twenty-ninth usually form the sacrum. The atlas ring is formed by the joined first neural arch and basiventral, while the second basiventral is attached to the cranial surface of the second centrum and produces, like the next few following, a sharp median ventral apophysis. The third has the indication of a transverse process; the fourth carries a short, straight, free rib. The fifth carries the first thoracic rib, so that the neck is remarkable for its shortness, having only four vertebrae. Intercentra occur throughout the trunk and chevrons in the tail. Those of the middle of the trunk may be of considerable size. The bulk of the axial portion of the vertebrae is formed by the true centrum, which in the Geckos seems to contain most or the whole of the original intervertebral cartilage, with however great preponderance of the interventral share. The vertebra would therefore be in the very important stage of being still holospondylous, potentially quadripartite, but on the way to the typical Amniotic gastrocentrous condition, reminding us of the unsettled condition of the Palaeozoic Stegocephali and Cricoti. Yet the Geckos, although bridging the gulf between Amphibia and gastrocentrous Reptiles, are neither, because their vertebral axial mass is not subdivided into equal anterior and posterior halves as in the Cricoti, and because their basiventrals are already converted into almost typical intervertebral pads.

The vertebrae of the trunk are of elongated hourglass shape. Each centrum is at first a cartilaginous tube which carries by high right and left wings the whole of the broad-based neural arch, as indicated by a typical neurocentral suture. Exactly in the middle of this calcifying and ossifying tube or shell is formed a cartilaginous septum, the so-called chordal cartilage; see discussion under Coeciliae, p. 136. When incomplete, as for instance in the trunk of *Phyllodactylus*, the chorda is only constricted, and is therefore capable of continuous growth throughout the column with the exception of the end of the tail, where the chorda is destroyed by the surrounding cartilage, as in *Chimaera*, Dipnoi, Urodeles and many Lizards. The septum may in time be invaded by calcification. Its large nucleated cells retain the appearance of embryonic cartilage, and it coincides exactly with the line of transverse division of most of the caudal vertebrae, when these break into an anterior and a posterior half and from thence reproduce the new tail.

Albrecht had the brilliant idea that this break indicated the junction of the cranial and the caudal half of the primordial skelerotome and that the line of division still retained "interprotovertebral tissue". Such tissue would be capable of reproducing anything of a skeletal nature, for instance, the continuous loss of part of the tail. The idea is suggestive and still favoured by some morphologists. It is precisely the level at which the tail breaks off and whence it is renewed. Nevertheless it is erroneous, owing to some inadvertent confusion. If the septum coincided with the original ends of the successive skelerotomes, then the basiventral mass would lie in front of it and would form the anterior half of the vertebra, while the centrum would be the posterior half; but it is this posterior portion, the centrum itself, which is split transversely, as can be seen in any well-macerated tail vertebra in any of those Reptiles which can renew their lost tail. These transverse splits are not reminiscences but later acquisitions. The cartilaginous septum is a necessary condition for the reproduction of the tail, but in the trunk, where of course this phenomenon of self-mutilation cannot occur, the preformed split of the centrum does not take place, although the septum is there. Hyrtl and Gegenbaur had already observed that the split begins on the outside of the shell, becomes deeper and gradually extends peripherally across the neurocentral suture and there even attacks and partly divides the neural arch. The whole process is postembryonic. Further, as the diagrams show, the split may pass behind or even right through the transverse process. Lastly, the split is entered by a blood vessel which is not Schauinsland's "Intersegmentalgefaess".

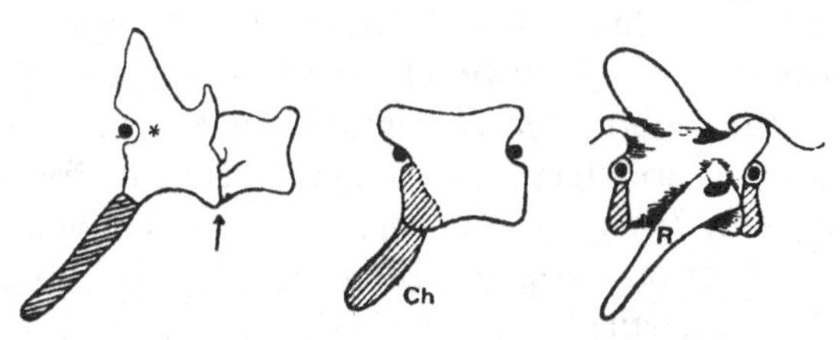

Fig. 97. *Gecko verus*, Camb. Mus. One mid-caudal, one anterior caudal and one lumbar vertebra, seen from the right side. The * in the left figure indicates the place of origin, the ● the exit of the spinal nerve. The whole mid-caudal vertebra is divided into a smaller anterior, and a larger posterior portion, ↑ indicates the line of cleavage, the former carrying the rudimentary transverse process, the latter the chevron, *Ch*. Basiventralia and chevrons are shaded obliquely. *R*, rib.

It enters at the only place where it can survive, namely where the basidorsals and centra produce and send inwards the so-called chordal cartilage.

The reproduced tail is only a sham, although sometimes a surprisingly close counterfeit, covered with skin which produces a caudal scaling more or less true to type. But the core of the whole is only a non-segmented rod or tube of fibro-cartilage without the chorda and without spinal cord. Even this new tail often gets broken and reproduced, for instance, the Tejid Lizard *Cnemidophorus*, of which I have collected not a few specimens with the originally long tail ending with three new instalments.

The split caudal vertebrae of Geckos and Lizards (Fig. 97) seem to have invited comparison with the "double" vertebrae of the Cricoti; quite wrongly, since the anterior and posterior halves together make the centrum of the Geckos and Lizards, while the two blocks of the Cricoti are the basiventral and the centrum. Nor are the chordal septa of these Reptiles homologous with those of the Urodela and Coeciliae, because in the former such septa are formed by the neural arch and the centrum and in the latter by the neural arch and the still large basiventral. Compression by the dorsal and ventral units has in both cases produced the same phenomenon. The change of the amphibian pseudocentrum is brought about by the extension backwards of the base of the neural arch upon the posterior mass, as it actually exists and takes place in *Cricotus*; this shift of the neural base will result in setting free the anterior end of the basiventrals. A similar shift of the neural arch also accounts, in some of the Gecko's caudal vertebrae, for the large spinous process which forms the posterior dorsal half of the whole vertebra; the anterior half of this split vertebra is then mistaken for the real basiventral.

Another feature characteristic of the Geckos, rather perplexing at first sight, is the intervertebral region. In the dried skeleton the osseous vertebrae are deeply biconcave. In the fresh condition, however, there is between the vertebrae a biconvex lens-shaped swelling of the chorda, peripherally bordered by a complete ring of the intervertebral cartilage, narrower dorsally and broadening ventrally, and in the tail carrying the chevrons. In the trunk this ventral swelling is just touched by the capitulum of the rib (Fig. 97). The whole complete ring is of course equivalent to the complete pair of basiventrals, which had formed first a semi-ring, the horns of which later growing upwards completed themselves into a meniscus, intervertebral pad, or intercentrum, acting like a washer or pad to this most imperfect of axial joints. It bears a great resemblance to the equally imperfect joint of the Urodela, before this ring is vertically split. But whilst in the Urodeles this ring is composed of the ID plus IV, i.e. the whole intervertebral mass, that of the Geckos and other Amniota is formed of the pair of basiventral units. Other changes in this unstable region have been discussed. Confusion of fundamental importance still exists, even in Schauinsland's elaborate and for the most part very careful account.

For some reason or other the chevrons with their rather large intercentral, i.e. basiventral cartilage, heal on to, or fuse by subsequent ossification on to, the posterior part of the vertebra next in front, that is to say, on to the wrong vertebra, a feature well known to occur in various Lizards. When in old specimens all the components of the vertebrae are co-ossified, and with the help of the skeletogenous mantle the sutures are abolished or at least hidden, such vertebrae bear a great resemblance to those of some Urodeles. In these, however, the chevrons are direct, legitimate outgrowths of the pseudocentra, and *eo ipso* form from the beginning part of the posterior continuation of their own vertebra. Whatever may be the mechanical advantage of the attachment of the ventral tail muscles to the preceding vertebra instead of to the next following, the necessary arrangement has been brought about by different means: either simply by elongating the outgrowth of the large pseudocentrum, basiventral, or, on the other hand, effecting a similar elongation by adding the whole chevron minus the vanished mother substance, namely the axial portion of the basiventral, to the hinder end of the vertebra in front, and this end naturally happens to be the centrum, the pseudocentrum having vanished. It is a clear case of isotely, the accomplishment of the same end but by different means, without regard to complete homology of material.

Chapter XXX

LACERTAE

THE LIZARDS AND CHAMAELEONS

The Lacertae possess procoelous vertebrae; the ribs have capitula, but reduced tubercula. Short paired or median haemapophyses in the neck; chevrons in the tail. Hypapophyses absent.

The group is cosmopolitan, with about 1800 recent species arranged in many families, and is apparently still actively developing. They are essentially terrestrial, although the Iguanid *Amblyrhynchus* of the Galapagos dives into the sea to feed upon algae, and the Central American arboreal *Basiliscus* runs paddling over the surface of brooks and lagoons. Some other Iguanids dive for safety, to hide at the bottom. Various Varanids also take freely to the water. The limbs vary extremely in the genera, and even in the species of some families, from the fully developed four to two or no limbs. Pre-Tertiary ancestors of present families are very scarce. Most of them are extinct. Some of these doubtful creatures date back to the Lower African Trias. For the very important Palaeagama and Palaeiguana of the Karroo Trias, see Eosuchia and Pseudosuchia, p. 293. The origin of the Lacertilia is still problematical. We have to fall back upon the Microsauri and Geckos. Their relationship to Pythonomorpha and Snakes will be discussed later. The number of vertebrae in the neck and trunk varies as follows:

Varanus giganteus; 9 cervical, sacrals nos. 30 and 31 left; 31 and 32 right.

Varanus giganteus; 8 cervical, sacrals nos. 30, 31.

Iguana tuberculata; 8 cervical, sacrals nos. 26, 27.

Uromastix spinipes; 7 cervical, sacrals nos. 25, 26.

Agama stellio; 7 cervical.

Lacerta viridis; 7 cervical, sacrals nos. 27, 28.

Cyclodus gigas; 7 cervical, sacrals nos. 37, 38.

Chamaeleon vulgaris; 6 cervical (4th and 5th with very long ribs). Sacrals nos. 23, 24, very loosely connected by short ribs.

Sphenodon (for comparison); 8 cervical, 3 sternal, 11 other thoracic, 3 lumbar, 2 sacral (26th and 27th).

The vertebrae of the Lizards and Chamaeleons are typically gastrocentrous. The embryonic chorda shows two constrictions to each vertebra. First, intravertebrally, due to the formation of the "chorda-cartilage"; secondly, owing to the basiventral ring. The last traces of the chorda vanish during adolescence. The neural arch tends to fuse with the centrum, abolishing the sutures, so that the adult vertebra becomes quite stereospondylous and solid. In the Iguanidae zygosphenes and zygantra secure an additional interlocking, besides the usual anterior and posterior zygapophyses (Fig. 98). The basiventrals are reduced to osseous unpaired nodules or wedges and persist as intercentra between most of the cervical vertebrae; they are absent in the thoracic and lumbar regions and reappear in the tail, either as little wedges or as chevrons; sometimes with vestigial menisci. BV I forms the ventral half of the atlas ring, with the neural half of which (BD I) it is connected by suture. BV 2 fuses usually with the cranial portion of the second centrum, and sometimes also with the caudal and ventral portion of the odontoid; in either case the first and second basiventrals are in full contact with each other (Fig. 99). The odontoid is firmly sutured or fused on to the second centrum. BV 3 either remains with the cranial end of the third vertebra, or it attaches itself to the caudal end of the second, so that this epistropheus carries two pairs of basiventrals, cf. also Chapter XIII. BV 4 and BV 5 are still more reduced in size to intercentral wedges, or nodules, as in most Lizards, or they anchylose with one or other of the adjoining vertebrae, e.g. in Anguids

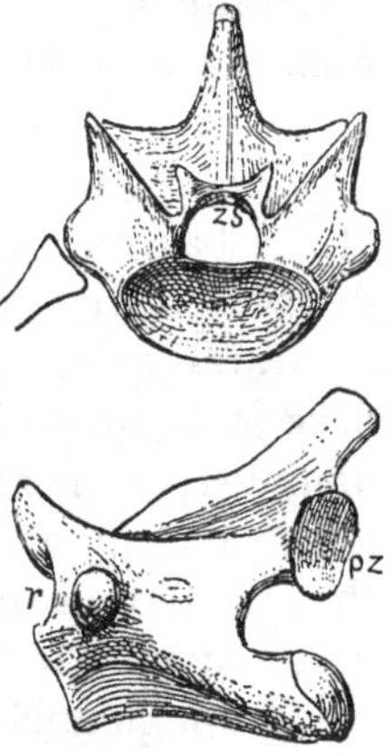

Fig. 98. *Iguana* (Lacertilia). Anterior dorsal vertebra, from the front and side. *zs*, zygosphene; *pz*, postzygapophysis; *r*, rib facet. (Williston.)

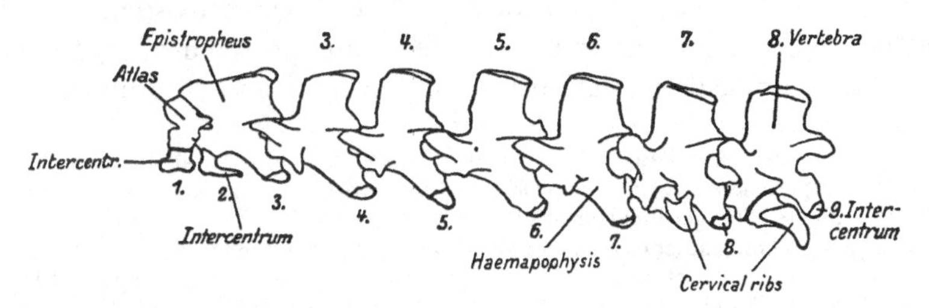

Fig. 99. Cervical vertebral column of *Varanus*, showing haemapophyses and intercentra. (Ihle.)

and Varanids. That these basiventral elements in the neck are homologous with the chevrons is indicated by the fact that, when in certain families the cervical wedges are fused on to the vertebrae, the same applies to the tail, and when they are free they are free also in the tail.[1] Further, in *Heloderma* the wedges of the anterior caudal vertebrae are sometimes still paired knobs, not yet elongated into the right and left shanks of the chevrons, retaining (as also in the BV 1) their original paired nature. Lastly, when the caudal vertebrae are strongly procoelous, the chevrons are attached to the neck of the pronounced posterior knob, so that they seem to belong to the posterior end of this vertebra, although as basiventral elements they belong of course to the next following. Even the axial portion of the basiventrals can persist as a fibrous disk which is interposed between the caudal end of the centrum and the knob which calcifies separately; and when, as in *Pseudopus*, the chevrons fuse with this neck in front of the knob, they look as if they were the exact serial homologues of the Newt's caudal haemapophyses. In reality this is a case of only general serial homology since, expressed numerically, the pseudocentrous Newt's ventral outgrowths would belong to vertebra 30, but the chevrons of the Lizard would belong to vertebra 31, although anchylosed to the caudal end of vertebra 30. The centra of Lizards develop no hypapophyses, differing thereby essentially from the Snakes. There are normally two sacral vertebrae and two pairs of sacral ribs. The caudal vertebrae, except for the first few in various families, are split in two like those of the Geckos, the split making its appearance with the ossification of the vertebral bodies. The resulting anterior half is usually the smaller, often reduced to a very narrow disk; farther back in the tail they are more equal. It is one of the indications that the tail usually retains more ancient features.

The atlas is always devoid of ribs. Their former existence on the second vertebra is sometimes (e.g. an adult specimen of *Hydrosaurus giganteus*, Cambridge Museum) indicated by a pair of unmistakable

1 This sympathy is a case of "homoeosis", a way in which serial homology asserts itself. The experiences or gains by one segment may affect its neighbours. The individual gain has affected the germinal mass of that respective series of organs, these behaving alike in neck and tail, either remaining independent or fusing on to the centrum. We cannot well assume that the neck and tail labour under the same condition as opposed to the trunk (in which the absence of wedges is clearly due to suppression) and that therefore the equal behaviour of cervical and caudal wedges is merely a survival of the original state.

capitular facets. In ripe embryos of *Lacerta vivipara* ribs are broadly attached from the fourth vertebra onwards, the attachment lying across the neurocentral region. The ribs of the trunk are often mentioned as single-headed, a convenient way of avoiding particulars. They have large, thick capitula articulating near or across the suture; the tubercula are much weaker, hook-like, extending obliquely dorsally and caudally without a direct bony attachment, but only a ligamentous one. In the tail the capitular portion is much reduced, while the tubercular is much stronger and lies behind, caudally, no longer above the capitulum, and is directly fused on to the body of the vertebra. In those Lizards whose vertebrae split transversely into an anterior and a posterior portion the ribs or "transverse processes" come to be carried by the posterior larger half. In short, the so-called transverse processes of the Lizard's tail (and the same applies to the Crocodiles and the Tortoises) are formed entirely by the elongated tubercular or dorsal half of the ribs. The neurocentral suture persists sometimes on the anterior caudals; on the posterior it is completely abolished, and the resulting transverse process behaves either like a normal diapophysis, where the rib was attached above the suture, or like a parapophysis, if the rib is attached to the middle of the centrum, so that the germ of the rib, amalgamated with the centrum, behaves as if it were an interventral instead of a basiventral element. This is an excellent instance of cenogenesis having produced a new kind of transverse process, initiated by the shifting of the rib, itself a cenogenetic act. The rib, genetically the distal portion of a basiventral, has been forced on to the centrum, and thence eventually with its capitulum and tuberculum on to a neural diapophysis, which in the lumbar region of many Amniota may then later lose this rib and remain as a large upper transverse process. In the tail, where the rib lies flat against the base of the vertebral bodies, this base (which is either the centrum, i.e. interventral, or the neural arch, i.e. basidorsal) is induced to form by proliferation an entirely new transverse process which mimics the ribless lumbar process. The tail seems to have returned to that primitive condition in which a lateral outgrowth or true apophysis separated off its distal portion and thus produced a rib.

The bases of the neural arch send out right and left thickenings of a cortical layer of partly cartilaginous or calcifying tissue, which like a mantle comes to surround most of the vertebra and soon abolishes every trace of sutures between the constituent units; this is a feature rather characteristic of the Lizard's tail and of most vertebrae of Snakes. In the Lizard embryos these thickenings of the pedicles of the neural arches cause the latter to extend downwards, grasping the centra, so that these offer a larger articular surface for the capitula. They further make it understandable how the new transverse process mentioned above may after all be formed only apparently out of the centrum (IV), but in reality out of the mantle which is of diapophysial nature, amalgamated with the rib.

Chapter XXXI

THE PYTHONOMORPHA AND OPHIDIA

THE EXTINCT AND RECENT SNAKES

PYTHONOMORPHA

The group consists of procoelous, mostly marine, long-tailed, extinct Reptiles, with more or less paddle-shaped fore- and hind-limbs. Cretaceous with the exception of the gigantic *Megalania* of the Pleistocene, Queensland.

The neural arches are fused, without suture, to the strongly procoelous centra. Chorda abolished. Condyle triple. Ribs capitular, with the tubercular portion reduced. Tail with movable or anchylosed chevrons. Basiventrals absent in the trunk and neck, with the exception of the unpaired ventral piece of the atlas, and an unpaired intercentrum between the immovable odontoid and the second centrum. Zygosphenes and zygantra occur in various genera, also typical cervical hypapophyses. The number of presacral vertebrae is at least 28.

The whole assembly of Pythonomorpha contains:

(1) *Aigialosauri*, i.e. strand or shore Reptiles. Lower Cretaceous. They have seven cervical, 21 thoraco-lumbar; i.e. 28 presacral vertebrae; two sacral. Girdles and limbs well developed; about six sternal ribs. Some reached or surpassed the size of large Monitors or Varanidae.

(2) *Dolichosauri*. Quite aquatic. Mostly less than 3 ft. total length. Lower and Upper Cretaceous. Thirteen cervical, total number of presacrals about 40. Whole body rather snake-shaped; long tail with strong chevrons. Cervicals with hypapophyses. Zygosphenes variable, present in *Dolichosaurus*. The presence of true central hypapophyses and chevrons or haemapophyses in the same creatures is remarkable as combining Lacertilian and Ophidian characters. The fore-limbs are much shorter than the hind-limbs and both are developing into paddles.

(3) *Mosasauri*. Upper Cretaceous. Mostly gigantic; marine. *Mosasaurus*, "the Meuse Reptile", 25 ft. long, with an enormous head of about 4 ft. *Liodon*, which roved from North America and Europe to New Zealand, has been computed vaguely to have reached the

length of 100 ft. The Mosasauri are well diagnosed by their underjaws. First, because they have a strong ligament instead of a symphysis. Secondly, because such a mandible has a compound joint between the dentary plus splenial and the other bones, so that the mouth, armed with terrific teeth, can grasp and swallow large prey. These features may have helped them to the name of Pythonomorpha. The vertebral column shows advance and also reduction in comparison with the Aigialosaurs. The number of cervicals remains seven, but that of the total presacral varies from about 30 to 46. It is significant that this considerable elongation of the trunk is correlated with the reduction to one sacral instead of two, the front or dorsal end of the ilium being scarcely connected with the sacral vertebra, or at most only by a ligament (Fig. 100). The sacrum is formed by number 30 in *Platecarpus* and *Liodon*, 39 in *Plioplatecarpus*, 41 in *Clidastes*, 47 in *Mosasaurus*. The number of caudal vertebrae seems thereby not affected, on the contrary the tail also increases in length.

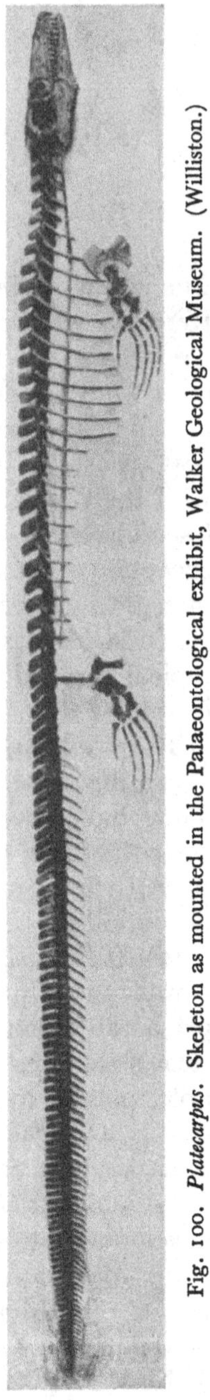

Fig. 100. *Platecarpus*. Skeleton as mounted in the Palaeontological exhibit, Walker Geological Museum. (Williston.)

Zygosphenes and zygantra vary from full development (*Clidastes*) to absence (*Mosasaurus*). With the exception of the atlas the cervicals carry large typical hypapophyses with an apparently independent epiphysis (Fig. 101). They are absent in the tail, which has typical chevrons. Although Pythonomorphs have both hypapophyses and chevrons, they show thereby no ancestral affinities to either Lizards or Snakes, since Lizards have only chevrons and Snakes only hypapophyses often all through the body and tail.

The sea has always attracted Tetrapods, and those who have taken to it have mostly ended as highly specialised marine predaceous rovers. They acquire such characters as the following: elongation of the body and tail; transformation of the limbs into paddles, generally with an increased number of phalanges and even rows additional to the pentadactyle type; weakening of the ilio-sacral connexion leading to loss of the girdles and limbs. A long tail becomes the principal propeller. Many attained enormous size, a sure sign of their doom.

Amphibia produced the Carboniferous *Aistopodes*; *Plesiosaurs* began with the Trias, *Ichthyosaurs* in the Middle Trias, and both ended overspecialised with the Cretaceous epoch. Marine *Crocodiles* in the wider sense are Jurassic and changed in the Cretaceous into semiterrestrial creatures. The *Pythonomorpha* began with the Lower Chalk as Aigialosauri and produced as side-shoots the Dolichosauri and later the Mosasauri, whilst the rest of these strand or shore Reptiles are supposed not to have gone to sea, but, via brackish estuaries, to have resumed terrestrial life and thus to have founded the Autosauri or Lacertilia and Snakes. Here utter confusion begins. Snakes may conceivably be derived from such "Strand Lizards" by great elongation of the whole body and loss of the limbs; as indeed has happened and is still happening in many recent Lizards. Baron Nopcsa, on the other hand, pleads for "Kreideschlangen". To wait for the transformation into terrestrial Snakes until their ancestors had nearly lost their limbs would be a *petitio principii*. To allude to the Hydrophinae or recent Seasnakes, who are simply specialised Cobras, would be futile. The extra joint in the jaws of the Mosasauri is likewise merely a case of isotely, although the dentary of the Colubrine genus *Polydontophis*, loosely balanced upon the elongated angulare, is a surprisingly close analogy. Something similar occurs in some Teleostei. On the other hand, the derivation of the Snakes from Lizards is possible, but far from proven; certainly not through the Varanidae, which are the least snake-like of all Lizards, although they have preserved various generalised old cranial and visceral characters.

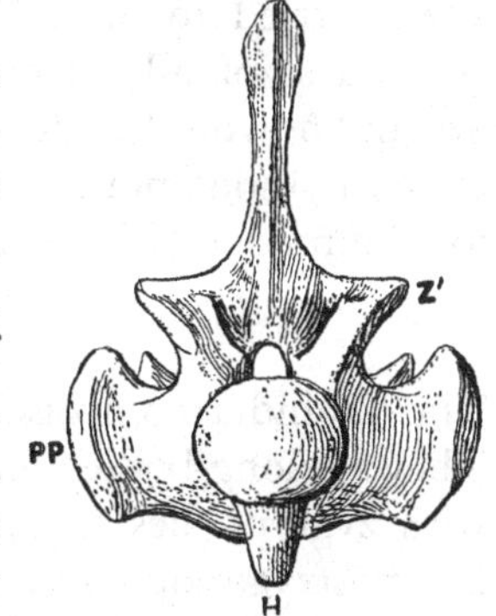

Fig. 101. *Clidastes* (Mosasauria). Posterior cervical vertebra, from behind. *Z'*, postzygapophysis; *PP*, parapophysis; *H*, hypapophysis. (Williston.)

The case of the Aigialosaurian change into Lacertilia is much more complicated and confused. When some authors, e.g. Broili, combine the Varanidae with the Dolichosauri and Aigialosauri into a superfamily Platynota; and the Lacertilia (inclusive of Geckos, but exclusive of Amphisbaenas and Chamaeleons) into a superfamily *Lacertilia vera* (Kionocrania), we are forced to look for the Varanidae as the link, at least as the last surviving offshoot of the Platynota, which has little in common with the Autosauri. The latter are much older than the Pythonomorpha, traceable through the Cretaceous into the Upper Jurassic, and possibly (Palaeiguana, Broom) into the Trias, not to mention the much debated position of *Lysorophus*. However, I have tried to show that the Geckos are the most primitive and the oldest of all Lacertilia, although fossils, even of recent kinds, are unknown. In short, the Mesozoic Lacertilia are the ancestors of the Pythonomorpha, a side-branch, while the main stem qualifies for the modern Lizards.

OPHIDIA

The Ophidia comprise the modern representatives of the Snakes. They are long-bodied, having numerous strongly procoelous vertebrae with zygosphenes and zygantra in addition to the prezygapophyses and postzygapophyses. Freely movable ribs which articulate by their capitula only. Sternum entirely absent. Hypapophyses found especially in the anterior half of the trunk and in the tail. Limbs and girdles are entirely absent in the vast majority, but vestigial pelves and hind-limbs still exist, tucked away on the sides of the vent in the older families, which are therefore termed Peropoda, or maimed-footed, and which form about 15 per cent. of all Snakes. Traces of the limb plexus still exist, easily demonstrated in the sacral region near the root of the tail, but vestiges of a brachial plexus are difficult to ascertain, only being identifiable by slight disturbance and delicate connexion of some spinal nerves in the region in which the shoulder girdle should lie.

The recent species of Snakes (about 1500) are as numerous and cosmopolitan as the Lacertae; these are the two most flourishing and still ascendant branches of the reptilian tree. Sparse fossil remains date from the Upper Chalk and are supposed to indicate large Boa-like creatures. The Mid-Eocene *Palaeophis* of Egypt seems to have been a monster, to judge from the enormous size of its well-preserved vertebrae and ribs.

The Snakes are generally supposed to be a highly specialised terminal side-branch of terrestrial Lacertilia. Another, rather more popular, view hailed the Pythonomorpha as the materialisation of the fabulous Sea Serpents. Recently Baron Nopcsa has suggested that the Ophidia have been evolved from aquatic Reptiles which have become apodous and were closely allied to the Dolichosauri.

For the present it seems advisable to look upon the three orders as certainly allied, of which the Pythonomorpha were the oldest branch. As has so often happened in evolution, the latter improved their particularly lucky hits rapidly, attained great size and wide distribution and died out with the Cretaceous epoch. Of the terrestrial Mesozoic Reptiles only the present Lizards and Snakes remained. Those of them which were the first to undergo apodism became Snakes, while the others retained their fore- and hind-limbs as normal features. They are even now still in the experimental stage, as shown by the fact that in several families (Skinks, Iguanas, Slow-worms) the loss of either fore- or hind-limbs, or both, is frequently a generic, even specific character. In the case of these Lizards the loss of the limbs and elongation of the body amounts to degradation, mostly correlated with a burrowing underground life. In the case of the Snakes, on the contrary, it is as clearly a case of higher specialisation, equally well efficient for swift terrestrial, arboreal or even marine life. It is worth remembering that underground life mostly spells degeneration, and that the Snakes with these habits, the Peropoda, are the ancient families. They must have taken to this mode of life so early that their still generalised structure was unable to progress much in the Ophidian direction. The only exception are the Boidae, which Boulenger considers rightly as ancestral to all modern Snakes.

We are still ignorant of the primary causes of the snake-shaped, many segmented, apodous body. The obvious assignment of an aquatic ancestry must be taken with caution, because an aquatic life, correlated with conversion of the pentadactyle limbs into paddles, is certainly inapplicable to the Anguidae, one of which, *Pseudopus*, the "Glass-snake", makes a fair approach to the true Snakes; and equally inapplicable to the peculiar Pygopod Lizards without fore-limbs, but with the hind-limbs in the shape of a pair of scaly flaps.

Lastly, we may consider the differential characters of Lizards and Snakes usually given in textbooks; namely, that the connexion of the mandibles is ligamentous and not symphysial; that the suspensorium of the mandibles is by means of a freely movable quadrate (streptostylic), which in its turn is quite loosely attached to the cranium by the horizontally elongated, very movable supratemporal; and that the attachment of the whole palatal apparatus is movable. These characters fail us in the burrowing Snakes. But there are three other characters not usually mentioned which may be of assistance.

(1) The Snakes have no intercentra in the neck and trunk, therefore also no chevrons in the tail; in the Lizards the former are of frequent occurrence in the neck and the tail chevrons are normal.

(2) Snakes have hypapophyses, outgrowths from the centra; in the tail they are substitutes for the chevrons.

(3) The cutaneous part of the skin of the Snakes is absolutely devoid of osteoderms, while such are of frequent occurrence and are often considerably developed in several of the chief families of the Lizards, e.g. Anguidae, Scincidae and Lacertidae; they are much reduced in others and are absent in the Agamidae, Iguanidae, Tejidae and Varanidae (with the exception of the Giant Komodo *Varanus*). There is thus a tendency in the Lizards to reduce these ancient osteoderms. In various Lizards these osteoderms are so numerous in the trunk as to remind one of a dermal armour, e.g. *Pseudopus* and the Slow-worm; in some Lacertidae and Skinks they form thick, large plates on the head, overlaying the parietals and frontals, which themselves are dermal bones, so that the osteoderms represent a second edition of bony sclerosis.

The vertebrae in the Ophidia range in numbers from about 200 to more than 350, including the tail, which may be very long or quite stunted. The precaudal numbers likewise vary considerably, not only within the families, but also to a lesser extent in individuals. Boulenger has found from examination of great numbers of the Common Viper, *Pelias verus*, that the mountain races have a distinctly shorter number than those which live in the lower plains. This is easily ascertained without the laborious preparation of the skeleton, since in the Snakes the number of transverse ventral shields agrees with that of the vertebrae.

The vertebrae lose their sutures at an early stage. They are strongly procoelous, with a very pronounced ball-shaped knob which is carried by a neck.

The zygosphenes arise as strong wedges from the anterior aspect of the united right and left halves of the neural arches, above the spinal canal; they fit with a pair of diverging flanges into a corresponding cavity, the zygantrum, on the posterior aspect of the preceding vertebra. The ventro-lateral roof of this cavity forms a right and a left facet, which, looking towards each other, underlap the down-looking articular facets of the sphenes. The mutual over- and under-lapping is therefore the reverse of that of the zygapophyses. Consequently the Snake vertebra carries four pairs of articular facets in addition to the axial cup-and-ball joint; total, ten facets, so arranged and interlocked as to allow right and left excursion but very little dorso-ventral bending of the column.

The hypapophyses are direct outgrowths from the ventro-lateral aspect of the vertebral centra, usually half-way down the middle and extending downwards. They vary extremely in the different families and in different regions of the column, from well-developed knobs, strong stalks, or sharp blades to mere rugosities or complete absence. Those of the trunk are best developed in the cervical region, from the atlas down to the region of the heart. They serve for the musculus longus capitis and its continuation the musculus longus colli; the tendinous bundles of these muscles, and their equivalents in the rest of the column, being inserted upon the prominent basis cranii and near the bases of the ribs. These hypapophyses are large and numerous in the Cobra's neck. In *Python molurus*, from about the fortieth vertebra they dwindle in size and become more and more restricted to the neck of the posterior axial ball as mere tuberosities, or they are absent. They reappear in the tail, increasing on the fifth to ninth caudals into unpaired pedicles, which bifurcate farther back with reduction of the pedicle itself until, on about the twenty-eighth vertebra and onwards, the two arms arise well separated directly from the centra and become almost equal in size to the transverse processes. Still farther back they are reduced to a widely separated pair of knobs. They are absent on the last 20 vertebrae.

All these modifications are intimately connected with the muscles of the tail, which in these Snakes is used as a comparatively short but very strong grasping prehensile organ.

The most peculiar use of the hypapophyses is met with in two widely separated kinds of Snakes, which seem to live entirely upon eggs, preferably those of Birds. The first species is *Dasypeltis scabra*, sole species of the South African Rhachiodontinae, a subfamily of aglyphous harmless Colubrines; the other species is the rare *Elachistodon westermanni* of Bengal, which is the only instance among the numerous Opisthoglypha or Snakes with furrowed poison fangs far back on the maxillae. In these two otherwise normal Snakes about half a dozen hypapophyses in the region of the thirtieth cervical vertebra permanently perforate the oesophagus with their very hard apices, erroneously stated to be covered with enamel. They crush the swallowed egg, the shell of which is then disgorged. Most of the teeth on the jaws are reduced.[1]

Ribs are absent on the atlas and usually also on the second and third cervical vertebrae, then present throughout the column and freely movable, except on the tail, where they are completely fused on to the centra as "transverse processes". Those of the trunk are more or less spongy, even partly hollow; their distal portion remains cartilaginous, the free ends fitting and fastening into the connective tissue of the sides of the ventral transverse scales, which, imbricating and sharp-edged, form the chief locomotory apparatus. They are freely movable in a forward or backward direction. The hood of the Cobras is spread out by the levator muscles of the much elongated ribs. The articulation of the trunk ribs much resembles that of the Lizards, the tubercular portion being represented by a small process, which arises from the posterior dorsal surface but does not articulate with the diapophysis, being only attached by ill-defined strands of ligamentous tissue. The capitulum is strong and lies either across the neurocentral suture or below it. Frequently it is divided into a dorsal and a ventral knob, corresponding with slight knobs on a pedicle which arises from the centrum with an oblique downward direction. In other cases the facets are almost flat, quite broad, with scarcely a pedicle. Unfortunately, the ontogeny of Snakes is still very imperfectly known. Possibly these pedicles represent the fused true diapophysial and parapophysial or central portion of the vertebra, overlaid and thus combined by a strong osseous mantle (cf. p. 276, Lizards). The combined oblique downward growth would account for the tubercular knob of the rib no longer reaching the diapophysis. The whole feature amounts to an unusually strong preponderance of the capitular portion of the rib, which has annexed the whole diapophysis and has ousted the tuberculum.[2]

1 The whole process of this specialisation is interesting. Placoid scales have become teeth by entering the service of the digestive apparatus. Ultra-Darwinians may think of spontaneous variation so that the weak toothed were driven to the egg diet. At any rate the teeth have lost their use. The perforation of the oesophagus by the preformed hypapophyses was originally an incident worth improving, and the "egg-cracking" "teeth" are as firmly protected by the now adherent oesophageal wall as our own teeth are by the gums. Such a permanent perforation by skeletal parts is very rare, the only analogous case is that of the sharply pointed ribs of the Newt Pleurodeles.

2 Huxley (*Manual*, p. 170) states that "the dorsal vertebrae of the snakes have double tubercles in the place of transverse processes (Perospondylia)". It would be premature for us to assume a change analogous to the formation of a new shoulder of the Urodele ribs, which have produced a new tuberculum at the expense of the vanishing capitular

Towards the tail the ribs diminish in size, often rapidly; the last frequently reduced to a vestigial condition and quite loosely attached. Then comes a sudden change. All the *bona fide* caudal vertebrae, freely movable at their procoelous joints, are completely anchylosed with their ribs, which are deeply bifurcated, the dorsal and ventral finger grasping the lymph heart of its side like a protecting claw (Fig. 102). The number of these vertebrae depends upon the length of the lymph hearts, which varies much and often asymmetrically. A few segments farther back comes another abrupt change. Instead of a pair of claws the vertebrae carry only a pair of straight, horizontal "transverse processes". Posterior lymph hearts at the beginning of the tail exist in Fishes, in the Amphibia and the Sauropsida, vanishing, however, in the young Birds. In Mammalia they seem to be absent. In the Reptiles they lie well fastened upon the dorsal side of one of the anterior caudal transverse processes, endowed with a muscular coating and showing rhythmical contractions, independent of those of the heart. The right and left ducts combine into a "cisterna", placed along the ventral side of the vertebral column. We can therefore conclude that the ventral finger of the claw is the rib itself, and that the dorsal finger is a secondary formation for the protection of the lymph heart, which is fastened to it by ossifying connective tissue. This interpretation is supported by the fact that, when there are several such clawed vertebrae, the upper finger of the foremost and the last show rudimentary or vestigial conditions.

The whole arrangement was first well described and figured by Panizza, who found a fair agreement between a Boa and *Coluber flavescens*, the Aesculap Snake. Hoffmann has given a very garbled and confused account of his own and other observations. He refers to Salle, whose apparently careful paper is inaccessible to me. He distinguishes between "lymphapophysen" (the mostly movable ribs) and "costo-transversarii" (anchylosed to the vertebrae), both kinds together varying in Snakes from two to eight pairs, and restricted to one pair in some apodous Lizards. Hoffmann calls the former kind presacral, the others postsacral, thereby treating a sacral vertebra as non-existing. This is unsatisfactory. It cannot reasonably be

arm (cf. p. 116, Urodeles) whilst in Snakes and Lizards the much stronger and curved ribs have retained the capitular at the expense of the arrested new tubercular portion. Possibly, however, the Snakes' tubercle may account for the *bifurcated ribs* to be mentioned presently.

doubted that the original Snakes, or at least their direct ancestors, had a limb-bearing pelvis and therefore at least one sacral vertebra, and this cannot have dropped out. Even now there are vestigial plexuses, but the problem of the position, composition and eventual shifting of the postulated Ophidian sacrum is still unsolved. The disturbance is obvious enough, and greatly enhanced by the enlargement of the lymph hearts and their mode of protection. The difficulty is the behaviour of the presacral segments. Fig. 102 shows the conditions in the Pythons and will save a long description. The Molurus series is drawn from a specimen in the Cambridge Museum, the other from the original description of Hoffmann. Salle's costal lymphapophyses or Hoffmann's presacrals are labelled *A*, *B*, *C*, *D*. The costo-transversarii are numbered I–IV.

The following summary may be made:

(1) It would be wrong to homologise the dorsal finger of the claw with the rib, the ventral finger with a true transverse process. Nor would it be admissible to take the two fingers as the materialisation of dorsal and ventral ribs.

(2) The dorsal fingers of *B* and *C* are incipient protectors of forwardly extending lymph hearts, the connective tissue mantle having thrown out anchorings.

(3) The vestigial condition of the whole rib of *A* indicates that these ribs are the last of the original trunk and have degenerated before subsequent forward extension of the lymph hearts.

So far we may feel fairly justified in concluding that number I is, or was, the first caudal, and *A* the first and only sacral vertebra, which has lost its pelvic-bearing function and therefore became reduced. The serial number of this *A* vertebra does not matter, considering the great amplitude of variation in even closely allied Snakes; in the two Pythons about 268; in Rattlesnakes 210; in Vipers 200 minus the tail, so that the problematic sacral would be far below the 200th. But the nerve plexus presents the greatest unsolved difficulty. In many Lizards the nerve issuing between the two sacral vertebrae is the last to form a part of the isciadic plexus. In many other Lizards, in the Crocodiles and Tortoises, two or three post-sacral nerves also contribute to this plexus. The respective muscles arise from the transverse processes of the tail. With our present ignorance of these plexuses it is safer to assume that the well-nigh complete reduction of the limbs and girdles, which have now arrived at the anal region, has brought about the complete reduction of the

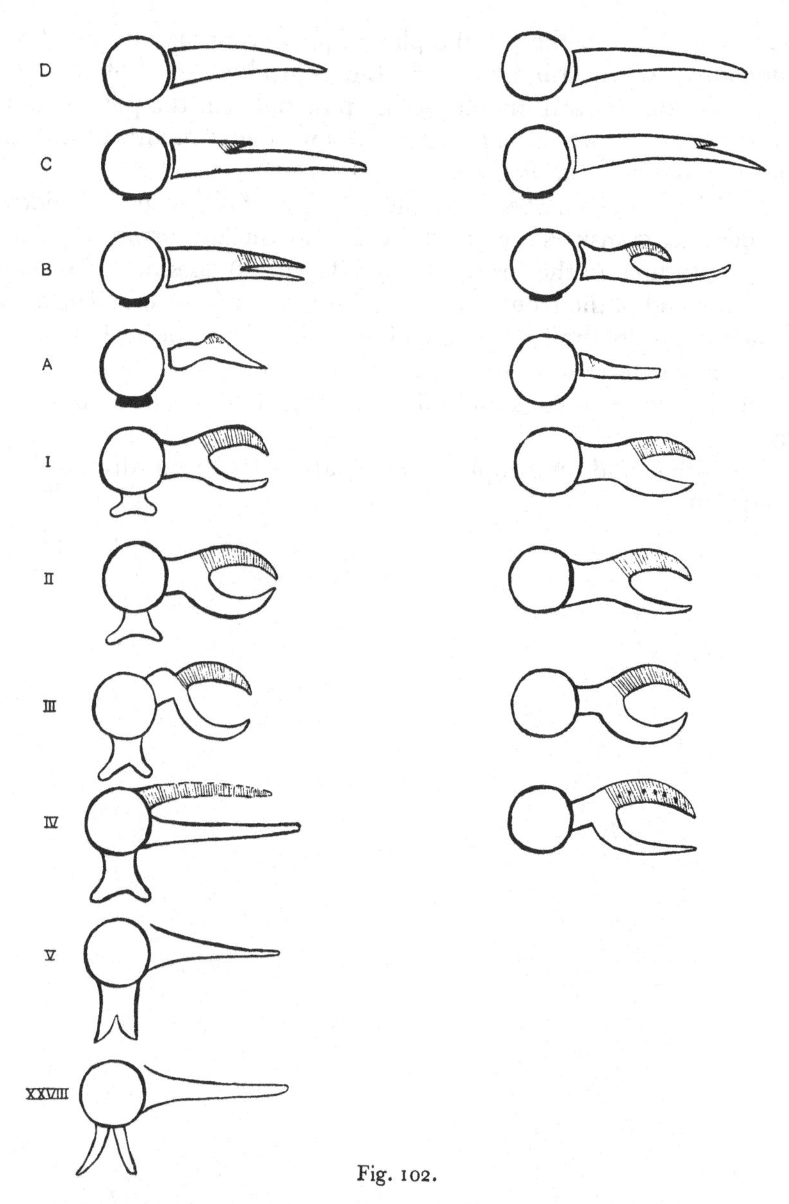

Fig. 102.

Shows the lymphapophyses and costo-transversarii. The former are labelled *A*, *B*, *C*, etc., and the latter I, II, III, etc. *A* is the only sacral vertebra, I the first caudal. XXVIII indicates the final form of the caudal vertebra which is reached far down in the tail. The series on the left are drawn from a specimen in the Cambridge Museum, that on the right from the original description of Hoffmann for comparison.

crural and most of the isciadic plexus, preserving mainly the plexus pudendus, which still innervates the ventral muscles of the anal sphincter, the ventral muscles of the postanal tail, the paired copulatory organs stowed away in the tail, and the remnants of the still movable limbs of the Peropoda.

We have to remember that the position of the vent (plagiotrematous, i.e. transverse in the Lizards and Snakes, orthotrematous, i.e. longitudinal in the Crocodiles and Tortoises) does not necessarily mark the end of the trunk, but may, for instance in several Tortoises, lie far back in the tail. The pelvis has in these cases been slung on to the trunk far in front of the anal region and has thus established the usual division into trunk and tail vertebrae quite independent of the viscera.

The atlas and epistropheus of Snakes has been discussed in Chap. XIII.

Chapter XXXII

"SQUAMATA"

THE VALUE OF THE TERM IN CLASSIFICATION

It has recently become the fashion to combine the Lizards and Snakes, which have been discussed in the last three chapters, as "Squamata". De Blainville (1816) distinguished between *Amphibiens squamifères* (i.e. Reptilia since Latreille, 1825, Joh. Mueller, and finally Gegenbaur) and *Amphibiens nudipellifères* (i.e. Amphibia). Merrem (1820) called the Reptiles Pholidota because of their pholis or horny outer covering, a name now by consent transferred to the Pangolins. Squamata comprised the Tortoises, Snakes, and Lizards in the widest sense, the latter being subdivided into *Sauria loricata* (Crocodiles), *S. squamata* and *S. annulata* (Amphisbaenidae).

Squamata therefore meant armoured as well as scaled Reptiles; but squama = Schuppe = scale = écaille, and this soon came to mean chiefly or solely the horny epidermal covering of the cutis. It was not appreciated until later that the complete scale is essentially a connective tissue product, a part of the cutis which frequently ossifies and is not shed like the cornified portion of the epiderm; and that the whole scale grows peripherally at its base. It is a detail whether the scale has a broad flat base, with or without an apex; or whether the neighbouring scales imbricate. It is likewise a detail and distinction of degree whether the sheath remains throughout life (e.g. the scutes of Crocodiles, or shields, the so-called "Tortoise-shell" of Tortoises), or is more or less periodically shed in flakes or entire as in some Lizards and in all Snakes. Nor is the size of importance, since the individual scales, or the shields, may be reduced to granules. It took a long time to appreciate the fact that the scales of many Lizards contain thick osteoderms, often so powerful as to enclose the whole body in such a compound armour, e.g. Skinks. In many other Lizards they are restricted to the head. Every intermediate condition is represented; even Tortoises have small, flat, reticulated scales on their soft parts like the neck, armpits and the groin, parts which in land Tortoises may peel in flakes. The same individuals may have osseous nodules on their legs, covered with conical pointed horny spikes.

A valid diagnosis of Squamata and non-Squamata is therefore impossible.

Lastly, since Lizard-like scales have been discovered in some Dinosaurs, these should also be classified with the Squamata; possibly also other fossil reptilian groups, for instance Pterosauri, unless they were naked; and Birds, which still have typical scales on their legs. Neither horny scales nor dermal ossifications, however, need indicate the descent or the affinity of their possessors; witness the Pangolins and Armadillos, in which they are certainly neomorphs so far as Mammalia are concerned. Independent development may apply to the various main groups or "Formenkreise" of the Reptiles. Certainly calcifications, osteoderms or dermatosts in the skin are now old-fashioned; their suppression means progress preparatory to a better coating of feathers or hair, thereby stabilising the blood temperature. Not the absurd inverted way suggested but a few years ago, that feathers caused the higher temperature!

The more we scrutinise the evolution and structure of the skin of the various orders of the Reptilia, the more the term Squamata justifies its original meaning as a diagnostic character applicable to all Reptilia, in opposition to the recent Amphibia, which have a naked surface, not horny and not scaly. This applies even to the Coeciliae with their hidden scales. How far back the scutes and scales date we do not know, nor do we know the outer covering of the armoured Labyrinthodonts. But it may reasonably be assumed that scutes and scales and certainly cornification of the epiderm have all resulted from terrestrial life. Nakedness would therefore be an aquatic adaptation, pseudoprimitive.

Chapter XXXIII

THE CROCODILIA AND THEIR PERMIAN AND TRIASSIC FORERUNNERS

The Crocodiles are powerful, more or less aquatic, rapacious, well-protected beasts, for the efficiency of which it would be difficult to suggest morphological improvements, although such are still possible and are no doubt proceeding, without turning these creatures into anything other than Crocodiles, Alligators and Gavials. We are certain that they had a greatly diverging number of Cretaceous and Jurassic ancestors, although the connexion with Triassic forms is still vague.

The term suchos is supposed to mean Crocodile. Huxley distinguished between Eusuchia, the Crocodiles proper, and Parasuchia, notably *Belodon*, or *Phytosaurus*, large, Triassic, rapacious beasts without descendants. Zittel added Pseudosuchia, comprising the likewise Triassic but still small and terrestrial *Aetosaurus*. Broom (1914) founded the Eosuchia upon *Youngina capensis*, a Permian Lizard-shaped creature, which with other recently discovered forms is still so generalised that he and other authorities have been able to discern in them one of the most important keys of Post-Permian reptilian evolution.

There was at those ancient periods already much terrestrial but still more aquatic life; the latter by its abundance and cosmopolitan distribution offering irresistible attraction to the terrestrial fauna. The main adaptational impetus and reactions were due to the change from terrestrial to more or less aquatic, even marine life, as predaceous creatures with a powerful skull, body armour plated, propelled by the tail when swimming, and, in the case of Crocodilia, retention of the full development of the four limbs for progress on land.

EOSUCHIA

Lizard-shaped creatures with the earliest satisfactorily known two-arched skull, but still without antorbital vacuities. Vertebrae amphicoelous; all the basiventrals still persisting as large intercentra or chevrons.

Youngina capensis, of the Upper Permian, is the type specimen.

Fig. 103. A nearly complete skeleton of an Eosuchian reptile, *Palaeagama vielhaueri* (Broom).

Palaeagama vielhaueri (Fig. 103), of the Karroo Trias, with nearly complete skeleton, is a near ally of *Palaeiguana*, shortly to be described, but is Eosuchian, being diapsid without antorbital vacuities. All the centra are much constricted in the waist and between each pair is a well-developed intercentrum. "Ribs apparently as in Lizards, i.e. with a single articulation near the middle of the vertebra", which means that the ribs have already been shifted back from the basidorsals. About 26 presacrals and two sacrals. Broom considers these later Triassic Eosuchians as intermediate ancestors of Lizards.

PSEUDOSUCHIA

Advanced beyond the Eosuchian stage by formation of antorbital fossae.

Palaeiguana whitei, Karroo, is according to Broom the oldest known Reptile with a Lacertilian skull and thereby more advanced in the direction of a suitable Lizard ancestor. The bridging of the still existing gap from the Trias to the Upper Jurassic is a question of time.

Euparkeria capensis, Upper Trias. Occipital condyle formed by the basioccipital and by part of the lateral occipitals. Proatlas doubtful. About nine cervical vertebrae with short, slightly procoelous centra. Ribs two-headed, with small uncinate processes. About 26 presacral and two sacral vertebrae; those of the long tail with strong chevrons.

Aetosaurus ferratus. Zittel's type of the Pseudosuchia. Kemper of Wuerttemberg. Completely encased by numerous flat dermal plates. About 25 presacral and two sacral vertebrae. Cervical ribs short, hatchet-shaped, with capitulum and tuberculum. Movable chevrons. Caudal transverse processes large (Fig. 104).

Ornithosuchus, Mid-Trias of Elgin.

Scleromochlus taylori, Mid-Trias of Elgin. Neck and trunk reduced to about 20 presacrals; sacrals correspondingly increased to three or four. Tail very long. Total length of this slender and long-limbed creature about 8 inches; obviously arboreal (Fig. 105). According to Broom it helps by analogy towards understanding the evolution of the Pterosaurs, just as *Galaeopithecus* suggests how Bats may have arisen.

The evolutionary importance of the Pseudosuchia has long been recognised. According to Boulenger they stand as near to the Parasuchia as to the Dinosaurs. Broom tersely states that the ancestor of the theropodous Dinosaurs was a Pseudosuchian. E. T. Newton indicated the relationship of *Ornithosuchus* to Crocodiles, Dinosaurs and *Aetosaurus*; he further pointed out the affinity of the Pseudosuchia and Ornithosauria. Von Huene likewise considers them as ancestral to Dinosaurs and Pterosaurs, Crocodiles and Birds, all of which he

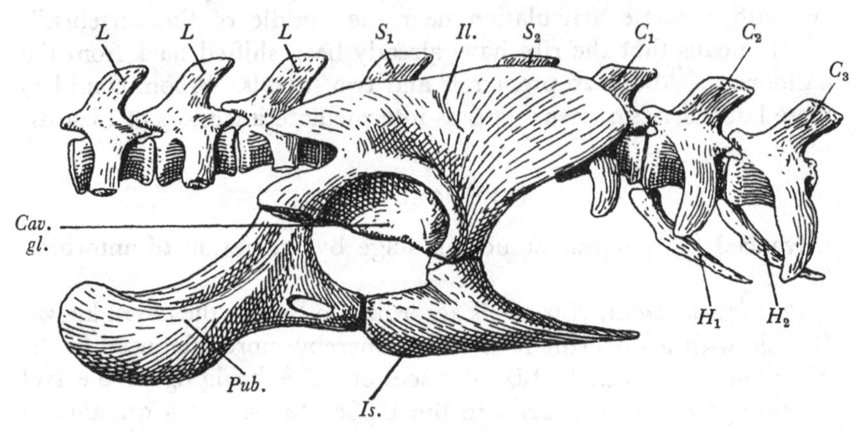

Fig. 104. The pelvis and the neighbouring vertebrae of *Aetosaurus crassicauda*; from the left side. *L*, lumbar vertebrae; S_1, S_2, sacral vertebrae; C_1, C_2, C_3, caudal vertebrae; *Cav. gl.*, acetabulum; *Il.*, ilium; *Pub.*, pubis; *Is.*, ischium; H_1, H_2, haemapophyses. (Fraas.)

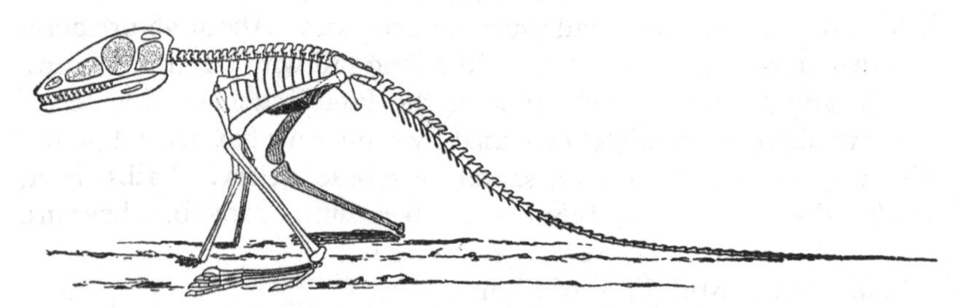

Fig. 105. Reconstruction of *Scleromochlus taylori*. (F. von Huene.)

grouped as Archosauri in opposition to the other Reptiles. It is another question how this "Formenkreis" is related to the still older Cotylosauria, which through the Theromorpha are reasonably supposed to contain the roots of the Mammalia. On the other hand, ancestral connexion of the Cotylosaurs with the remaining third Formenkreis, the Tocosauria or Lepidosauria, is beginning to be solved by Broom's Eosuchia.

PARASUCHIA

From the Lower to the Upper Trias, with wide distribution. Large, long-snouted, aquatic creatures, in many respects resembling the Crocodilia proper without direct connexion. They form an early

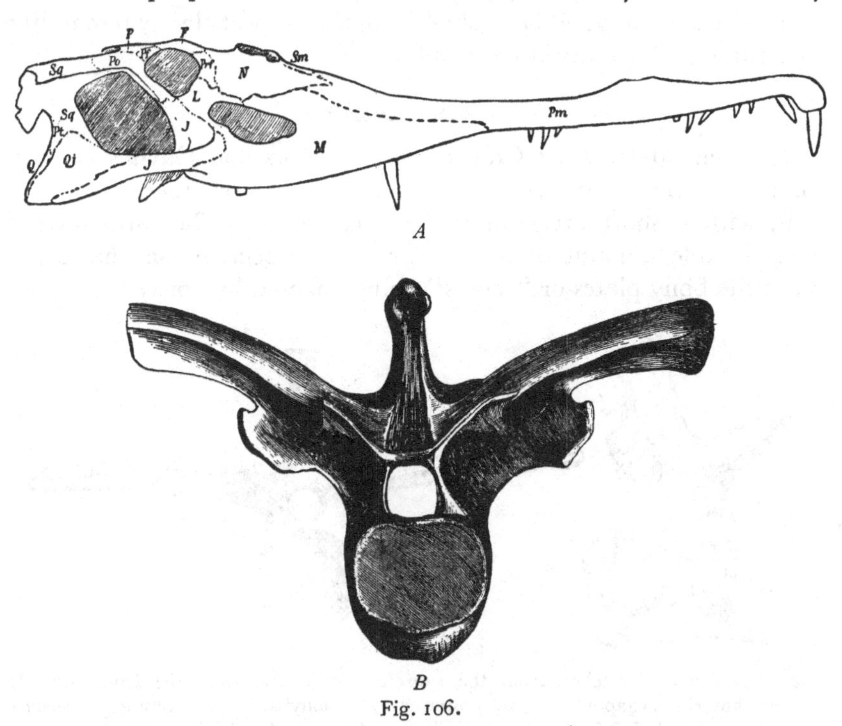

Fig. 106.

A. *Phytosaurus plieningeri* H. v. Meyer. Skull from right side. *Pm*, praemaxilla; *M*, maxilla; *Sm*, septomaxilla; *N*, nasal; *Prf*, prefrontal; *L*, lacrymal; *J*, jugal; *Qj*, quadratojugal; *Q*, quadrate; *Sq*, squamosum; *Pt*, pterygoid, which is also seen projecting down below the jugal; *Po*, postorbital; *Pf*, postfrontal; *P*, parietal; *F*, frontal. The crater-like opening above the letter *N*, behind the *Sm*, is the nasal opening. (F. von Huene.)

B. *Belodon*. Dorsal vertebra. Reduced one-half. (Zittel.)

specialised group which would be collateral if they had not died out before the modern Crocodiles had been evolved from as yet unknown direct ancestors.

The best known is *Belodon* with antorbital foramina between the lacrymals and maxillae, reminding one of the very variable vacuities of Dinosaurs (Fig. 106). The vertebrae are slightly amphicoelous, platycoelous or procoelous. Atlas and axis without ribs. The other cervicals and some thoracics carry separate diapophyses and parapophyses for the attachment of the ribs, which on the rest of the trunk are carried by the long diapophyses, as in the modern Crocodiles. While *Belodon* is devoid of dermal armour, the closely allied and partly contemporary *Stagonolepis* was protected by elaborate dorsal and ventral compound bony shields; in this respect closely resembling the Jurassic Teleosaurid Crocodiles.

CROCODILIA

The recent and extinct Crocodiles, Alligators and Gavials are distinguished by the possession of complete ribs from the atlas to the tail, with a short break in the lumbar region. They still have a considerable amount of dermal armour, especially on the dorsal side; the bony plates or "scutes" being covered by horny "shields".

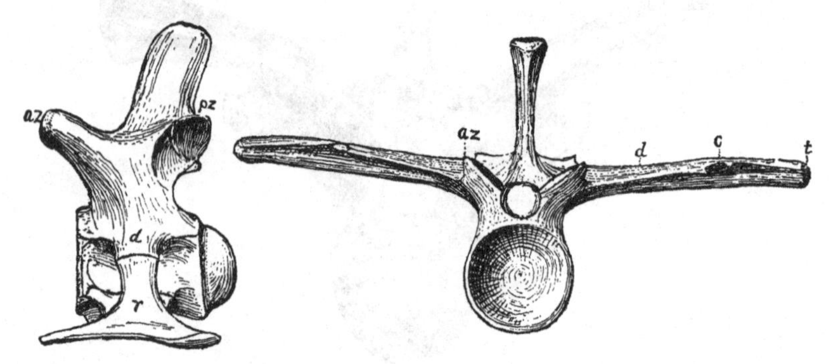

Fig. 107. *Gavial.* Vertebra, from the side (cervical), and from the front (dorsal). *az*, anterior zygapophysis; *pz*, posterior zygapophysis; *d*, diapophysis; *r*, cervical rib; *c*, articulation for head; *t*, for tubercle of dorsal rib. (Williston.)

While they are reduced to 20 recent species, they were much more diversified and numerous in the Mesozoic age, traceable into the Lias, and Lower Jurassic; most of these earlier forms were marine, with weaker and shorter fore- than hind-limbs. Those which shortened and broadened their fore-limbs to incipient paddles died out with the Jurassic. It was only with the approach of the Tertiary epoch that the others overcame the discrepancy between fore- and hind-limbs and recovered their status as more terrestrial walking, even occasionally running, creatures. Some can assume a surprisingly straight-limbed gait, reminding one of the restored theropodous Dinosaurs. The strongly developed ankle joint and the typical pentadactyle limbs of the Liassic forms show that these must have sprung from strictly terrestrial ancestors, semiplantigrade, with a tendency to shortened fore-limbs. Steiner goes further, and on the strength of Pseudosuchian and Eosuchian descent pleads for an arboreal common ancestor for the whole Archosaurian assembly.

The usual numbers of the vertebrae are nine cervicals, two of which are transitional; 8–9 carrying sternal ribs, about three carrying floating ribs, and three ribless trunk vertebrae; two sacral and 30–40 caudal. Total about 60. The vertebrae consist of the centrum, dorsal arch and the basiventrals with their derivatives the meniscus, ribs and chevrons (Fig. 107). The vertebrae are solid, but remnants of the chorda persist for years in the middle of the centra, which were still amphicoelous or at least slightly biconcave in the Jurassic types, with probably considerable intervertebral portions of the chorda. From the Cretaceous onwards they became procoelous with the exception of the first caudal, which has a prominent knob at either end, so that naturally the posterior of the two sacrals is opisthocoelous. The same applies to some Dinosaurs. Cartilaginous, thin, meniscus-like rings are found regularly throughout the column, unless they are abolished, squeezed out of existence by the growing and ossifying centra. Ventrally these ring-disks are continuous with the chevrons. In well-macerated and disarticulated specimens the first chevron, with its still cartilaginous ring, can be lifted off the neck or base of the posterior knob of the first caudal to which it does not genetically belong. The ribs of any complete immature specimen afford an ideal demonstration, a summary of all the changes which the ribs of the Tetrapoda can possibly undergo (Fig. 108). The first rib is a broad blade extending back to the level of the third vertebra. It articulates by its capitulum with the posterior corner of the united pair of first basiventrals; its tuberculum is vestigial, continued by a ligamentous string to the odontoid. In the Mesozoic *Metriorhynchus* it was still complete, as shown by a facet. The second rib is still larger and broader; capitulum articulating with the second basiventral, which in the young extends as a cartilaginous disk upwards between the first and second centra; the tuberculum is well marked, but tapers to a point which either touches, or is otherwise connected with, a small protuberance of the odontoid: in *Metriorhynchus* it seems also to have reached the second neural arch. The third to seventh cervical ribs are complete, adze-shaped; capitulum articulating with the centrum, tuberculum with the diapophysis well above the neurocentral suture; the resulting foramen transversarium encloses the vertebral artery and the deep sympathetic nerve strand. The next ribs are transitional, slender and pointed, looking like the dorsal portions of formerly complete thoracic ribs. The tenth and eleventh have been shifted dorsally; the still

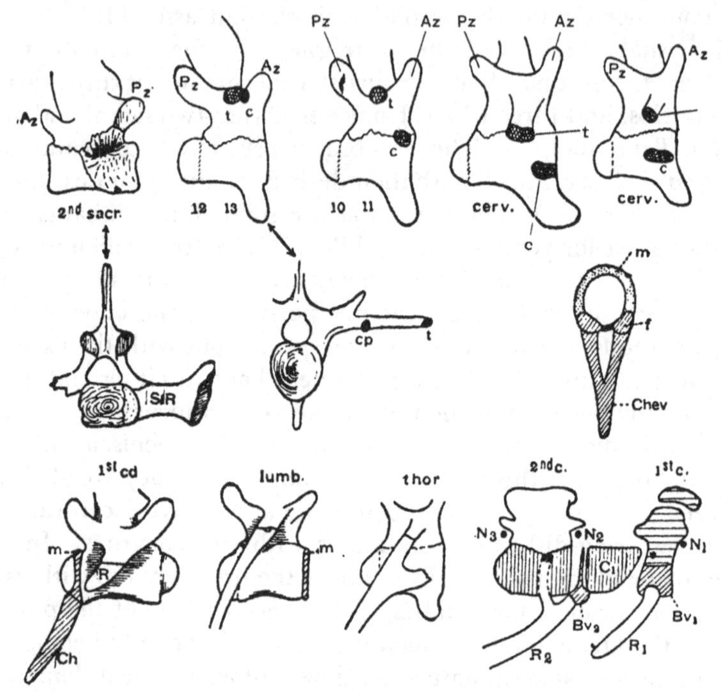

Fig. 108. *Crocodilus acutus.*

Second sacral vertebra seen from the left side after removal of the sacral rib, the broad attachment of which is indicated.

The same vertebra seen from behind, with the sacral rib, *SR*, attached; observe the extent of the central articular facet, which is formed by the centrum proper and by part of the rib.

The twelfth and thirteenth, tenth and eleventh, a posterior and a middle cervical vertebra, seen from the right, show the changing positions of the capitular (*c*) and tubercular (*t*) attachments.

The right figure in the middle series shows a front view of a pair of chevrons; *f*, facets articulating with the posterior end of the centrum of the next previous vertebra; *m*, cartilaginous meniscal ring.

The lower series shows the first caudal, a lumbar and thoracic vertebra, and the analysis of the first and second cervical vertebrae. *Ch*, chevron; *R*, rib; *N*, position of nerve exit. *Az*, prezygapophysis; *Pz*, postzygapophysis.

well-developed capitulum and tuberculum enclose the foramen transversarium, but the capitulum lies right across the neurocentral suture, and the tubercular diapophysis starts from the horizontal level of the base of the spinous process in a transverse plane well behind that of the capitulum. The twelfth marks a sudden change which it would be interesting to follow up in early fossils; both capitulum and tuberculum are carried by the long diapophysis, which looks like, but is not, a transverse process. The tuberculum is attached to the elongated distal end or tip of the process, the capitulum half-way along the caudal side. The foramen transversarium has ceased and the proximal end of the rib has undergone a rotation, capitulum and tuberculum now lying in the same horizontal level. This arrangement is continued until, in the lumbar region, the reduced ribs are attached to the tip of the diapophysis.

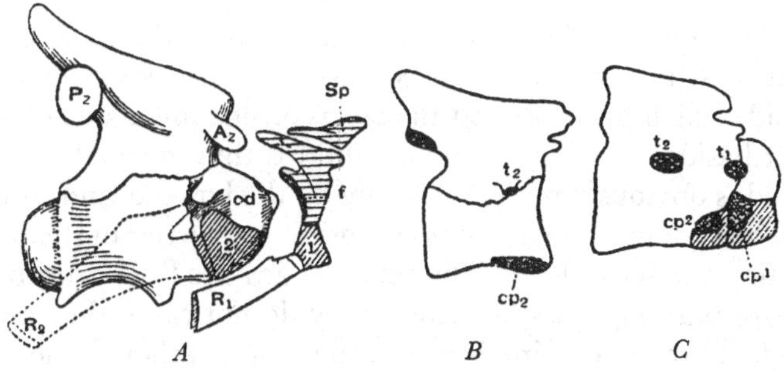

Fig. 109.

A. Crocodilus vulgaris. Adult. Camb. Mus. First and second cervical vertebrae. The second rib is dotted in outline. *f*, articulating facet of the neural arch of the atlas, playing upon the odontoid process *od*; *Sp*, proatlas; 1, 2, first and second basiventrals; *Az*, prezygapophysis; *Pz*, postzygapophysis.

B. Steneosaurus sp., Jurassic. England. Brit. Mus. Specimen described by Hulke. Second cervical vertebra, showing the neurocentral suture, the capitular (cp_2) and tubercular (t_2) articulating facets of the second rib.

C. Metriorhynchus sp., Jurassic. England. Brit. Mus. Specimen described by Hulke. Atlanto-epistropheal complex. The articular facets of the first and second rib are carried by the first and second right basiventralia respectively. The first tubercular facet is slightly shifted backwards, and lies across the suture between the first and second vertebrae. cp^1, cp^2, capitular facets; t_1, t_2, tubercular facets.

The two sacral ribs are very stout and suturally wedged between the flange-like base of the neural arch and the corresponding centrum, which is laterally so much compressed that it is almost severed from its basidorsal; as a result of this a considerable part of the sacral rib forms the articular cup for the knob of the biconvex first caudal. The ribs of the other caudal vertebrae are vestigial and soon fuse with their transverse supports. The neural arches of the trunk carry well-developed spinous processes and zygapophyses, but no traces of zygosphenes and zygantra.

The atlas is very exceptional (Fig. 109). Its right and left basidorsals are not joined by a spinous process. The postzygapophyses are large and functional, but the anterior zygapophyses are decidedly reduced. Above them, and filling the occipito-cervical gap, lies the shield-shaped rather large remnant of the proatlas. The first nerve passes in front of the arch. The very stout basidorsal pair is loosely sutured to the first pair of basiventrals, which are fused into one thick plate, the upper portion of which forms a large but shallow concavity for the occipital condyle. Ventrally the plate sends out a pair of stout processes for the capitula of the first ribs. The caudal or posterior aspect of the plate is almost flat, resting against the second likewise united basiventrals. These are firmly fused with the first centrum so that, at least in the adult, this compound mass looks like one huge odontoid, which plays against the facets on the inner side of the right and left basidorsals. Since the odontoid is thus grasped by the atlas ring, and is obviously pushed dorsally by the large second basiventral mass, it has got into the anomalous position of carrier of the tubercula for the first and second ribs. The centrum has two little protuberances, which are parapophyses, although they do not carry the capitula.

The first trochoid joint is also anomalous, although not unique. Ventrally it is intervertebral because the first and second basiventrals are in movable contact; dorsally it is intravertebral as in the Mammalia, the joint being formed between the first centrum and the first neural arch. The almost entirely basioccipital condyle moves upon the basiventral plate as mentioned above, but also plays against facets of the pedicles of the first arch, without, however, reaching the small prominent "dens" of the odontoid.

Chapter XXXIV

DINOSAURIA AND PTEROSAURIA

DOMINANT MESOZOIC REPTILES

DINOSAURIA

Mesozoic Reptiles with more than ten cervical vertebrae, long tail, and terrestrial limbs. Without osseous remnants of the basiventrals, excepting the chevrons. All have numerous teeth. Extending from the Lower Trias to the end of the Cretaceous, they became cosmopolitan, carnivorous and herbivorous, as the dominant land vertebrates.

The whole order comprising numerous families with almost endless modifications used to be divided according to the formation of the limbs; this was not very satisfactory, since the respective purely morphological terms, as usual, cannot express the evolution of such a large diversely flourishing assembly. Their classification on this basis is as follows:

1. *Theropoda*, with digitigrade hind-limbs and Saurian pelvis; e.g. *Hallopus*, *Coelurus*, *Compsognathus*, *Ornithomimus*, *Anchisaurus*, *Ceratosaurus*, *Tyrannosaurus*, *Spinosaurus*.

2. *Sauropoda*, with plantigrade limbs and Saurian pelvis; quadrupedal: e.g. *Atlantosaurus*, *Brontosaurus*, *Diplodocus*.

3. *Orthopoda*,[1] with plantigrade or digitigrade hind-limbs, but with Bird-like pelvis; e.g. the bipedal *Laosaurus*, *Hypsilophodon*, *Iguanodon*, *Trachodon* (*Diclonius*); the quadrupedal *Stegosaurus*, *Scelidosaurus*, *Ankylosaurus*, *Triceratops*.

Meanwhile, because these Orthopoda differ by their very avian ischial and postpubic combination from the Theropoda and the Sauropoda, the two latter have been combined as Saurischia in opposition to the Ornithischia. These form the two main suborders.

1 This term has come into general use from a confusion and abundance of synonyms. *Ornithopoda* (Marsh), bird-footed, with reference to the bipedal, essentially three-toed Iguanadonts. *Orthopoda* (Cope), straight limbed; with reference to the straight, not bent, position of the quadrupedal gait of *Stegosaurus*, *Triceratops*, in opposition to their bipedal relations. *Predentata*, comprising the whole third group, with reference to the predental bone which carries no teeth. Pre-dentate must mean creatures with front teeth; cf. the correctly termed E-dentata!

Lastly, thanks to von Huene, Broili and Abel, the series *Hallopus* to *Ornithomimus* has been separated as Coeluria, as the perhaps most fundamental third suborder; all digitigrade, bipedal jumpers or runners, dating also from the Lower Trias and culminating in the Upper Cretaceous *Ornithomimus* with the most astonishingly Bird-like foot, but with the Saurischian pelvis. It was sagaciously suggested that these mostly small, slender creatures were originally arboreal. This has been proved by the elaborate study of the hands and feet of not only Dinosaurs, but also Crocodiles, Birds and other groups, so that we arrive at the low level of the Pseudosuchia and maybe the Eosuchia as the roots of a great Archosaurian "Formenkreis", cf. p. 294. It may be noted that "arboreal" life, itself an episode of originally terrestrial creatures, can easily lead to bipedal running or hopping, but the reverse conclusion from such a gait to former arboreal descent is often fallacious; cf. jumping Rodents and Insectivores.

The order as such is morphologically rather disappointing, because of the marked indecision of progressive evolution, notable in almost every region of the skeleton, particularly in the skull, hind-limbs, pelvis, gait and vertebral column. Some begin well, orthogenetically improving certain parts, then stopping long before the morphological terminus is reached, but transferring active evolution to other organs without necessarily producing high-class specialisation; or in the middle of their career they revert to an obviously ancestral condition. For instance, the three suborders had perhaps started with a tendency to an upright gait, the fore-limbs being shortened, the hind-limbs becoming stronger and longer with correlated strength of the pelvic region. Quite upright gait with almost shrivelled fore-limbs was reached in the Lower Cretaceous epoch by the Theropodous *Ceratosaurus*. In the same epoch the Sauropodous *Apatosaurus* had sunk down with its terrestrial fore-limbs, a giant in length, with about 14 cervicals and an enormous tail. At the same time the Orthopodous *Iguanodon* stood and sat down upright; and *Stegosaurus* had sunk down upon his short but stout fore-limbs with five stunted fingers; short necked and long tailed, a monster of gait and figure.

Perhaps all this vague "planless" evolution happened and continued so long thanks to the absence of dangerous competition, except such as appeared within the whole group itself, and many tried to overcome this danger by force, which to them meant size, certainly not brain. The Dinosaurs came upon the scene when the

Labyrinthodonts vanished. The clumsy type of Permian Reptiles had already failed, and only the small-sized members of the remaining stock of the terrestrials proper continued, and made the most of that fortuitous combination of characters which turned some of them from potential into actual Sauro-Mammalia; too small, too late, perhaps also too local for serious competition with the rapidly rising Dinosaurs. The latter, however, after an enormously long innings, seem to have died out before the Tertiary epoch.

Vertebral column. Besides a proatlas remnant, the presacrals seem to range from about 23 to 28. At least 10 of these are cervical. Increase of length of neck is effected by lengthening of the vertebrae themselves, and at the expense of the thoraco-lumbar. This reciprocate condition is shown clearly by *Iguanodon*: 10 cervicals, 18 thoraco-lumbar, = 28. *Diplodocus*: 15 cervical, only 11 other presacrals, = 26 (Fig. 110). Some Therapods 10 plus 13 = 23. It is sometimes difficult to ascertain the true number, because of the variable number in the lumbar region converted into sacrals; the sacrum itself consisting of about four to six.

A characteristic feature of the vertebrae is the suppression of the basiventral elements, notably the intercentra, traces of which occur only on the atlas and as usually well-developed chevrons on the tail. The centra, amphicoelous in the earlier forms, show a tendency to lose the cavity at the anterior end, resulting in opisthocoelous joints, especially in the neck. Those of the thoracic and lumbar region are sometimes but slightly bi-

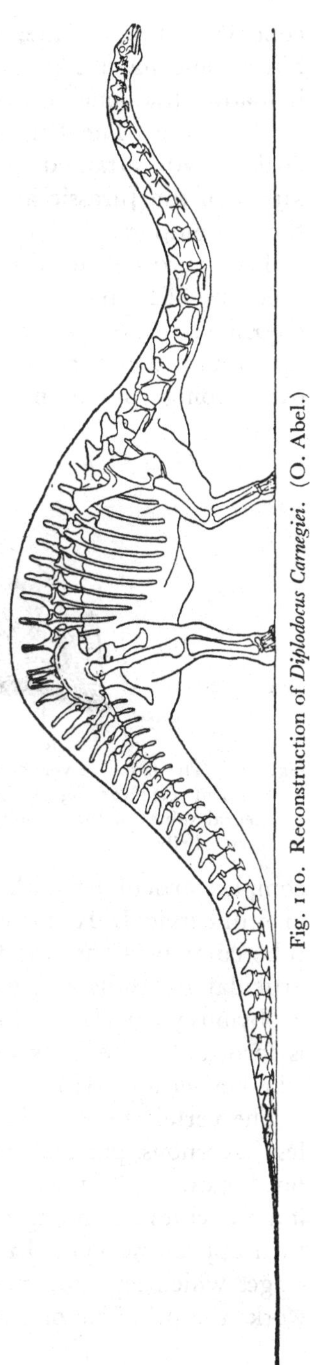

Fig. 110. Reconstruction of *Diplodocus Carnegiei*. (O. Abel.)

concave, almost plane, especially in heavy quadrupedal genera. Since some have, like Crocodiles, a biconvex first caudal vertebra, it follows that one procoelous vertebra, the second caudal, occurs in this order. The Dinosaurs have, however, not risen to the procoelous type attained by the Crocodiles of the Lower Cretaceous, although the Jurassic ancestors of the latter were amphicoelous like the contemporary Dinosaurs.

The neurocentral sutures frequently give way to fusion. A remarkable case of fusion of vertebrae, unique in its extent, is the complete co-ossification of the centra of the atlas and the second, third and apparently the fourth vertebra, together with a partial fusion of the overlapping spinous processes in *Triceratops* (Fig. 111). This fused

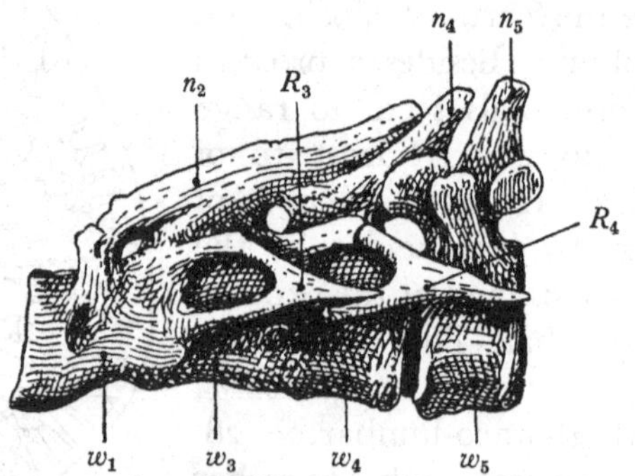

Fig. 111. The first five vertebrae of *Triceratops prorsus* (Marsh). w_1, atlas and second cervical vertebrae; w_3–w_5, bodies of the third, fourth and fifth vertebrae; n_2, n_4, n_5, neural spines of the second, fourth and fifth vertebrae; R_3, R_4, third and fourth cervical ribs.

complex articulates with the occipital condyle and with the fourth or fifth cervical. It carries two pairs of ribs with long capitular and tubercular portions; the first is fused on to its vertebra, the second articulates with its centre and diapophysis. This unique arrangement is obviously correlated with the huge hood-shaped neck shield which is formed by the back extension of the squamoso-parietal region; whether with or without additional dermal bone is doubtful.

The vertebrae of various Sauropoda and Theropoda are more or less cavernous, particularly in the tail of *Coelurus*. They are, however, not hollow in the pneumatic sense, that is to say, with one or more large cavities extending into the sides of the centrum, which would then appear swollen. In reality the feature is produced by flanges, ridges which expand, meet and fuse peripherally like fancy faience work; the principle of basket structure, which allows a much larger and yet not too heavy a bone to be produced (Fig. 112). Especially the neural arches, with their various processes, often present a very cavernous, gnawed out, intricate appearance. It is not necessary to think of the pneumaticity of Birds; they may be marrow bones which, although otherwise restricted to Mammals, have been thus anticipated by some of the many least Reptile-like looking Dinosaurs. Some writers have even suggested their being warm-blooded.

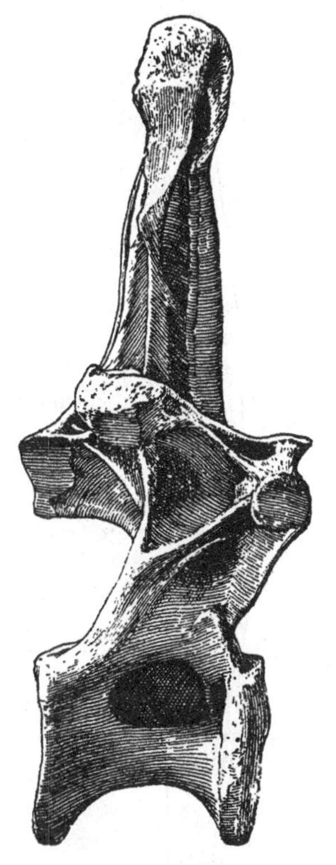

Fig. 112. *Diplodocus* (Saurischia). Dorsal vertebra. One-tenth natural size. After Hatcher. (Williston.)

Ribs are absent on the first and second vertebrae, otherwise they are two-headed. In the cervical region the capitulum is attached to the centrum, the tuberculum to its "diapophysis". Farther back on the trunk the ribs are moved dorsally on to the horizontally elongated diapophyses of the neural arches. This dorsal shift, recalling the Crocodilian condition, results in an increased capacity of the dorsal portion of the body cavity. Sometimes, e.g. *Apatosaurus*, the diapophyses and parapophyses of the neck are very long, fuse distally, enclosing a wide space, and end in two knobs for the rib. Usually the capitulum and tuberculum of the ribs are fairly long; the body of the movable rib is hatchet- or adze-shaped as in Crocodiles.

The pelvis is carried by three or four to ten vertebrae, which, when the rib-containing processes are laterally anchylosed, give the sacrum a very mammalian appearance. In *Ankylosaurus*, of the Upper Cretaceous, the ribs of the trunk are so strong, closely packed and partly synostosed as to form a box-like armour. The tail is completely encased in thick dermal armour and some buckles exist on the skull. Lastly, osteodermal plates form a weak armour in *Scelidosaurus*; in *Stegosaurus* this is strong; thick bony plates, supposed to be covered with horny sheaths, make a series of high defensive crests upon the back from head to tail. While osteoderms are apparently restricted

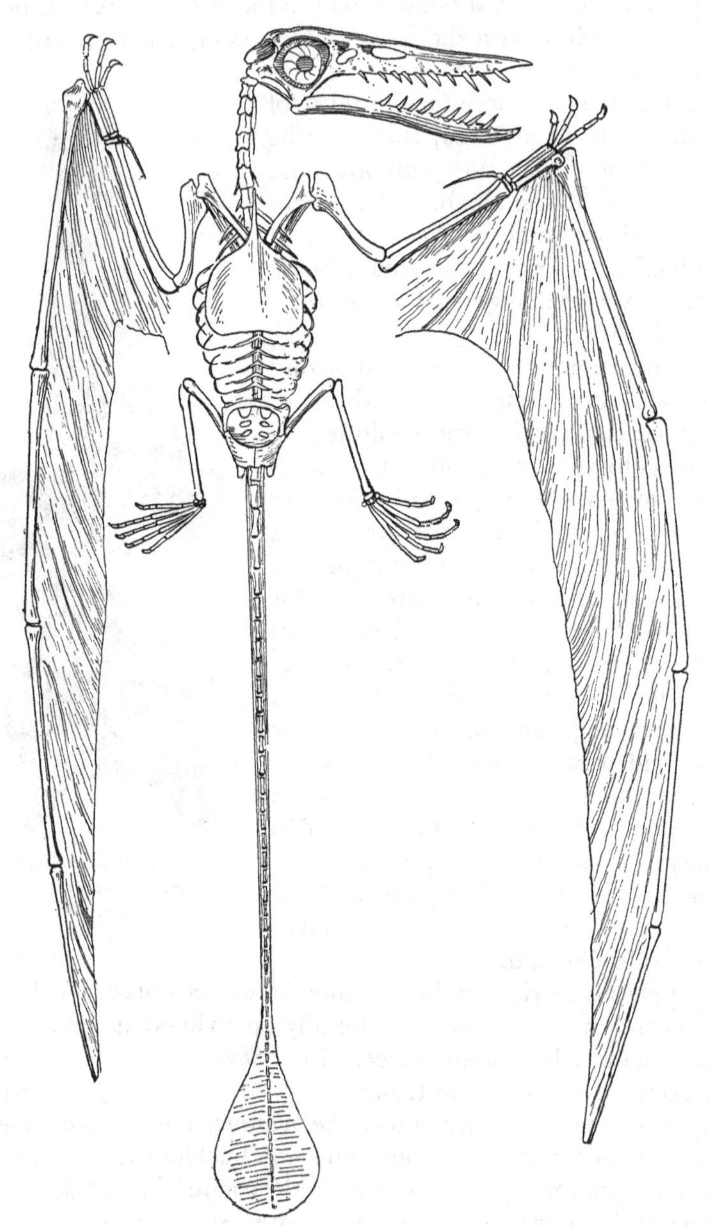

Fig. 113. *Rhamphorhynchus* (Pterosauria). Skeleton, one-third natural size. (Williston.)

to the quadrupedal Ornithischia, but are absent in the majority of the whole order, a strong distinction from the Crocodilia, all the Theromorpha and Sauromorpha have numerous well-developed abdominal ribs.

The sporadic way in which agreements and disagreements are distributed between Dinosaurs and other groups (Abel's Spezialisations-Kreuzungen) must not be used as indicating the very remoteness of the common root of the two main Dinosaurian suborders, nor is it a safe guide to the relationship with other orders. They are instances of the indecisive character of Dinosaurian evolution, for instance, the abdominal ribs, osteoderms, evolution of the pelvis independent of that of the limbs, and other features mentioned in the text.

PTEROSAURIA

Mesozoic flying Reptiles. They appear already in the Lias, Lower Jurassic, and died out, highly specialised, in the Upper Cretaceous. Examples are the long-tailed *Rhamphorhynchus* (Fig. 113), the short-tailed *Pterodactylus* (Fig. 114), and *Pteranodon*, a giant, and one of the latest.

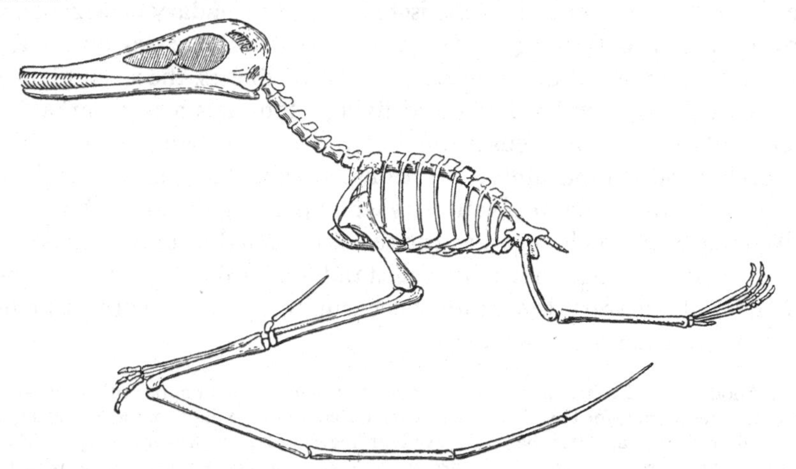

Fig. 114. *Pterodactylus*. Skeleton, enlarged one-third. (Williston.)

The vertebral column consists, rather variably, of 7–9 cervicals; 10–15 thoracics; 4–10 partly fused sacrals and a very long or very short tail. The caudal vertebrae are amphicoelous, the others procoelous; the first and the second are said to be anchylosed, a feature paralleled by the Hornbills among the Birds. The neurocentral suture is abolished. The centra are more or less pneumatic, which is not due to basket-work-like structures, but indicates air sacs (Fig. 115).

The cervical ribs are very short and thin, directed tailwards as in Birds; the thoracic ribs are long, with capitulum and tuberculum, some are attached to the keeled sternum. The scapulae are long, sabre-shaped and turned back as in Birds. In some, especially *Pteranodon*, an unique feature has arisen: the scapulae are attached by ligamentous strands to several of the spinous processes, and these ossify into a thin horizontal plate which renders the thoracic vertebrae thus involved immovable and anchylosed. A similar tendinous, horizontal, ossifying sheet covers the expanded iliac bones and causes the formation of a stiff sacrum, composed of from four to as many as ten metameres partly fused together. This reminds us of the sacrum of the *Cheiroptera* with 3–4 sacrals, and *Galaeopithecus* with 4–5 sacrals, which in the short-tailed genera coalesce with the caudals and thus increase the sacral region. In *Pteropus* the long transverse processes of these pseudo-sacrals join the ischia. Such secondary enlargements or additions to this region exist also in most Edentates and a few other Mammals, for different reasons. Nevertheless, the resemblances between Pterodactyles, Birds and flying Mammals are remarkable.[1] The tail, when long, is ensheathed by the long ossifying tendons. The well-developed hind-limbs, the strong sacrum, the long tail and the modified hands point to an ancestry with upright gait; some Dinosaur-like creatures which began as runners and ended as parachuters. In their incipient stage they may well stand for Baron Nopcsa's Proavis, their contemporary but more successful competitor, feathers being better than naked patagia.

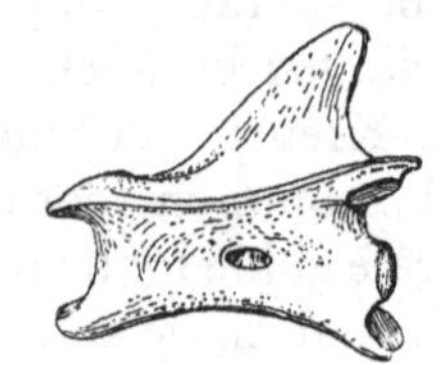

Fig. 115. *Pteranodon* (Pterosauria). Cervical vertebra, from the side. (Eaton.)

1 Another bird-like feature concerns the hand skeleton, a point apparently overlooked and thus responsible for the occasional statement that the large finger of the Pterodactyles is the fourth, or at any rate that they have four fingers only. The first metacarpal is short and separated from the others, directed inwards and forming an acute angle with the radius; it served as a stay in the preaxial patagium. The other metacarpals are much elongated and carry the second to fifth fingers, the outer of which is the strongest and largest and alone carries the patagium. The resemblance to Birds rests with the abduction of the short pollex and elongation of the remaining metacarpals.

Chapter XXXV

AVES

The ontogeny of the cervical vertebrae of the Chick was first and well studied by Froriep in 1883. Important stages in their development are found from the fifth day onwards.

Middle of the fifth day: A perichordal ring, consisting of bundles of longitudinal fibres, extends dorsally as the future basidorsal, and ventrally is thickened into a "hypochordal clasp" (the basiventral = hypocentrum = intercentrum). Both the dorsal and ventral arches are still in a precartilaginous state.

End of the fifth day: The perichordal ring forms the rudiment (Anlage)[1] of the future "intervertebral ligament". The right and left dorsal arches, still separate in the middle, extend slightly tailward from the ring. It is connected by indifferent tissue with the hypochordal clasp, which is now cartilaginous, horseshoe-shaped, with the arms looking upwards. A new thickening cluster has appeared in the skeletogenous layer, slightly behind the hypochordal clasp. The new cluster is unpaired, thickest laterally and ventrally. During its growth it grasps round the chorda from below upwards, behaving in fact like a second horseshoe.

Middle of the sixth day: The ventral cluster is transformed into a cartilaginous ring surrounding the chorda almost completely, being

1 There is still the search for a scientific rendering of *Anlage*. Many years ago Allman honoured me with that question. I could tell him only what this amoeboid term can mean. The planning, laying out of a future park, is an "Anlage", and often remains known as "die Anlage" when the park has grown up. By investing money in shares I am making an "Anlage", by "anlegen" that money, hoping that it may produce growing dividends. The money is "angelegt", not "beigelegt", i.e. laid by or laid up. The musician has an inborn "Anlage", at least an "ear for music", without which it may be difficult to steer clear between the *Wacht am Rhein* and the *Marseillaise*! In German morphology it is still customary to mix up rudiment and remnant. For instance "Rudimentaerorgane" do not mean the latent germs but the more or less degenerated remnants of the former, or elsewhere normal organs. It is often difficult to divine which category an author means. Thanks to G. B. Howes' proposal of "vestiges" or "vestigial", we can now distinguish between this and something else which should be "rudiments" or "rudimentary", and would cover "Anlage" without a rest. "Rudis indigestaque moles", the "Anfangsgruende", "die letzten organischen Bestandtheile", the three Rs are the "Anlagen" or Rudiments of something promising, capable of further development. If a building is badly "angelegt", it may break up into *rudera* and become a vestige.

the foundation or beginning of the centrum or body of its vertebra. Towards the end of this day the chorda is surrounded completely, but lies still nearer the dorsal side. It is only later that the chorda is turned into a central position. We conclude that the same process accounts for the variable position of the chorda in the vertebrae of other Amniotes, recent and fossil.

Later: Fusion of the central cartilage with the anterior mass (BD plus meniscus plus IV) produces the complete vertebra. The ventro-lateral corners of the clasp show a thickening of their perichondrium, which extends laterally into the intermuscular septa as dense fibrous tissue and indicates the future ribs.

The spinal nerve issues originally behind the neural arch, but later it comes to lie at the level of the posterior portion of the centrum and in front of the hypochordal cartilage of the vertebra next behind. This apparent shifting of the nerve is caused by the changes in the relative growth of the skeletal parts. The centrum grows chiefly headwards in length, and comes to lie with its somewhat conical apex between the arches, in the same way as the odontoid process, which, as the centrum of the atlas, is grasped by the arches. The whole reads like an epitome of the phyletic evolution of the Amniotic vertebra, especially as to the origin of the centrum, which Howes found in *Sphenodon* in the still more primitive paired condition.

Although it amounts to one of the best proofs of the interventral origin of the centrum, Froriep's elaborate account (more condensed in *Phil. Trans.* 1896) is usually ignored by those who will not, or cannot, appreciate the difference between Notocentrum and Gastrocentrum.

The basiventral masses of cartilage are gradually reduced and disappear, with the exception of the first and second, and those of some in the tail, which are there known as intercentra.

Concerning the atlas and axis: the first centrum, or odontoid, does not fuse directly with the second centrum, there being interposed the whole of the meniscus of the second basiventral as an intervertebral or intercentral pad. The first pair of fused basiventrals articulates, when ossified, solely with the cranial surface of the second basiventral, but not with the second centrum. The neural arch of the second vertebra is shifted backwards upon centrum 2, with which it forms the first neurocentral suture. The innermost portions of the ventral half of the atlas ring do not ossify, but remain cartilaginous or fibrous; they form in the adult the ligamentum transversum atlantis. It is serially homologous with the pad between odontoid and centrum of the axis and farther back with the pads known as menisci or annuli fibrosi. Even these last traces of the axial portion of the basiventrals may disappear when successive vertebrae become synostosed. The centre of each meniscus is of course perforated by the chorda, the last trace of which remains as the ligamentum suspensorium of the vertebral bodies, the first or foremost of these being the better known ligamentum suspensorium dentis, which passes into the occipital condyle. Since in all complete menisci the peripheral ring is thicker than the middle of the disk, this is, in amphicoelous vertebrae, bordered in front and behind by chorda so far as this is not constricted. Farther, the first pad between centrum 1 and centrum 2 firmly fuses them together and is moreover so large, or rather high vertically and broad transversely, that it prevents the BV 1 from coming into articulation with centrum 2. Thus it comes to pass that in the Birds the first highly movable joint within the whole column is an intervertebral joint. It is formed between the homologous units of two successive vertebrae. Different from the joint which is physiologically the same in the Mammalia, since the latter is intravertebral, being formed between BV 1 or BD 1 and centrum 1, which are units of the same vertebra. Cf. chapter on atlas joints.

The embryonic vertebrae of all Birds are at first amphicoelous, then they change through opisthocoelous into the heterocoelous or saddle-shaped type, which is a combination of procoelous and opisthocoelous and represents the highest stage of interaxial joint, allowing of most excursion. It is best developed in the neck and thorax, and is restricted to and characteristic of Birds. The anterior facet of the heterocoelous type is concave in the horizontal plane, convex vertically. The posterior facet shows the reverse; horizontally convex, vertically concave. Seen from the ventral aspect the whole series looks procoelous, seen from the lateral aspect all look opisthocoelous (Fig. 116).

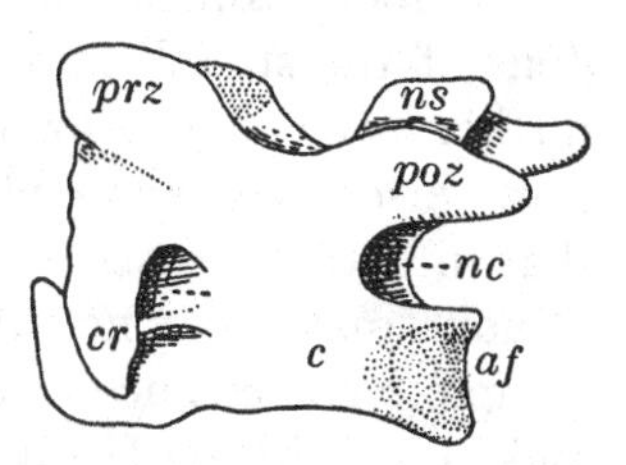

Fig. 116. Cervical vertebra of a Bird showing the saddle-shaped articular surface *af* on the centrum *c*; *cr*, cervical rib; *nc*, neural canal; *ns*, neural spine; *poz*, *prz*, post- and pre-zygapophyses. (Kingsley.)

The present ontogeny is a recapitulation of phyletic stages:

Amphicoelous were *Archaeopteryx*, and the cervico-dorsals of the well-flying Cretaceous *Ichthyornis*, while some of the neck vertebrae indicate transition towards the heterocoelous condition. Lastly, the tail of recent Birds often retains biconcave joints.

Opisthocoelous are those of the thorax of many recent Birds; most pronounced in the Penguins, to a lesser degree in the Cormorants, Gannets, some Gulls, Auks and Shorebirds, Parrots, etc.

Procoelous is of course the atlas. W. K. Parker found that in the Nidifugous birds the concavity is a perfect half-cup; that of the more advanced Nidicolae or Nestsitters is kidney-shaped, grasping part of the odontoid.

Heterocoelous was already the swimming, flightless Cretaceous *Hesperornis*. The majority of existing Birds possess such vertebrae.

Paired articulation between the neural arches is restricted to the usual prezygapophyses and postzygapophyses. Thoracic vertebrae of various Birds exhibit ventral outgrowths from the centra, true hypapophyses. They may be simple vertical blades, or ⊥-shaped and long, or paired knobs, such modifications often occurring in the same Bird; and they serve for the thoracic origin of the musculus longus colli anticus, reaching their greatest development in Penguins and Divers. In various Birds the carotid arteries, ascending on the ventral side of the neck vertebrae, are partly protected by low, paired flanges or ridges, which are also hypapophysial growths of the centra. Frequently some thoracic vertebrae show a tendency to a more rigid junction in various stages. In the simplest case this occurs in old Birds by the mere ossification of the ligaments, or even by that of the tendons of the spinal muscles. In other cases consolidation is carried further by co-ossification not only of the centra, but also of the spinous, transverse and zygapophysial processes of adjoining vertebrae, so that in extreme cases the whole dorsal region may become one continuous mass of bone. A feature quite unique among Birds is the complete coalescence of the atlas and axis so as to form one physiological unit in the Hornbills, e.g. *Buceros* and *Bucorvus*.

All these are instances of a process which, starting pathologically, frequently becomes a normal feature, ontogenetically hurried on so as to be ready in the immature as a character of species, genera or whole families as the case may be. The acquired character has become an inherited institution. For instance, in most Pigeons, including Dodos, the fifteenth, sixteenth and seventeenth vertebrae, being generally the three middle thoracics, are thus consolidated. In the *Tinamus* and most Gallinae, in the Flamingoes and Sandgrouse, the last cervical and the first three or four thoracics coalesce, and in many

Birds of Prey the first four thoracics are fused together. The sacrum, for which in its entirety the hybrid "synsacrum" has been invented, is very complicated, the large forward and backward extension of the pelvis having pressed a very great number of presacral and postsacral vertebrae into its service.

The whole sacral region may be subdivided into four successive subregions, each very variable in its extent and detail (Fig. 117).

(1) Anterior or crural portion. The vertebrae are united with the ilium by strong diapophyses and parapophyses. The first of these often bears a complete thoracic rib; the others bear more or less aborted ribs, with a tendency to lose their capitulum and tuberculum, the remaining shaft fusing with the surface of the iliac expansions. The last pair of these fused shafts often forms very strong bridges. The nerves which pass out of this crural portion form the crural plexus.

(2) The second or ischiadic portion. Parapophyses and ribs are absent; the diapophyses are reduced to thin blades extending obliquely, sometimes almost vertically, upwards to the dorsal median rim of the iliac bones. In the large fovea thus formed is embedded part of the kidneys. The number of vertebrae in this region cannot be ascertained other than by the exits of the nerves forming the ischiadic plexus; the whole series of mostly 3–5 vertebrae being fused together.

(3) The third portion or sacral proper is connected with the dorsi-median rim of each ilium by transverse diapophysial and by ventral bridges. The first two vertebrae are the primary or true sacrals, and they lie just behind a line drawn from one acetabulum to the other. Consequently, the original ilio-sacral connexion is postacetabular as in Reptiles, in contradistinction to the preacetabular Anura and Mammalia. Their ventral transverse buttresses are not outgrowths of the centra, but are ribs, as can still be ascertained in very young Birds.

(4) The postsacral portion consists of vertebrae, which in many Birds (*Pavo* for example) behave partly as do the primary sacrals and partly pass into normal caudals. Their diapophyses and parapophyses are always present, but, fusing with each other into transverse processes, abolish the foramen transversarium, while they abut upon the dorsal rim of the postacetabular part of the ilia. The first postsacral sometimes retains a pair of rib elements which either abort, or form a third sacral pair, while, on the other hand, only one sacral pair may exist. The general tendency of modern Birds seems to be

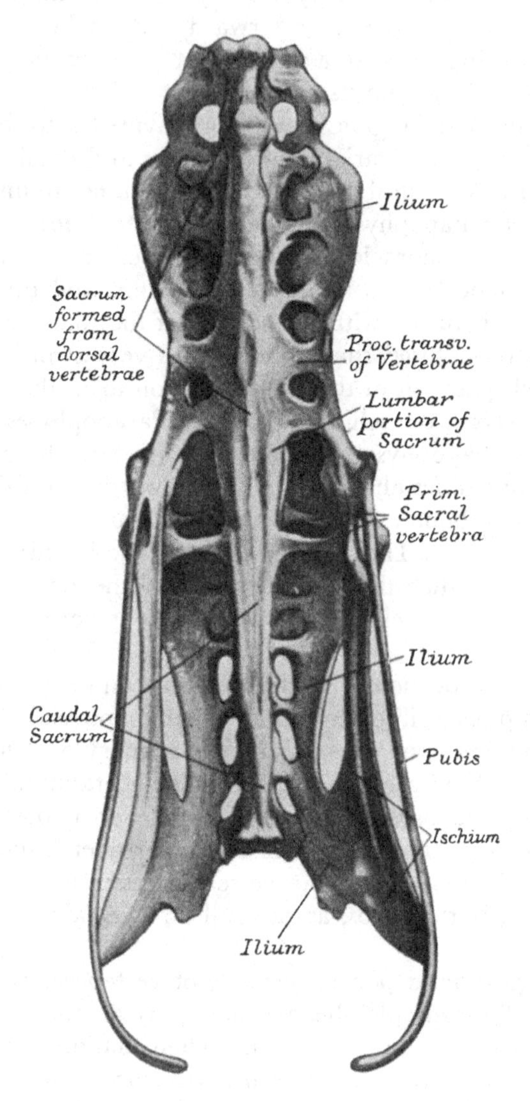

Fig. 117. Sacrum and pelvic girdle of a Bird (*Sula*), seen from the ventral aspect. (After Mivart.)

towards an increase in the number of postsacrals, compensated often by a smaller number of presacrals, and especially of those of the ischiadic portion. The whole complicated building up of the sacrum is due to several agencies and requirements. Forward elongation of the preacetabular iliacs catches all the existing ribs, converting them into transverse struts, e.g. the last thoracic, and reduces those of the ischiadic region eventually to nought. This procedure amounts in Mammals to a forward creeping of the whole pelvis, which presses one presacral segment after another into its service, and at the same time releasing others converts them into more or less reduced and synostosed caudals.

In Birds this forward shifting has been stopped by the formation of the strong buttresses in front of the ischiadic foveae. In Birds, further, the whole postacetabular half of the pelvis has become greatly elongated and commandeers postsacrals which were originally free caudals, converting them into additional sacrals. In a certain specimen of *Pavo*, the fifth postsacral is actually in the transitional condition, still free but just touching the ilia, which are here amalgamated with the ischia. In *Colymbus*, a Diver, the conversion has extended to the seventh.

In fact, the whole postsacral sacrum of Birds is a secondary acquisition, not a derelict as in Mammals. The most unexpected freak, an unique modification, is typical of Rhea, the three-toed American Ostrich: the two extremely elongated ischia fuse in the midline into a long ischiadic symphysis which lies above the gut, separating it from the kidneys. In adults the distal ends of the ilia fuse with these united ischia leaving foramina, and herewith is correlated a still more striking feature, namely the gradual absorption as maturity approaches of nearly the whole postsacral vertebral column, so that the caudal vertebrae seem to be attached to the united ischiadic and iliac ossifications.

The caudal vertebrae, namely all those which are not converted into postsacrals, have strong transverse processes, with scarcely any vestiges of ribs. The spinous processes often show a slight bifurcation at the end. They articulate almost entirely by the centra, which are usually slightly amphicoelous, with the interposition of the fibro-cartilaginous disk, which frequently displays a median osseous nodule, the last remnant of the intercentrum. Small hypapophyses, whether double or single, are mostly restricted to the last of the free vertebrae and to the first of those which form the *pygostyle* (Fig. 118 *A*). The latter is composed, as ontogeny shows, of six to seven vertebrae, which in almost all Carinatae coalesce into a terminal bony expansion, an upright plate which carries the rectrices, of which there are usually six pairs, but which are in some cases increased to 10 or 11,

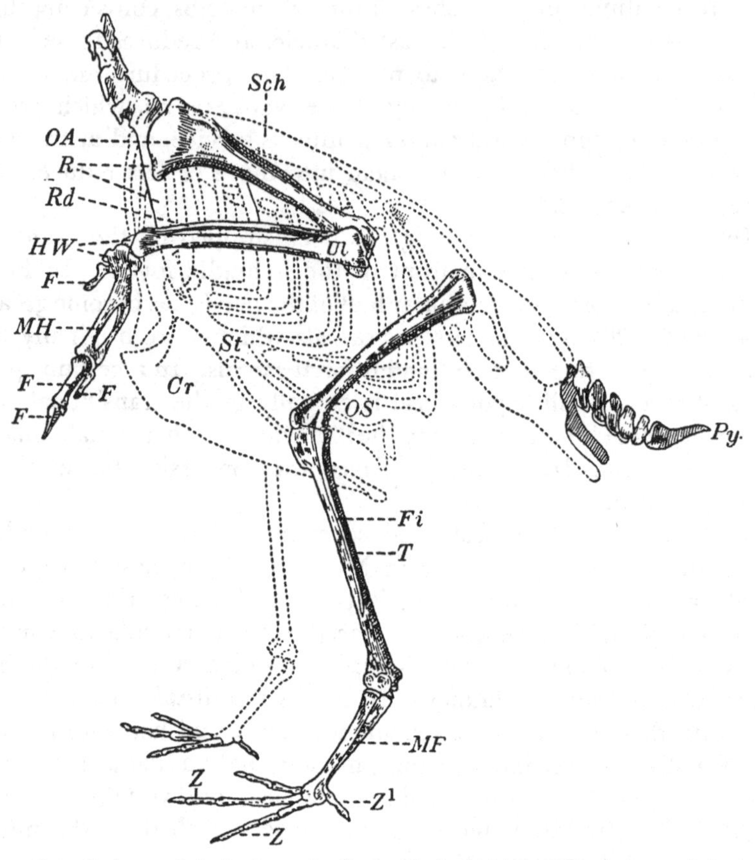

Fig. 118 *A*. Skeleton of the limbs and tail of a Carinate bird. (The skeleton of the body is indicated by dotted lines.) *F*, digits; *Fi*, fibula; *HW*, carpus; *MF*, tarsometatarsus; *MH*, carpometacarpus; *OA*, humerus; *OS*, femur; *Py*, pygostyle; *R*, coracoid; *Ul*, ulna; *Sch*, scapula; *St*, sternum, with its keel (*Cr*); *T*, tibiotarsus; *Rd*, radius; *Z*, Z^1, digits of foot. (Wiedersheim.)

in others decreased to four. These variations are brought about by the secondary addition, or suppression, of the outer pairs, and are in no correlation with the pygostyle, although this owes its origin to the paired enlarged rectrices and necessary concentration upon a shortened axis. The whole tail of *Archaeopteryx* is ideally primitive

(Fig. 118*B*): it possessed 21 (perhaps one or two more very weak ones at the tip) free and separate vertebrae behind the sacrum, which carry 12 pairs of well-developed rectrices; without a trace of pygostyle.

Fig. 118*B*. *Archaeopteryx lithographica*. From the Berlin specimen. *c*, carpal; *cl*, furcula; *co*, coracoid; *h*, humerus; *r*, radius; *sc*, scapula; *u*, ulna; 1, 2, 3, 4, digits. (Parker and Haswell, *Zoology*.)

The last caudals are far too small and weak to have supported the rectrices, which therefore cannot have been strictly terminal. Hence Haeckel's term Sauriurae (Lizard-tails) for *Archaeopteryx* as sole representative; all the other Birds he called Ornithurae, with the typical fan-shaped tail of feathers supported by a pygostyle. The conception of Ornithurae may be said to have lapsed; it has not been used from Huxley onwards in any of the more seriously constructed systems. It has been rightly abandoned, because it was felt that the difference implied is not fundamental, only one of degree. The tail of all Ratitae is devoid of a pygostyle with the individual exception of *Struthio*, in which sometimes a fair-sized terminal complex is developed. In a specimen of an old Ostrich in the Cambridge Museum it is some 2 in. high and nearly an inch and a half long. Other cases have since been mentioned. I. Parker has found indications in *Apteryx*. In the Cretaceous *Ichthyornis*, one of the toothed, well-flying Birds, it is very small. Its absence in these Birds is most probably a pseudoprimitive feature; the same applies probably to the Neotropical *Tinamus*, which can fly well. But the Cretaceous, toothed *Hesperornis*, a large marine Bird without a keel to the sternum and with extremely reduced fore-limbs, has as many as 14 sacrals connected with the very long ilium, and 12 free caudals which end in a point without trace of fusion. It is not impossible that his Mesornithic ancestors could fly, flutter or parachute like *Archaeopteryx*, and became thoroughly aquatic before pygostyles were "invented".

Ribs. The atlas shows no trace of ribs. The first existing is that of the axis. The ribs are at first carried entirely by the basiventrals, but they are soon transferred upon the centra, which produce parapophyses for the capitula, and the neural arches furnish the diapophyses for the tubercula. In the neck the vertebral arteries pass through the foramina transversaria, including those of the axis. In all Birds each rib is carried by the centrum and neural arch of the vertebra to which it belongs. Intervertebral articulation of the capitulum does not occur. The cervical ribs soon become anchylosed immovably to the vertebrae. Their body is directed backwards, parallel with the column, tapering to a sharp point. The last few have a longer body and, not reaching the sternum, are in a transitional, variable condition. Those of the thorax are well articulated and are sharply divided into a dorsal and ventral or sternal half (costalia and sternalia), both of which undergo proper chondral ossification, not calcification. Their number shows great amplitude of variation in the different groups. The same applies to the posterior thoracics, which gradually lose their share in building up the sternum, lose their sternalia and become short lumbar ribs which gradually lose their independence, helping to build up the sacrum (Fig. 119).

Uncinate processes are present in all Birds with the exception of the Neotropical Screamers, *Palamedea* and *Chauna*. The flat, at first independent and entirely cartilaginous processes arise from the posterior rim of the costalia (Fig. 119). They are best developed on the thoracic

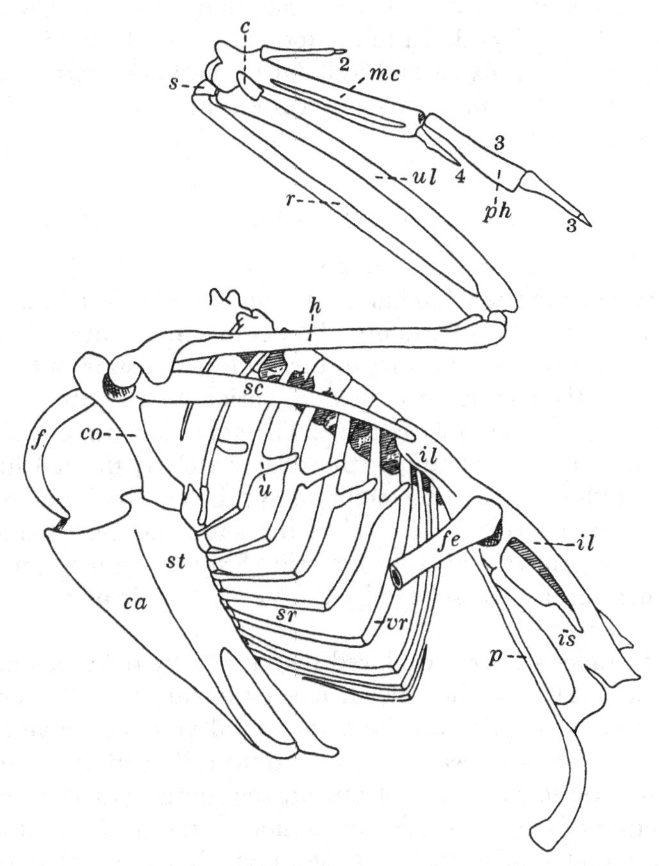

Fig. 119. Skeleton of trunk of Common Goose, *Anser domesticus*. *c*, cuneiform; *ca*, carina; *co*, coracoid; *f*, furcula (clavicle); *fe*, femur; *h*, humerus; *il*, ilium; *is*, ischium; *mc*, metacarpals; *p*, pubis; *ph*, phalanges; *r*, radius; *s*, scaphoid; *sc*, scapula; *sr*, sternal rib; *st*, sternum; *u*, uncinate process; *ul*, ulna; *vr*, vertebral rib; 2, 3, 4, digits. (Kingsley.)

ribs, beginning as small outgrowths sometimes on the last cervicals, continued back to the floating ribs, which consist only of their costalia. But they vary much in numbers, length and width. Extending over and even beyond the next following rib, with which they are connected by fasciae, they make the whole thorax more compact and yet yielding. But they are not intimately correlated with flight, being strong and numerous in the Ratitae, altogether absent in the Screamers, which are extremely good flyers and soar for hours for sheer enjoyment, trumpeting until almost out of sight. But these large Birds are so pervaded with air sacs that they can blow up the loose skin of their legs down to the toes, and the whole skin of body and neck feels then like a multicellular air cushion. This is unique, and so is the absence of uncinate processes, the morphology of which has been discussed on p. 265.

SUMMARY OF SERIAL AND REGIONAL HOMOLOGY

As has already been stated, the usual division of the whole column into cervical, thoracic, lumbar, sacral and caudal vertebrae is one of convenience, rule of thumb. Proper diagnoses are often very difficult to apply, partly owing to the transitional state of these five regions with their frequent individual variation, and above all due to the absence of a zero datum line. The only fixed zero, applicable to all the higher vertebrata, is the atlas vertebra; the tail literally tapers to nothing even in the embryo. A primary division into trunk and tail is indicated by the level of the anus, but even the always present vent can be shifted far forward (Fishes; postanal gut), or it can, as in some Tortoises, be lodged in a considerable portion of what is now *bona fide* tail.

The Tetrapoda have introduced the sacrum by the attachment of the pelvic girdle, and henceforth it is customary to call caudal all those vertebrae which lie behind the sacral vertebra or vertebrae. From the Amphibia onwards begins also the differentiation into neck and trunk, quite gradual and undefinable until with the Amniota the meeting of elongated ribs and fusion in the medio-ventral line lays the foundation of the costal sternum. But the first sternal or thoracic rib is by no means a reliable fixed point, since there are many instances of the first sternal not reaching the sternum, the articulation being lost, the connexion being reduced to a more or less complete, short or ligamentous strand, the sternale being reduced to a short vestige, or altogether absent. What is individually a variation (a case of "epistasis", arrested development in one species) reveals itself almost certainly as the normal condition in an allied species. The reverse process is likewise observable in this way; for instance, the rib of the fifteenth vertebra, normally being reduced to a cervico-dorsal condition in all but one species of a *bona fide* group, reveals the tendency in that single species of improving itself into a complete thoracic rib; this, of course, amounts to a shortening of the neck and a forward increase of the thorax. Instances of this process account for the really short-necked Birds like Pigeons, Oscines and Swifts, all of which are terminal groups of Birds. It is quite out of the question that these short necks represent an ancient, older condition, as can be proved by circumstantial evidence, comparison with their nearest kindred groups. Again, the long-necked Birds have acquired their feature clearly by progressive conversion of thoracic into cervico-dorsal and these into cervical vertebrae. This process of floating forwards and backwards almost forces us to assume that the "germ" of these ribs persists in a latent condition, to be revived when change of habits (environment) requires this reversion of evolution. If by long-continued disuse the "germ" has become too vestigial, the potentiality has lapsed. There is, for instance, a limit to shortening the Pigeon's neck beyond the thirteenth vertebra, because the twelfth and eleventh have already lost completely the normal backward extensions of the ribs which are present in the middle cervicals. It is also worth noting that the domesticated Pigeons, all descended from *Columba livia*, are in a state of flux, there being individuals with 13 and 15 instead of the normal 14 cervical vertebrae. The domestic fowls are likewise much more liable to such variations than wild species of Birds, which have been and can be studied in sufficient numbers. The variation in the behaviour of the posterior thoracic ribs is still more extraordinary. Sometimes the last thoracics have very long and slender sternal portions, which individually meet or do not meet the sternum for articulation. Sometimes two of these sternalia fuse together, the joined portion perhaps reaching the sternum asymmetrically, or the posterior sternale has lost continuity with its costal portion and is itself reduced to a literally floating sternale, loosely attached to the sternale of the previous vertebra. Normally the thoracics farther back are reduced to their longer or shorter costal portions, and when these are overlaid by the forward extension of the ilium and fuse on to it, their respective vertebrae are transformed from thoracic through lumbar into lumbosacral vertebrae. These and their ribs vary considerably in number in the different Birds. The last are those which form the strong buttresses for the ilia in front of the ischiadic recess. But these last buttresses themselves are not always serially homologous, the hindmost frequently being in a state of reduction. As a rule the nervus furcalis issues between the last pairs of buttresses; this nerve forms the connexion between the crural and the ischiadic plexus, and its stem is continued as the nervus obturator, passing through the foramen obturatum between os pubis and os ischii; it supplies the obturator muscles and the musculi pubi-ischio-femoralis (adductor longus). This nervus furcalis is of very limited help in unravelling the composition of this part of the sacrum.

More important and serious proved the search for a zero datum for the true primary sacral nerve. Gegenbaur concluded, from examination of the domestic Fowl, Bustard, Pigeon and Buzzard, that the true sacral nerve always issues between the two primary sacral vertebrae, and is the last nerve which sends a branch to the ischiadic plexus and at the same time a branch to the plexus pudendus. But I found from an examination of now more than 100 Birds representative of every group, that the above statement is weakened by about 40 per cent. of exceptions. One explanation is that there are in Birds, as sometimes in Reptiles, three primary sacral vertebrae, sometimes only one, because either the first or the first and second have been assimilated to those of the ischiadic recess, which thereby has been increased in extent, or the last of the three is in a more or less complete state of reduction. Lastly, for instance in certain examples of *Colymbus* (*Cygnus*), the sacral ribs or struts have vanished without traces, so that the ischiadic recess is increased to four vertebrae (numbers 29–32), as indicated by the nerve holes, and is bordered behind by the first of the nine typical postsacrals.

The conversion of the column following behind the primary sacrals into postsacrals (all those connected with the postacetabular half of the ilium), free caudals and the pygostyle is graphically expressed in Table III.

The whole question of regional homology, especially the composition of the Bird's sacrum, has to some extent been discussed with many illustrations and tables in Bronn's *Thierreich*. The problem involves the pelvis, in particular the extension of the ilium to support and fix the hind-limbs; this involves the muscles and, since several of the thigh muscles range from excessive size to complete absence, the nerve plexuses are correspondingly affected and in their turn greatly modify the sacrum. As the economy of a Bird and the gross structure of these parts are easily ascertained, the above correlations suggest a fruitful field for physiologico-anatomical examination.

AVES

It is a peculiar fact that none of the abundant textbooks, manuals, or primers contain more than a general account of the whole vertebral column of Birds; notably, this applies also to the common Fowl, and the Pigeon, which ranks as one of the standard "Types". The reason for this deficiency is that there happens to be no textbook from which to copy, and yet in a boiled Spring Chicken or young Pigeon or Duck the halves of the pelvis come off easily and show the whole sacrum, the only difficult region, like a diagram.

Table III shows in diagrammatic form the sacral and postsacral subregions in various Birds.

THE DESCENT OF BIRDS

Numerous footprints of three-toed creatures with upright bipedal gait, some of gigantic size, were discovered in 1835 in the Triassic Sandstone of Connecticut, and were described as belonging to Birds. The spoors themselves were called Ornithichnites.

The first Bird as yet known is *Archaeopteryx*, discovered in the year 1861 in the lithographic Slate, Upper Jurassic, of Solenhofen in Bavaria. It was soon recognised as a true Bird with various important reptilian characters. The close affinities of Birds and Reptiles were finally settled by Huxley, who combined them as Sauropsida, and gave in his celebrated Manual of 1871 lists of their numerous homologous and differential characters. It is all the more surprising that in Gegenbaur's textbook (1898, p. 137) *Archaeopteryx* is mentioned as "der den Voegeln am naechsten stehende Saurier"; that Reptile which stands nearest the Birds!

The discovery of *Hesperornis* (splendidly monographed by Marsh, *Odontornithes*, 1873), in the Cretaceous deposits, bridged the gulf within the Birds, but unfortunately Marsh saw in the absence of a sternal keel in *Hesperornis* close relationship with the Ratitae. Dollo took up this idea and called *Hesperornis* an aquatic carnivorous Ostrich, which was soon emphasised in German and English literature by naming it "the swimming ostrich". C. Vogt, and also Wiedersheim (1884), perhaps further misled by the Triassic footspoors, declared that these keel-less birds were directly derived from Dinosaurs. Albrecht helped the confusion by declaring the recently discovered Iguanodonts of Belgium as swimmers, like the Ducks, on the strength of their possessing a fourth trochanter, of which he gave a sketch with the muscles attached. To lay this ghost was not easy,

Table III

Vertebrae	27	28	29	30	31	32	33	34	35	36	37	38	39	40	41	42	43	44	45	46	Total Vertebrae
Columba		1	2	3	4	5	1	2	3	4	5	6	Large Pygostyle →								44 or 45
Corvus			1	2	3	4 Trans.	1	2	3	4	5	6	Large Pygostyle →								44 or 45
Hesperornis				1	2	3	4	5	6	7	1	2	3	4	5	6	7	8	9	10–12 All free	47 or 48
Archaeopteryx					1	2	3 All post sacrals are free and continued by free caudals	4	5	6	7	8	9	10	11	12	13	14	15 – 20 or 21		50 or 51
Pavo					1	2	3	4	1	2	3	4	Pygostyle →								46 or 47
Gallus						1	2	3	4	1	2	3	4	5	Pygostyle →						46 or 47
Cormorants							1	2	3	4	5	6	7	1	2	3	4	5	Pygostyle to 52		52
Colymbus							1	2	3	4	5	6	7	8	9	1	2	3	4	5 +Pygostyle	57
Tinamus								1	2	3	4	5	6	1	2	3	4	5	6	7–9 All free	48
Rhea									1	2	3	4	5	6	1	2	3	4	5	6–7 All free	47
Apteryx											1	2	3	4	1	2	3	4	5	Pygostyle to 50	50
Struthio												1	2	3	4	5	6	7	1 – 12 All free		56
Casuarius												1	2	3	4	5	6	7	8	1–12 All free	58
Cygnus																	1 – 9		1–6 free & Pygostyle		64

Scheme of the sacral (red), postsacral (green) and caudal (blue) vertebrae of different birds, to shew their great variability in position and number. In some all caudal vertebrae are free, in others the terminal ones are fused into the pygostyle.

and it took time, because it appealed to the multitude and obtained serious revivals when Baur and others examined the legs of the Ornithopoda and the pelvis of the Ornithischia, and showed their surprising resemblance to the conditions in young Birds. I raised objections in a paper published in *Proc. Camb. Phil. Soc.*, where it has remained hidden, seen by few zoologists, referred to by none. The latest serious contribution is by Steiner who, by an elaborate ontogenetic and phylogenetic study of the fore-limbs of Birds and Reptiles, came to the satisfactory conclusion that the relationship of Birds and Dinosaurs is rather remote, direct descent from any known group of the latter being out of the question.[1] There is no objection to representing them in a general diagram of Sauropsidan pedigree by parallel lines, both arising from the unknown.

1 The origin of Birds has aroused popular unabated interest; witness the book by G. Heilmann, *The Origin of Birds*, London, 1926; 208 pages and hundreds of excellent detailed illustrations. The author does not profess to be more than interested in the question, not posing as an authority. He has from extensive reading compiled many data, morphological, etc., concerning Birds and notably Dinosaurs, making the most of the agreements. He ends with the inevitable picture of the *Proavis*. In connexion with this I beg to state that in 1893 (Bronn, Voegel, pt. II) I proposed (p. 86) for the still unknown transitory Reptile-Birds (the *Proavis* of up-to-date writers) the term *Herpetornithes*, and on p. 90 the term *Mesornithes*, for those likewise unknown Birds which must have bridged the gap between *Archaeopteryx* (solitary representative of the first "Birds" or *Archaeornithes* of the Upper Jurassic) and the *Odontornithes* of the Upper Cretaceous, which were already *Neornithes*. That such intermediate forms must have existed is certain. It is but a question of time and luck for them to be found.

Chapter XXXVI

ICHTHYOSAURIA AND SAUROPTERYGIA

THE MESOZOIC MARINE ROVERS

ICHTHYOSAURIA

These aquatic Reptiles characteristically possessed a short neck and a long tail. Without cervical ribs but with many trunk ribs which, although long, are restricted to the costal portion, and there is no sternum. Very numerous abdominal ribs form a complicated ventral protection; they exceed the metameres of the respective region of the trunk in number, and are so closely packed that this belly-shield seems to be due to a telescoping of neighbouring regions. Fore- and hind-limbs modified into broad, hyperphalangeal, sometimes also hyperdactyle paddles. From Mid-Trias to Upper Cretaceous with the widest distribution.

Occipital condyle large, entirely basioccipital, carrying the lateral occipitals with normal hypoglossal foramina.

The vertebral column shows many peculiar features. It is composed of many, up to 150, vertebrae, about two-thirds of which are caudal.

The atlas. Centrum concave in front for reception of the condyle, scarcely fused with centre 2 except in older specimens of Post-Triassic Ichthyosaurs; it carries the but partly united first pair of neural arches (Fig. 120). The first pair of basiventrals, equivalent to the ventral half of the atlas ring of most other Reptiles, thus becomes an intercentral, unpaired wedge between the occipital condyle and centre 1. Smaller intercentra belong to the second and third vertebrae; farther back they are absent; even chevrons are restricted to the Triassic Mixosauri.

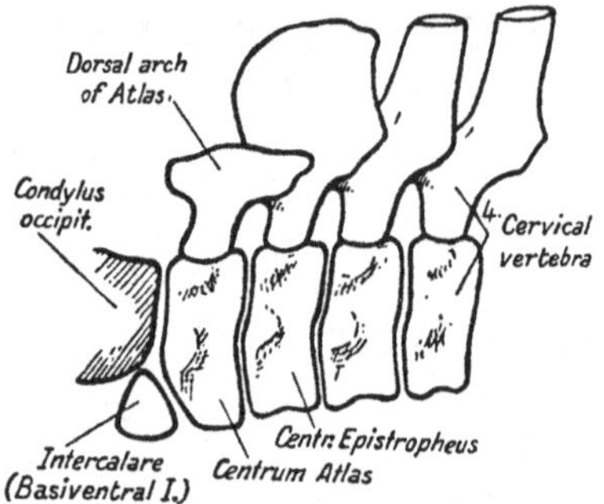

Fig. 120. Cervical vertebrae of an Ichthyosaur seen from the left side. The first basiventral is separate and the centrum of the atlas is not fused to the second vertebra. (After Jaekel.)

Cervical ribs are entirely absent. Owing to severe suppression of the basiventral elements, of the intercentra and even of the menisci, the vertebrae of at least the neck and trunk consist of only the centra and the neural arches, which are loosely united together. The largest and most characteristic part is the centrum, which itself is shortened (hence the short neck) to the shape of a disk, deeply biconcave, perforated in the exact middle by the much constricted chorda. The disk itself is much higher than it is long, and it carries laterally or more ventrally short processes for the thoracic ribs (Fig. 121). These processes and the ribs have undergone a steady change. In the Mid-Triassic older forms like *Mixosaurus*, only a few of the anterior trunk ribs are clearly two-headed, capitulum and tuberculum being

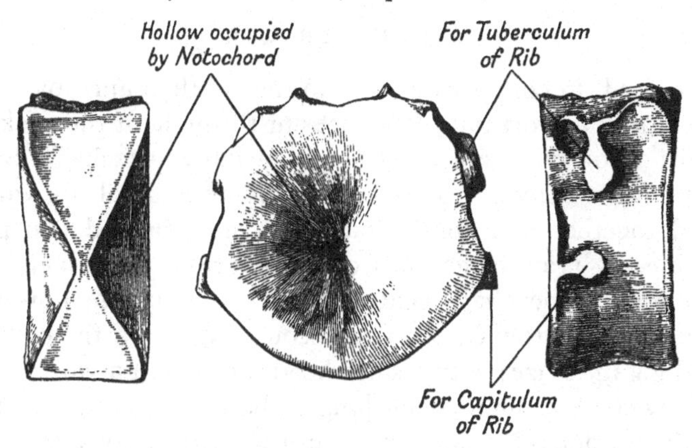

Fig. 121. *Ichthyosaurus.* The centrum of an anterior dorsal vertebra, viewed in section, and from the anterior and left lateral aspects. (After Owen.)

carried by well-separated facets, which both lie in the same transverse vertical plane. The other ribs and those of the tail are one-headed. This condition prevails also in some genera of the Upper Trias. But in others a change has by this time taken place, which lasts throughout the Jurassic to the Upper Cretaceous, e.g. in *Ichthyosaurus*, the longest lived genus of all. In them the trunk vertebrae carry two-headed ribs; the bearing processes steadily approach each other until they are in contact, and at the same time are shifted towards the ventro-lateral side of the centrum. The caudal ribs seem always to be single-headed. If these trunk vertebrae co-ossified with their neural arches they would become stereospondylous and resemble closely those of the Labyrinthodonts, although of fundamentally different composition. The ruling mechanical principle is the same.

The sacrum as such is abolished, the ilia being much reduced and separated from the column. The tail always contains many vertebrae. That of the Mixosaurs is still straight. During the Jurassic the dorsal tail fin (adipose as shown by its imprint in the matrix) becomes so preponderantly larger than the ventral fin that the column is bent down at an angle and the tail ends in a large bifurcated, vertical, semilunar fin. This is composed almost entirely of the original dorsal fin, the upper blade. The original ventral, hypaxial caudal fin is completely reduced and the lower blade is made of the enlarged integument on the dorsal side of the bent-down column, the vertebrae of which are reduced to nodules.

SAUROPTERYGIA

Characterised, in contradistinction to the Ichthyosauri, by having a long neck and a short tail. Ribs present throughout the neck, trunk and tail; two- or one-headed, and variably attached. Sternum absent; but the belly is protected by many abdominal ribs which are crowded together and consist each of a median and two pairs of lateral pieces. Transformation from the normal pentadactyle limbs into enlarged hyperphalangeal paddles begins in the Lower Trias and finishes with the Upper Cretaceous. Many of these creatures attained gigantic size; with a stout head more than a yard in length, carried by a neck often much longer than the robust trunk. Tail, although considerably shorter than the trunk, consists of many shortened and dwindling vertebrae, and was not used for locomotion.

They occur from the Lower Triassic Sandstone to the end of the Cretaceous and ultimately had the widest distribution.

Ichthyosauri and Plesiosauri used to be combined as Eualiosauri. Owen separated them as Ichthyopterygia and Sauropterygia. Later it was recognised that at least *Nothosaurus* and, later, *Lariosaurus* are allied to the Sauropterygia, but the Mesosauri, until then vaguely grouped with the Rhynchocephalia, or linked with *Protorosaurus* as Progonosauria, have only recently received their proper place in the system as members of the Sauropterygia, which thereby have at least a plausible ancestry.

But the Sauropterygia have produced a great many forms which, as usual with such a great group, have not kept step in the specialisation of skull, limbs, girdles and vertebral column, so that according to personal bias concerning taxonomic importance some authorities discern for instance in such forms as *Nothosaurus*, *Lariosaurus* and *Trachelosaurus*, various indications of relationship with Ichthyosaurs; while *Placochelys*, an ally of *Placodus*, suggests not only obvious affinity with the Sauropterygia, but, through the unusual presence of osteodermal buckles on the skull, more than mere convergence with the contemporary *Toxochelys* and the much older *Triassochelys*. Appeal to convergence is a polite way of dismissing inconvenient facts, but it does not do away with the many observed instances of ancient structures which, lost for many generations, crop up again sporadically "because they are in the blood". Perhaps the Mesosaurs stood as near the other Triassic Sauropterygia, whose widely divergent descendants may have been the Placodonts and perhaps Chelonia, as did the Ichthyosaurian Mixosaurs. Most of them specialised for marine life, the latest to do this being some of the Crocodiles and Chelonians. This may be a dream, but nothing is gained by the honest confession of our ignorance when treating them all as so many highly specialised ancient separate orders of, say, a widened "Formenkreis" of Archaeosauri, with which they will fit in just as suitably as with Lizards, Snakes and Birds.

The following list of the more important genera may serve as a basis for a general account of the trend of evolution of the Sauropterygia:

Stereosternum and *Mesosaurus*; size of small Varanids, semiaquatic, Lower Trias; the first from Brazil; the second from S. America and S.W. Africa, therefore strongly suggesting the existence of a continuous coast across the Atlantic.

Trachelosaurus, Lower Trias. *Nothosaurus*, whole Trias; size of large *Varanus*. *Lariosaurus*, Mid-Trias. All three European; hands and feet of incipient aquatic type, but not yet paddles.

Pliosaurus, *Plesiosaurus*, *Elasmosaurus*; beginning with the Lower Jurassic, ending with the Cretaceous.

The vertebral column may contain from about 80 to more than a 100 units. The vertebrae are rather short, progressing from deeply amphicoelous, through slightly biconcave to plane centra, with or without a neurocentral suture (Fig. 122). The cervical ribs are short, attached to their centra, becoming gradually longer towards the thoracic region. Their number varies greatly; 16 to 20 in the Triassic genera. The Plesiosauridae proper, from the Jurassic onwards (Fig. 123), mark an increase which reaches the record of more than 70 cervicals in the Upper Cretaceous monster *Elasmosaurus* with a neck much longer than trunk and tail together. The ribs of the trunk, described as one-headed, are transferred on to the diapophyses of the neural arches, unless these stout, double faceted processes are in reality transverse processes. The Plesiosaurian ribs are generally strong, sometimes stout, and their dorsalwards shifted attachment (in opposition to Ichthyosauri) means a widened chest.

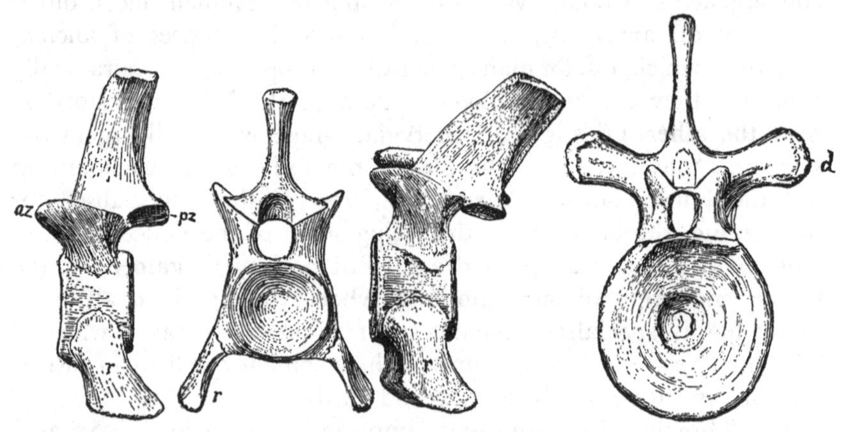

Fig. 122. *Polycotylus* (Plesiosaur). Cervical vertebrae from the side and behind, and dorsal vertebra from in front: *az*, anterior zygapophysis; *pz*, posterior zygapophysis; *r*, *r*, *r*, cervical ribs; *d*, diapophysis. (Williston.)

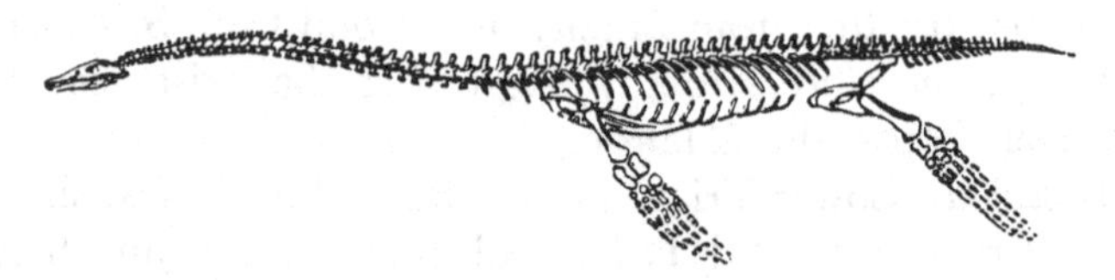

Fig. 123. Skeleton of *Plesiosaurus dolichodirus*. (Conybeare.)

The short, dwindling ribs of the tail are attached to the centra well below the suture. The chevrons are small, right and left separated, movably attached.

Zygapophysial articulations are well developed only in the neck and sometimes zygantra and zygosphenes are indicated; all, with the retained ribs, are in correlation with the muscles of the long, flexible neck. In the trunk, even the zygapophyses have become vestigial or absent, while the diapophyses are prominent (Fig. 122). A strong ventral armour of abdominal ribs is present in all.

The sacral region is in a remarkably variable condition; difficult to decide whether incipient or vestigial. In the Triassic smaller genera, still with at least semi-terrestrial hind-limbs, there are four to five or even six lumbar-like vertebrae. In *Lariosaurus* five such rather long sacral ribs converge towards the ilium, touching each other without fusion. *Stereosternum* has four, two carrying like typical sacrals the posterior acetabular half of the broadened ilia, the others, more lumbar-like, converge and touch the anterior end of the pelvis. The allied *Proneusticosaurus* is in an intermediate condition, the two foremost ribs are free. In the Plesiosaurs the ilia are decidedly reduced, connected with only two or even one pair of ribs, cf. chapter on the Sacrum.

BIBLIOGRAPHY

To find a reference: The page number in the last column is that on which the author is mentioned in this book. Opposite will be found the particular reference with its page number in the column in italics. If two or more separate papers by a single author are referred to on the same page, the references are indicated by inferior figures, e.g. 75_1, 75_2.

ABEL, O.	Die Stämme der Wirbelthiere. 1919.	*22*	9
		28	29
		28	31
		308	139
		333	176
		524	180 n.
		308	182
		Fig. 229, p. 308	191 F.
		395	209
		451	257
		580	302
		Fig. 478, p. 610	303 F.
			307
ADOLPHI, H.	Über Variationen der Spinalnerven und der Wirbelsäule anurer Amphibien. Morph. Jahrb. XIX.	*313*	91
	Ibid. XXII.	*449*	
ALBRECHT, P.	Sur les copulae intercostoïdales. Bruxelles. 1883.		268
			323
BALFOUR, F. M.	Monograph on the development of the Elasmobranch fishes, 1878.		
	Journ. Anat. and Physiol. x, 1876.	*377*	
		517	
		672	66
	Ibid. XI, 1877.	*128*	
		406	
		674	
	Ibid. XII, 1878.	*177*	
BAUR, G.	Ueber die systematische Stellung der Microsaurier. Anat. Anz. XIV, 1896.	*148*	191
	Dermochelys, *Dermatochelys* oder *Sphargis*. Zool. Anz. No. 270, 1888.	*1*	208
	Bemerkungen über das Becken der Vögel und Dinosaurier. Morph. Jahrb. x.	*613*	325

BELL, J.	Amphibia. Todd's Cyclopaedia of Anatomy and Physiology, I, 1835–6.	*90*	134
BEMMELEN, J. F. VAN	Bemerkungen zur Phylogenie der Schildkröten. Compte-Rendu des séances du troisième congrès international de zoologie. 1896.	*322*	208
	Bemerkungen über den Schädelbau von *Dermochelys coriacea*. Festschr. zum 70sten Geburtstag von C. Gegenbaur.	*279*	
DE BLAINVILLE, H. M.	Prodrome d'une nouvelle distribution du règne animal. Bull. Soc. Philom. 1816.	*113*	289
BOULENGER, G. A.	Les Batrachiens. Paris, 1910.	*6*	
	Catalogue of Chelonians in the British Museum. 1889.		166
	Remarks on a Note by Dr G. Bauer on the Pleuradiran Chelonians. Ann. Mag. Nat. Hist. VI, ii, 1888.		208
	Catalogue of Snakes in the British Museum. 1893–6.		281 282
	Catalogue of Chelonians, Rhynchocephalians and Crocodiles in the British Museum. 1889.		293
BRANSON, E. B.	Structure and relationships of American Labyrinthodontidae. Journ. Geol. XIII, No. 7, 1905.	*568*	9 155
BROILI, F.	Permische Stegocephalen und Reptilien aus Texas. Bd. LI, 1904.	*1*	170 176 180 193
	Die Rhachitomenwirbel der Stegocephalen. Zeitschr. Deutsch. Geol. Ges. LX, 1908.	*235*	51 279 302
BRONN, H. G.	Klassen und Ordnungen des Thierreichs. Aves, Bd. VI.	*4*	322
BROOM, R.	On a new Labyrinthodont (*Rhinesuchus Whaitsi*) from the Permian beds of South Africa. Ann. S. Afric. Mus. IV, 1908.	*373*	158
	A comparison of the Permian Reptiles of North America with those of South Africa. Bull. Amer. Mus. Nat. Hist. XXVIII, 1910.	*232*	180
	On the origin of Mammals. Rep. Brit. and S. Afr. Ass. III, 1907.	*1*	222
	A further comparison of the South African Dinocephalians with the American Pelycosaurs. Bull. Amer. Mus. Nat. Hist. XXXIII.	*135*	223

BROOM, R. (*cont.*)	On the structure of the skull in Cynodont Reptiles. Proc. Zool. Soc. 1911.	*893*	225
	Permian, Triassic and Jurassic Reptiles of South Africa. Bull. Amer. Mus. Nat. Hist. xxxv, ii, 1915.	*161*	291 293
	On the structure and affinities of the Multituberculata. Bull. Amer. Mus. Nat. Hist. XXXIII.	*115*	231 F.
	On a nearly complete skeleton of a new Eosuchian Reptile. Proc. Zool. Soc. 1926.	*486*	292 F.
BUDGETT, J. S.	On the breeding habits of some West African fishes, with an account of the external features in development of *Protopterus annectens*, and a description of the larva of *Polypterus lapradii*. Trans. Zool. Soc. XVI, 1901. On the structure of the larval Polypterus. *Ibid.* 1902.		113
BULMAN, O. M. B. and WHITTARD, W. F.	On *Branchiosaurus* and allied genera (Amphibia). Proc. Zool. Soc. 1926.	*533*	124
BÜTSCHLI, O.	Vorlesungen über vergleichende Anatomie. I. Lieferung. Berlin, 1910.	*200* *Fig. 124* *221*	114 204 F.
CASE, E. C.	Revision of the Pelycosauria of North America. 1907.		51 218 219 220
	Revision of the Amphibia and Pisces of the Permian of North America. Carnegie Institution of Washington Pub. No. 146, 1911.		98 142 153 F. 155 F. 161 194
	On the osteology and relationships of Protostega. Journ. Morph. XIV, 1897.	*21*	208
CHIARUGI, G.	Lo sviluppo dei nervi vago, accessorio, ipoglosso e primi cervicali nei Sauropsidi e nei Mammiferi. Atti Soc. Toscana d. sc. nat. x, 1889.	*149*	80
	Sur les myotomes et sur les nerfs de la tête postérieure et de la région proximale du tronc dans les embryons des Amphibiens anoures. Arch. Ital. de Biol. xv, 1891.	*229*	
CLAUS, C. and SEDGWICK, A.	Elementary Text-book of Zoology, vol. II, 1885.	*226* *Fig. 646*	202 F.

Conybeare, W. D.	Additional notices on the fossil genera *Ichthyosaurus* and *Plesiosaurus*. Trans. Geol. Soc. Lond. (2), 1824.	*103*	330 F.
Cope, E. D.	On the fossil remains of the Reptilia and Fishes from Illinois. Proc. Acad. Nat. Sci. Phila. 1875.	*404*	12, 37, 43, 149, 161, 170_1
	Fifth contribution to the knowledge of the Fauna of the Permian formation of Texas and the Indian Territory. Proc. Amer. Phil. Soc. xxii, 1884.	*28*	
	The Batrachia of the Permian period of North America. Amer. Nat. xviii, 1884.	*26*	
	Descriptions of extinct Vertebrata from the Permian and Triassic formation of the United States. Proc. Amer. Phil. Soc. xvii, 1878.	*188*, *526*	38, 98, 155, 156, 166
	The Rhachitomous Stegocephali. Amer. Nat. xvi, 1882.	*335*	45
	The Batrachian intercentrum. Amer. Nat. xx, 1886.	*76*	53
	On the intercentrum of the Terrestrial Vertebrata. Trans. Amer. Phil. Soc. xvi, 1888.	*243*	
	Synopsis of the Extinct Batrachia of North America. Proc. Acad. Nat. Sci. Phila. 1868.	*209*	134
	A Batrachian Armadillo. Amer. Nat. xxix, 1895.	*998*	170_2
	The skull of Empedocles. Amer. Nat. xiv, 1880.	*304*	175, 176, 200
	Second contribution to the history of the Vertebrata of the Permian formation of Texas. Proc. Amer. Phil. Soc. xix, 1880.	*38*	175, 196
	A description of the Genus *Protostega*, a form of the extinct Testudinata. Proc. Amer. Phil. Soc. xii, 1872.	*422*	208
	The long-spined Theromorpha of the Permian epoch. Amer. Nat. xx, 1886.	*544*	220 F.
Credner, H.	Die Stegocephalen und Saurier, aus dem Rothliegenden des Planenschen Grundes bei Dresden. Zeitschr. Deutsch. Geol. Ges. xxxiii–xlv, 1881–93.		74
	Entwicklungsgeschichte von *Branchiosaurus amblystomus*. Zeitschr. Deutsch. Geol. Ges. xxxviii, 1886.	*576*	124

CREDNER, H. (*cont.*)	Die Urvierfüssler (Eotetrapoda) des Sächsischen Rothliegenden. Naturw. Woch. Allgem. verständ. Naturw. Abhand. XV, 1891.	*1*	172 181
CUVIER, G.	Leçons d'anatomie comparée. 2nd ed. Tome I. Paris, 1835.	*188*	51
DAWSON, J. W.	On the air-breathers of the Coal Period: a descriptive account of the remains of land animals found in the coal formation of Nova Scotia. Montreal, 1863.		173 181
	On a terrestrial mollusk, a chilognathous myriapod, and some new species of reptiles from the coal measures of Nova Scotia. Quart. Journ. Geol. Soc. XVI, 1860.	*268*	189
	On the results of recent explorations of erect trees containing animal remains in the coal formation of Nova Scotia. Phil. Trans. Roy. Soc. CLXXIII, 1882.	*637*	
DOLLO, L.	Première note sur les chéloniens oligocènes et néogènes de la Belgique. Bull. du Musée Roy. d'Hist. nat. de Belgique, V, 1888.	*59*	208
	Sur l'origine de la Tortue Luth. Bull. Soc. Roy. Sciences médicales et naturelles de Bruxelles, 1901.		209
	Note sur la présence chez les oiseaux du "troisième trochanter" des Dinosauriens et sur la fonction de celui-ci. Bull. du Musée Roy. d'Hist. nat. de Belgique, II, 1883.	*13*	323
DUGÉS, A.	Recherches sur l'ostéologie et la myologie des Batrachiens à leurs différents âges. Mém. Instit. France, Sciences Math. et Phys. VI, 18 plates, 1835.	*216*	167
EATON, G. F.	Osteology of *Pteranodon*. Memoirs Conn. Academy of Arts and Sciences, II. New Haven, 1910.		308 F.
ECKER, A. and GAUPP, E.	Anatomie des Frosches von A. Ecker und R. Wiedersheim. 2te Abteilung. Auf Grund einiger Untersuchungen durchaus neu bearbeitet von E. Gaupp. 1896.	*165* *170*	91_1 91_2
FLOWER, W. H.	An introduction to the osteology of the Mammalia. 3rd edition.	*62* *39* *Fig. 15* *Figs. 20, 21*	50 61 63 107 238 F. 241 F.

Fraas, E.	Die Labyrinthodonten der schwäbischen Trias. Paläontographica, xxxvi, 1889.	*1*	158
	Neue Labyrinthodonten aus der schwäbischen Trias. Paläontographica, lx, 1913.	*275*	
	Aetosaurus crassicauda n.sp. nebst Beobachtungen über das Becken der Aetosaurier. Jahresh. der Ver. f. Vaterl. Naturk. in Württemberg. 1907.	*101*	294 F.
Fritsch, A.	Fauna der Gaskohle und der Kalksteine der Permformation Boehmens. Prag, i, 1883.	*68*	122
		129 *142*	126 F.
		108	145
		142	183
		159	189
		171	191
	Ibid. ii, 1889.	*24*	45
			151
		11	161
Froriep, A.	Zur Entwickelungsgeschichte der Wirbelsaeule, insbesondere des Atlas und Epistropheus und der Occipitalregion. I. Birds. Arch. f. Anat. u. Phys. pls. vii–ix. 1883.	*177*	309
	Ibid. II. Mammals. Arch. f. Anat. u. Phys. pls. i–iii. 1886.	*69*	106
			234
			235
			236
Fuerbringer, M.	Vergleichende Anatomie der Brustschulterapparatus. Jenaische Zeitschrift, vii, 1873.		75
	Ibid. viii, 1874.	*175* *280* *489*	75 87
	Ibid. xxxiv, 1900.		75
	Ibid. xxxvi, 1902.		75
	Zur vergleichenden Anatomie der Schultermuskeln. Morph. Jahrb. i, 1875.		75
	Ueber die Spinooccipitalen Nerven der Selachier und Holocephalen und ihre vergleichende Anatomie. Festschr. für Gegenbauer, iii, 1896.	*349*	75
		Figs. 1–6	78
		Pl. 1	
		Fig. 12	80
		Pl. 7	87
		Fig. 9	91_1
		Pl. 8	
		730	91_2
		Fig. 16	96
		Pl. 7	
	Untersuchungen zur Morphologie und		

FUERBRINGER, M. (*cont.*)	Systematik der Vögel, I, II. Amsterdam und Jena, 1888.		247
	Beiträge zur Systematik und Genealogie der Reptilien. Jena. Zeitschr. f. Naturwissenschaft, XXXIV, N.F. XXVII, 1900.	*597*	179 267
GADOW, H.	A classification of Vertebrata, recent and extinct. 1898.	*17*	12
	On the evolution of the vertebral column of Amphibia and Amniota. Phil. Trans. Roy. Soc. (B), CLXXXVII, 1896.	*29* *14* *26*	96 121 136
	On the modifications of the first and second visceral arches with special reference to the homologies of the auditory ossicles. Phil. Trans. Roy. Soc. CLXXIX, 1888. The evolution of the auditory ossicles. Anat. Anz. XIX, 1901.	*451* *396*	222
	Remarks on the supposed relationship of birds and dinosaurs. Proc. Camb. Phil. Soc. IX, 1897.	*204*	325
GADOW, H. and ABBOTT, E.	On the evolution of the vertebral column of fishes. Phil. Trans. Roy. Soc. CLXXXVI, 1895.	*163*	9 14 22 30 130
GASKELL, W. H.	The structure, distribution and function of the nerves which innervate the visceral and vascular systems. Journ. Physiol. VII, 1886. The origin of vertebrates. Longmans, Green and Co. 1908.	*1*	91
GAUDRY, A.	Les reptiles de l'époque permienne aux environs d'Autun. Bull. Soc. géol. France. 16th Dec. 1878.		68
	Les enchaînements du monde animal, dans les temps géologiques. Fossiles primaires, I. Paris, 1883.	*263*	195
GAUPP, E.	Die Entwickelung des Kopfskelettes. In O. Hertwig's Handbuch der vergleichenden und experimentellen Entwickelungslehre der Wirbelthiere, III, ii. Jena, 1906.	*597*	75
GEGENBAUR, C.	Lehrbuch. Vergleichende Anatomie der Wirbelthiere mit Berücksichtigung der Wirbellosen. Leipzig, 1898.	*249* *325* *241* *240* *257* *247* *256*	67 68 128 184 232 233 234

Author	Work		
GEGENBAUR, C. (*cont.*)	Lehrbuch (*cont.*).	*256*	235
		256	236
		258	245_1
		262	245_2
		461	244
		259	250
		255	269
		840	322
		137	323
	Untersuchungen zur vergleichenden Anatomie der Wirbelsäule bei Amphibien und Reptilien. 4 Taf. Leipzig, 1862.		30, 136, 167
	Zur Phylogenese der Zunge. Morph. Jahrb. XXI, 1894.		84
	Untersuchungen zur vergleichenden Anatomie der Wirbelthiere. Das Kopfskelet der Selachier, III. Leipzig, 1872.		66, 74, 75_3, 76, 87, 111
	Ueber die Occipitalregion und die ihr benachbarten Wirbel der Fische. Festschrift zu A. v. Koelliker's 70. Geburtstag. Leipzig, 1887.		75_1
	Die Metamerie des Kopfes und die Wirbeltheorie des Kopfskeletes. Morph. Jahrb. XIII, 1888.		75_2
GOEPPERT, E.	Zur Kenntniss der Amphibienrippen. Vorläufige Mittheilung. Morph. Jahrb. XXII, 1895.	*441*	117, 130, 147 (both works)
	Die Morphologie der Amphibienrippen. Festschr. zum 70sten Geburtstag von C. Gegenbaur, I, 1896.	*395*	
GOETTE, A.	Beiträge zur vergleichenden Morphologie des Skeletsystems der Wirbelthiere. II. Die Wirbelsäule und ihre Anhänge. Archiv Mikr. Anat. XV, 1878.	*442*	22, 117, 133, 164
	Ueber die Entwicklung des knöchernen Rückenschildes (Carapax) der Schildkröten. Zeitschr. wissensch. Zool. LXVI, 1899.	*407*	208, 234
GOODRICH, E. S.	On the dermal fin-rays of fishes. Quart. Journ. Micr. Sci. XLVII, 1903.		4
	Structure and development of vertebrates. London, 1930.	*Figs. 23, 24 b*	19 F.
		Fig. 79	57 F.
		Fig. 52	160 F.
GREGORY, W. K.	The skeleton of *Moschops capensis* (Broom), a Dinocephalian reptile from the		

GREGORY, W. K. (*cont.*)	Permian of South Africa. Bull. Amer. Mus. Nat. Hist. LVI, Art. 3, 1926.	*179*	225
	Studies in comparative myology and osteology. No. III (with C. L. Camp). Part IV. A reconstruction of the skeleton of *Cynognathus*. Bull. Amer. Mus. Nat. Hist. XXXVIII, Art. XV, 1918.	*447*	225 F.
GÜNTHER, A. C.	Contribution to the anatomy of *Hatteria* (*Rhynchocephalus*, Owen). Phil. Trans. Roy. Soc. London, CLVII, 1867.	*595*	179
HAECKEL, E.	Systematische Phylogenie der Wirbelthiere. Berlin, 1895. (Vertebrata.) Dritter Theil des Entwurfs einer systematischen Phylogenie.		36 72 170 174 176 179 317
	Generelle Morphologie der Organismen. 1866.		134
HASSE, C.	Das natürliche System der Elasmobranchier auf Grundlage des Baues und der Entwicklung ihrer Wirbelsäule. Eine morphologische und paläontologische Studie. Jena, 1879–85.		30 31 167
HAY, O. P.	On *Protostega*, the systematic position of *Dermochelys* and the morphology of the chelonian carapace and plastron. Amer. Natural. XXXII, 1898.	*929*	208
HEILMANN, G.	The origin of birds. London, 1926.		325 n.
HERTWIG, O.	Lehrbuch der Entwicklungsgeschichte. 1888.	*442*	14
HOFFMAN, C. K.	In H. G. Bronn's Thierreich, VI, iii. Schlangen und Entwicklungsgeschichte der Reptilien. 1890.	*1422*	285
HOWES, G. B.	Concerning the Proatlas. Proc. Zool. Soc. 1890.	*257*	67
HOWES, G. B. and SWINNERTON, H. H.	On the development of the skeleton of the Tuatara *Sphenodon punctatus*; with remarks on the egg, on the hatching, and on the hatched young. Trans. Zool. Soc. XVI, 1901.	*1*	8 39 44 80 136 165 172 173 236 261 263 265 309

Huene, F. v.	The skull elements of the Permian Terapoda in the American Museum of Natural History, New York. Bull. Amer. Nat. Hist. N.Y. xxxiii, 1913.		75
	Some additions to the knowledge of *Procolophon, Lystrosaurus, Notesaurus* and *Cistecephalus*. Trans. Roy. Soc. S.A. xiii, 1925.	*140*	228
	Beiträge zur Geschichte der Archosaurier. Geol. und Paläont. Abhandl. (N.F. Bd. xiii), G.R. xvii, 1914.		176 294
	Beiträge zur Kenntnis und Beurteilung der Parasuchier. Geol. und Paläont. Abhandl. (N.F. Bd. x), G.R. xiv. Jena, 1911.	*109*	295 F.
	Die Dinosaurier der europäischen Triasformation mit Berücksichtigung der aussereuropäischen Vorkommnisse. Geol. und Paläont. Abhandl. Supplementalband i, 1907–8.		302
Hulke, J. W.	Contribution to the skeletal anatomy of the Mesosuchia based on fossil remains from the clays near Peterborough in the collection of A. Leeds, Esq. Proc. Zool. Soc. 1888.	*417*	299 F.
Huxley, T. H.	A manual of the anatomy of vertebrated animals. London, 1871.	*Fig. 71* *170* *193*	58 F. 72 170 284 323
Hyrtl, J.	Ueber normale Quertheilung des Saurierwirbels. Wiener Sitzungsber. Math.-Naturw. Cl. x, 1853.	*185*	269
Ihle, J. E. W. and others	Vergleichende Anatomie der Wirbeltiere. Berlin, 1927.	*98* *Fig. 109*	273 F.
Jaekel, O.	Ueber *Ceraterpeton, Diceratosaurus* und *Diplocaulus*. Neues Jahrb. f. Min. Geol. u. Pal. Heft i, 1903.	*109*	140
	Die Wirbeltierfunde aus dem Keuper von Halberstadt. II. Teil: Testudinata. Paläontologische Zeitschr. ii. Berlin, 1916.	*88*	208
	Die ersten Halswirbel. Anat. Anz. xl, 1912.	*614* *Fig. 4*	326 F.
Kingsley, J. S.	Outlines of comparative anatomy of vertebrates. London, 1917.	*59* *Fig. 55* *62* *Fig. 60*	311 F. 319 F.

Klaatsch, H.	Beiträge z. vergl. Anat. der Wirbelsäule: I. Über den Urzustand der Fischwirbelsäule. Morph. Jahrb. xix, 1893.		2
	Ibid. II. Über die Bildung knorpeliger Wirbelkörper bei Fischen. Morph. Jahrb. xx, 1893.		30
	Ibid. III. Zur Phylognese der Chordascheiden. Morph. Jahrb. xxii, 1895.		30
Koelliker, A.	Ueber die Beziehung der Chorda Dorsalis zur Bildung der Wirbel der Selachier und einiger anderer Fische. Verhandl. d. Phys. med. Gesell. in Würzburg, x, 1860.		30
Krukenberg, C. F. W.	Ueber die chemische Beschaffenheit der sogenannte Hornfäden von *Mustelus* und ueber die Zusammensetzung der keratinösen Hüllen um den Eiern von *Scyllium Stellare.* Mitt. a. d. Zool. Stat. zu Neapel, vi, 1886.	*286*	4
Kuppfer, v. C.	Studien zur vergl. Entw.-Gesch. d. Kranioten. II. Heft. Die Entwicklung des Kopfes von *Petromyzon planeri*, 1894.		2
Lamarck, J. B.	Philosophie zoologique. 1809.		54
Latreille, P. A.	Familles naturelles du règne animal. 1825.		289
Leydig, Fr.	Die in Deutschland lebenden Arten der Saurier. Tübingen, 1872.		53
Lvoff, B.	Vergleichend-anatomische Studien ueber die Chorda und die Chordascheide. Bull. Soc. Imp. Nat. de Moscou, No. 2, 1887.		136
Macalister, A.	The development and varieties of the second cervical vertebra. Journ. Anat. and Physiol. xxviii, 1894.	*257*	105
	Notes on the development and variations of the Atlas. Journ. Anat. and Physiol. xxvii, 1893.	*519*	234 237
Männer, H.	Die Entwicklung der Wirbelsäule bei Reptilien. Zeit. f. wiss. Zool. lxvi, 1899.		97
Marcus, H. and Blume, W.	Wirbel und Rippen bei *Hypogeophis.* Zeitschr. f. Anat. u. Entw. lxxx, 1926.		8 31 42 44 135 136
Marsh, O. C.	Amphibian footprints from the Devonian. Amer. Journ. Sci. (4), ii, 1896.	*374*	158

MARSH, O. C. (*cont.*)	The Dinosaurs of North America. Sixteenth Annual Report U.S. Geol. Surv. 1894–5. Part I, 1896.	*206*	304 F.
	On a new subclass of fossil birds (*Odontornithes*). Amer. Journ. Sci. (3), V, 1873.	*161*	323
MATTHEW, W. D.	A revision of the Puerco fauna. Bull. Amer. Mus. Nat. Hist. IX, 1897.	*259*	38
MERREM, B.	Tentamen systematis Amphibiorum. 1820.		289
MEYER, H. v.	Reptilien aus der Steinkohlenformation in Deutschland. Palaeontographica, VI. Kassel, 1857.	*59*	159
MIVART, ST G.	The axial skeleton of the Pelecanidae. Trans. Zool. Soc. London, X, 1879.	*Pl. 60, Fig. 3*	314 F.
MOODIE, R. L.	The Microsauria as ancestors of the Reptilia. Geol. Mag. (5th series), VI, 1909.	*216*	170
MÜLLER, JOHANNES	Vergleich. Anat. der Myxinoiden. I. Abh. d. K. Akademie d. N. in Berlin, 1836.	*56*	51 52
	Beiträge zur Anatomie und Naturgeschichte der Amphibien. Treviranus Zeitschrift für Physiologie, IV, 1831.	*190*	134 169
NEWTON, E. T.	Reptiles from the Elgin Sandstone. Phil. Trans. Roy. Soc. London (B), CLXXXV, 1894.	*573*	294
NOBLE, G. K.	The phylogeny of the Salientia. I. The osteology and the thigh musculature: their bearing on classification and phylogeny. Bull. Amer. Mus. Nat. Hist. XLVI, 1922.	*1*	166
NOPCSA, F. v.	Zur Kenntnis der fossilen Eidechsen. Beiträge zur Paläont. und Geol. Österreich-Ungarns und des Orients, XXI, 1908.	*45*	279 280
	Über Dinosaurier. Centralblatt f. Mineralogie u.s.w. 1917.		308
OSBORN, H. F.	Origin of the Mammalia. III. Occipital condyles of reptilian tripartite type. Amer. Nat. XXXIV, 1900.	*943* *Fig. 6*	71 260 F.
OWEN, R.	On the anatomy of vertebrates, I, II. London, 1866.	*394* Vol. II, *448* Vol. I, *50, 51*	46 242 169 328
	History of British fossil reptiles. 1847–65.		327 F.
PALMER, R. W.	Note on the lower jaw and ear ossicles of a foetal Perameles. Anat. Anz. XLIII, 1913.		222

Panizza, B.	Sopra il systema linfatico dei rettili, ricerche zootomiche. Pavia, 1833.		285
Parker, T. J.	Observations on the anatomy and development of *Apteryx*. Phil. Trans. Roy. Soc. (B), CLXXXII, 1891.	*80* *83*	265 318
Parker, W. K.	On the morphology of birds. Proc. Roy. Soc. XLII, 1887.	*52*	312
Pearson, H. S.	A Dicynodont reptile reconstructed. Proc. Zool. Soc. 1924.	*827*	225
Peter, K.	Die Entwicklung und funktionelle Gestaltung des Schädels von *Ichthyophis glutinosus*. Morph. Jahrb. XXV, 1896.		67 75
Piiper, J.	On the evolution of the vertebral column in birds, illustrated by its development in *Larus* and *Struthio*. Phil. Trans. Roy. Soc. London (B), CCXVI, 1928.	*285*	7 9 15 24 42 44
Piveteau, J.	Paléontologie de Madagascar. XIII. Amphibiens et Reptiles Permiens. Annales de Paléontologie, XV, 1926.	*55*	226
Platt, Julia	Ectodermic origin of the cartilages of the head. Anat. Anz. VIII, 1893.	*506*	2
Quenstedt, Fr. A.	Handbuch der Petrefactenkunde. Dritte Auflage. Tübingen, 1885.		169
Rathke, H.	Vorträge zur vergleichenden Anatomie der Wirbelthiere. Leipzig, 1862.	*20*	53
Ridewood, W. G.	Development of the vertebral column in *Pipa* and *Xenopus*. Anat. Anz. XIII, 1897.	*359*	31 68 165 166 167 168
Rosenberg, E.	Untersuchungen über die Occipitalregion des Cranium und den proximalen Teil der Wirbelsäule einiger Selachier. Dorpat, 1884.		74
	Ueber die Entwicklung der Wirbelsaeule und des Centrale carpi des Menschen. Morph. Jahrb. I, 1876.	*131*	246
Salle, O.	Untersuchungen über die Lymphapophysen der Schlangen und schlangenähnlichen Saurier. Dissert. inaug. Göttingen, 1881.		285 286
Schauinsland, H.	Die Entwicklung der Wirbelsaeule nebst Rippen und Brustbein. Handbuch der vergleichenden und experimentellen Entwicklungslehre der Wirbelthiere (O. Hertwig), III. Jena, 1906.	*526* *525* *527* *420* *526*	7 8_1 8_2 9 21

SCHAUINSLAND, H. (*cont.*)	Die Entwicklung der Wirbelsaeule nebst Rippen und Brustbein (*cont.*).	*401*	22
		409	32
		516	33
		541	67
		536 Fig. *309*	95
		497 Fig. *284*	133
		526	165
		503	166
		502	167
		525	233
		549	234
		552	236
		514–542	261
		520	269
		339	270
	Weitere Beiträge zur Entwicklungsgeschichte der Hatteria. Arch. f. Mik. Anat. LVI, 1900.		14 31 40 42
SCHWARZ, H.	Über die Wirbelsäule und die Rippen holospondyler Stegocephalen. Beiträge zur Paläont. und Geol. Österreich-Ungarns und des Orients, XXI, 1908.		9 127 F. 166
SEELEY, H. G.	II. On *Pareiasaurus bombidens* (Owen) and the significance of its affinities to Amphibians, Reptiles and Mammals. Phil. Trans. Roy. Soc. CLXXIX, 1888.	*100*	71
	On *Protorosaurus Speneri*. Phil. Trans. Roy. Soc. CLXXVIII, 1887.	*187*	257
	Researches on the structure, organisation, and classification of the fossil Reptilia. VII. Further observations on *Pareiasaurus*. Phil. Trans. Roy. Soc. CLXXXIII, 1892.	*Pl. 17*	199 F.
SOEMMERRING, S. T. VON	Vom Baue des menschlichen Körpers, II. Leipzig, 1839.		51
STANNIUS, H.	Handbuch der Anatomie der Wirbelthiere. 2. Teil, 2. Aufl. 1854.	*16*	166
STEINER, H.	Die Ontogenetische und Phylogenetische Entwicklung des Vogelflügelskelettes. Acta Zoologica, Jg. 3, 1922.	*356* *307*	297 325
THÉVENIN, A.	Les plus anciens Quadrupèdes de la France. Annales de Paléontologie, V, 1910.	*43*	186
THOMAS, O.	On the homologies and succession of the teeth in the Dasyuridae, with an		

THOMAS, O. (*cont.*)	attempt to trace the history of the evolution of mammalian teeth in general. Phil. Trans. Roy. Soc. Lond. CLXXVIII, 1888.	*443*	251
VERSLUYS, J.	Über die Phylogenie des Panzers der Schildkröten und über die Verwandtschaft der Lederschildkröte (*Dermochelys coriacea*). Paläontologische Zeitschr. I. Berlin, 1914.	*321*	208
VOGT, C.	*Archaeopteryx macrura*, an intermediate form between birds and reptiles. The Ibis (4), IV, 1880.	*434*	323
WALDSCHMIDT, J.	Zur Anatomie des Nervensystems der Gymnophionen. Jena. Zeitschr. XX, 1887.	*461*	75 80
WATSON, D. M. S.	The evolution and origin of the Amphibia. Phil. Trans. Roy. Soc. London (B), CCXIV, 1926.	*189*	57 F. 67 68 76 80 153 156 158 159 163 178
	On *Seymouria*, the most primitive known reptile. Proc. Zool. Soc. London, 1919.	*267*	194
	The skeleton of *Lystrosaurus*. Rec. Alb. Mus. II, 1912.	*287*	226
WIEDERSHEIM, R.	Die Anatomie der Gymnophionen. Jena, 1879.		51 54 135
	Labyrinthodon rütimeyeri. Ein Beitrag zur Anatomie von Gesammtskelet und Gehirn der triassischen Labyrinthodonten. Abhandl. schweizer. paläont. Gesellsch. V, 1878.	*1*	228
	Die Stammesentwicklung der Vögel. Biolog. Centralbl. III, 1884.	*654*	323
	Vergleichende Anatomie der Wirbelthiere. Jena, 1906.	*Fig. 145* *71*	316 F. 117
WIJHE, J. W. v.	Ueber die Mesodermsegmente und die Entwickelung der Nerven des Selachierkopfes. Verh. Konink. Akad. Wet. Amsterdam, XXII, 1882.		74 77
WILLEY, A.	Orthogenetic variations in the shells of *Chelonia*. Zool. Results, Pt III, 1899.		210

WILLISTON, S. W.	*Cacops. Desmospondylus.* New genera of Permian vertebrates. Bull. Geol. Soc. Amer. XXI, 1910.	*283*	154 F.
		Pl. 9	
		272	187_2
		280	193
			195
			197
	New Permian vertebrates. Journ. of Geol. XVII, 1909.	*636*	155 F.
		Fig. 5	
	Primitive reptiles. Journ. Morphology, XXIII, 1912.	*655*	176
			182_2
		647	257
	Familial relations of early Vertebrata. Journ. of Geol. 1909.	*389*	182_1
	The oldest known reptile, *Isodectes Punctulatus* (Cope). Journ. of Geol. XVI, 1908.	*395*	186
			187_1
	The osteology of the reptiles. (Edited by W. K. Gregory.) Harvard Univ. Press, 1925.	*Fig. 161*	188 F.
		Fig. 74	197 F.
		Fig. 164	198 F.
		Fig. 80	203 F.
		Fig. 84	213 F.
		Fig. 167	214 F.
		Fig. 168, *Fig. 77*, *Fig. 78*	216 F.
		Fig. 165	217 F.
		Fig. 171	227 F.
		Fig. 73	260 F.
		Fig. 73	273 F.
		Fig. 184	278 F.
		Fig. 86	279 F.
		Fig. 90	296 F.
		Fig. 75	305 F.
		Fig. 188	306 F.
		Fig. 189	307 F.
		Fig. 89	330 F.
WORTMAN, J. L.	Studies of Eocene Mammalia in the Marsh collection, Peabody Museum. Amer. Journ. Sci. (4) XVI.	*345*	222
ZITTEL, K. v.	Handbuch der Palaeontologie, III, Munich and Leipsic, 1887–1890.	*370*	125
		397	157
		369	181
		502	205
		644	291
			293
		641	295 F.
ZYKOFF, W.	Ueber das Verhaeltniss des Knorpels der Chorda bei *Siredon pisciformis*. Bull. Soc. Nat. de Moscou, Pl. II, 1893.		136

INDEX

INDEX

INDEX

INDEX

INDEX

INDEX

INDEX

Printed in the United States
By Bookmasters